MARKET SEGMENTATION
Conceptual and Methodological Foundations

Second Edition

INTERNATIONAL SERIES IN QUANTITATIVE MARKETING

Series Editor:

Jehoshua Eliashberg
The Wharton School
University of Pennsylvania
Philadelphia, Pennsylvania USA

Other books in the series:

Cooper, L. and Nakanishi, M.:
Market Share Analysis

Hanssens, D., Parsons, L., and Schultz, R.:
Market Response Models: Econometric
and Time Series Analysis

McCann, J. and Gallagher, J.:
Expert Systems for Scanner Data Environments

Erickson, G.:
Dynamic Models of Advertising Competition

Laurent, G., Lilien, G.L., Pras, B.:
Research Traditions in Marketing

Nguyen, D.:
Marketing Decisions Under Uncertainty

Wedel, M. and Kamakura, W.G.
Market Segmentation

MARKET SEGMENTATION
Conceptual and Methodological Foundations

Second Edition

Michel Wedel
University of Groningen

Wagner A. Kamakura
University of Iowa

Kluwer Academic Publishers
BOSTON DORDRECHT LONDON

Distributors for North, Central and South America:
Kluwer Academic Publishers
101 Philip Drive
Assinippi Park
Norwell, Massachusetts 02061 USA
Telephone (781) 871-6600
Fax (781) 681-9045
E-Mail <kluwer@wkap.com>

Distributors for all other countries:
Kluwer Academic Publishers Group
Distribution Centre
Post Office Box 322
3300 AH Dordrecht, THE NETHERLANDS
Telephone 31 78 6576 000
Fax 31 78 6576 474
E-Mail <services@wkap.nl>

 Electronic Services <http://www.wkap.nl>

Library of Congress Cataloging-in-Publication Data

Wedel, Michel.
  Market segmentation: conceptual and methodological foundations/
  Michel Wedel, Wagner A. Kamakura.—2nd ed.
    p.cm.
  Includes bibliographical references and index.
  ISBN 0-7923-8635-3 (alk.paper)
  1. Market segmentation. 2. Market segmentation—Statistical methods.
    I.Kamakura, Wagner A. (Wagner Antonio)
HF5415.127 .W43 1999
658.8'02—dc21   99-049246

Copyright © 2000 by Kluwer Academic Publishers. Third Printing 2003.

All rights reserved. No part of this publication may be reproduced, stored in a retrieval system or transmitted in any form or by any means, mechanical, photo-copying, recording, or otherwise, without the prior written permission of the publisher, Kluwer Academic Publishers, 101 Philip Drive, Assinippi Park, Norwell, Massachusetts 02061

Printed on acid-free paper.

Printed in the United States of America

To our parents:

Hans & Joan

Antonio & Sumiko

Contents

PART 1 – INTRODUCTION   1

1   The Historical Development of the Market Segmentation Concept   3

2   Segmentation Bases   7
- Observable General Bases   8
- Observable Product-Specific Base   10
- Unobservable General Bases   11
- Unobservable Product-Specific Bases   14
- Conclusion   16

3   Segmentation Methods   17
- A-Priori Descriptive Methods   18
- Post-Hoc Descriptive Methods   19
- A-Priori Predictive Methods   22
- Post-Hoc Predictive Methods   23
- Normative Segmentation Methods   26
- Conclusion   28

4   Tools for Market Segmentation   31

PART 2 – SEGMENTATION METHODOLOGY   37

5   Clustering Methods   39
- Example of the Clustering Approach to Market Segmentation   42
- Nonoverlapping Hierarchical Methods   43
  - Similarity Measures   44
  - Agglomerative Cluster Algorithms   48
  - Divisive Cluster Algorithms   50
  - Ultrametric and Additive Trees   50
  - Hierarchical Clusterwise Regression   51
- Nonoverlapping Nonhierarchical Methods   52
  - Nonhierarchical Algorithms   53
  - Determining the number of Clusters   54
  - Nonhierarchical Clusterwise Regression   55
- Miscellaneous Issues in Nonoverlapping Clustering   56
  - Variable Weighting, Standardization and Selection   56
  - Outliers and Missing Values   58

	Non-uniqueness and Inversions	59
	Cluster Validation	59
	Cluster Analysis Under Various Sampling Strategies	60
	Stratified samples	60
	Cluster samples	62
	Two-stage samples	63
	Overlapping and Fuzzy Methods	64
	Overlapping Clustering	64
	Overlapping Clusterwise Regression	65
	Fuzzy Clustering	65
	Market Segmentation Applications of Clustering	69
6	**Mixture Models**	**75**
	Mixture Model Examples	75
	Example 1: Purchase Frequency of Candy	75
	Example 2: Adoption of Innovation	76
	Mixture Distributions (MIX)	77
	Maximum Likelihood Estimation	80
	The EM Algorithm	84
	EM Example	86
	Limitations of the EM Algorithm	88
	Local maxima	88
	Standard errors	88
	Identification	90
	Determining the Number of Segments	91
	Some Consequences of Complex Sampling Strategies for the Mixture Approach	94
	Marketing Applications of Mixtures	96
	Conclusion	99
7	**Mixture Regression Models**	**101**
	Examples of the Mixture Regression Approach	102
	Example 1: Trade Show Performance	102
	Example 2: Nested Logit Analysis of Scanner Data	103
	A Generalized Mixture Regression Model (GLIMMIX)	106
	EM Estimation	108
	EM Example	108
	Standard Errors and Residuals	109
	Identification	109
	Monté Carlo Study of the GLIMMIX Algorithm	110
	Study Design	110
	Results	112
	Marketing Applications of Mixture Regression Models	112
	Normal Data	113
	Binary Data	113

Multichotomous Choice Data	115
Count Data	116
Choice and Count Data	116
Response-Time Data	117
Conjoint Analysis	117
Conclusion	119
Appendix A1 The EM Algorithm for the GLIMMIX Model	120
The EM Algorithm	120
The E-Step	121
The M-Step	121

8 Mixture Unfolding Models — 125

Examples of Stochastic Mixture Unfolding Models	127
Example 1: Television Viewing	127
Example 2: Mobile Telephone Judgements	128
A General Family of Stochastic Mixture Unfolding Models	131
EM Estimation	133
Some Limitations	133
Issues in Identification	134
Model Selection	134
Synthetic Data Analysis	136
Marketing Applications	138
Normal Data	138
Binomial Data	140
Poisson, Multinomial and Dirichlet Data	140
Conclusion	140
Appendix A2 The EM Algorithm for the STUNMIX Model	142
The E-Step	142
The M-step	142

9 Profiling Segments — 145

Profiling Segments with Demographic Variables	145
Examples of Concomitant Variable Mixture Models	146
Example 1: Paired Comparisons of Food Preferences	146
Example 2: Consumer Choice Behavior with Respect to Ketchup	147
The Concomitant Variable Mixture Model	150
Estimation	152
Model Selection and Identification	152
Monté Carlo Study	152
Alternative Mixture Models with Concomitant Variables	153
Marketing Applications	156
Conclusions	156

10	**Dynamic Segmentation**	**159**
	Models for Manifest Change	160
	Example 1: The Mixed Markov Model for Brand Switching	161
	Example 2: Mixture Hazard Model for Segment Change	162
	Models for Latent Change	167
	Dynamic Concomitant Variable Mixture Regression Models	167
	Latent Markov Mixture Regression Models	168
	Estimation	169
	Examples of the Latent Change Approach	170
	Example 1: The Latent Markov Model for Brand Switching	170
	Example 2: Evolutionary Segmentation of Brand Switching	171
	Example 3: Latent Change in Recurrent Choice	175
	Marketing Applications	176
	Conclusion	176
	Appendix A3 Computer Software for Mixture models	178
	PANMARK	178
	LEM	179
	GLIMMIX	181

PART 3 – SPECIAL TOPICS IN MARKET SEGMENTATION   187

11	**Joint Segmentation**	**189**
	Joint Segmentation	189
	The Joint Segmentation Model	189
	Synthetic Data Illustration	191
	Banking Services	192
	Conclusion	194
12	**Market Segmentation with Tailored Interviewing**	**195**
	Tailored Interviewing	195
	Tailored Interviewing for Market Segmentation	198
	Model Calibration	199
	Prior Membership Probabilities	200
	Revising the Segment Membership Probabilities	201
	Item Selection	202
	Stopping Rule	202
	Application to Life-Style Segmentation	203
	Life-Style Segmentation	203
	Data Description	203
	Model Calibration	203
	Profile of the Segments	204
	The Tailored Interviewing Procedure	209
	Characteristics of the Tailored Interview	209
	Quality of the Classification	211

	Conclusion	214
13	**Model-Based Segmentation Using Structural Equation Models**	**217**
	Introduction to Structural Equation Models	217
	A-Priori Segmentation Approach	222
	Post Hoc Segmentation Approach	223
	Application to Customer Satisfaction	223
	The Mixture of Structural Equations Model	225
	Special Cases of the Model	226
	Analysis of Synthetic Data	227
	Conclusion	229
14	**Segmentation Based on Product Dissimilarity Judgements**	**231**
	Spatial Models	231
	Tree Models	232
	Mixtures of Spaces and Mixtures of Trees	235
	Mixture of Spaces and Trees	238
	Conclusion	238

PART 4 – APPLIED MARKET SEGMENTATION   239

15	**General Observable Bases: Geo-demographics**	**241**
	Applications of Geo-demographic Segmentation	242
	Commercial Geo-demographic Systems	244
	PRIZM™ (Potential Rating Index for ZIP Markets)	244
	ACORN™ (A Classification of Residential Neighborhoods)	247
	The Geo-demographic System of Geo-Marktprofiel	248
	Methodology	254
	Linkages and Datafusion	256
	Conclusion	257
16	**General Unobservable Bases: Values and Lifestyles**	**259**
	Activities, Interests and Opinions	260
	Values and Lifestyles	261
	Rokeach's Value Survey	261
	The List of Values (LOV) Scale	265
	The Values and Lifestyles (VALS™) Survey	266
	Applications of Lifestyle Segmentation	268
	Conclusion	276
17	**Product-specific observable Bases: Response-based Segmentation**	**277**
	The Information Revolution and Marketing Research	277
	Diffusion of Information Technology	277
	Early Approaches to Heterogeneity	278
	Household-Level Single-Source Data	279

	Consumer Heterogeneity in Response to Marketing Stimuli	282
	Models with Exogenous Indicators of Preferences	283
	Fixed-Effects Models	283
	Random-Intercepts and Random Coefficients Models	284
	Response-Based Segmentation	285
	Example of Response-Based Segmentation with Single Source Scanner Data	286
	Extensions	288
	Conclusion	292
18	**Product-Specific Unobservable Bases: Conjoint Analysis**	**295**
	Conjoint Analysis in Marketing	295
	Choice of the Attributes and Levels	296
	Types of Attributes	296
	Number of Attributes	297
	Attribute Levels	298
	Stimulus Set Construction	298
	Stimulus Presentation	299
	Data Collection and Measurement Scales	300
	Preference Models and Estimation Methods	301
	Choice Simulations	302
	Market Segmentation with Conjoint Analysis	303
	Application of Conjoint Segmentation with Constant Sum Response Data	303
	Market Segmentation with Metric Conjoint Analysis	305
	A-Priori and Post-Hoc Methods Based on Demographics	306
	Componential Segmentation	306
	Two-Stage Procedures	306
	Hagerty's Method	307
	Hierarchical and Non-Hierarchical Clusterwise Regression	307
	Mixture Regression Approach	308
	A Monté Carlo Comparison of Metric Conjoint Segmentation Approaches	310
	The Monté Carlo Study	310
	Results	312
	Predictive Accuracy	313
	Segmentation for Rank-Order and Choice Data	314
	A-Priori and Post-Hoc Approaches to Segmentation	315
	Two-Stage Procedures	315
	Hierarchical and Non-hierarchical Clusterwise Regression	316
	The Mixture Regression Approach for Rank-Order and Choice Data	316
	Application of Mixture Logit Regression to Conjoint Segmentation	318
	Results	319
	Conclusion	320

PART 5 – CONCLUSIONS AND DIRECTIONS FOR FUTURE RESEARCH 323

19 Conclusions: Representations of Heterogeneity 325
   Continuous Distribution of Heterogeneity versus Market Segments 325
      Continuous or Discrete 326
      ML or MCMC 327
      Managerial relevance 329
   Individual Level versus Segment Level Analysis 331

20 Directions for Future Research 335
   The Past 335
   Segmentation Strategy 336
   Agenda for Future Research 341

   References 345

   Index 371

List of Tables and Figures

Table	2.1:	Evaluation of Segmentation Bases	16
Table	3.1:	Evaluation of Segmentation Methods	29
Table	5.1:	The Most Important Similarity Coefficients	46
Table	5.2:	Definitions of Cluster Distance for several Types of Hierarchical Algorithms	49
Table	5.3:	Optimization Criteria in Nonhierarchical Clustering	52
Table	5.4:	Fuzzy Clustering Algorithm Estimating Equations	70
Table	5.5:	Cluster Analysis Applications to Segmentation	72
Table	6.1:	Results of the Green et al. Three-Segment Model	78
Table	6.2:	Some Distributions from the Univariate Exponential Family	82
Table	6.3:	Application of the EM Algorithm to Synthetic Mixture Poisson Data	87
Table	6.4:	Mixture Model Applications in Marketing	97
Table	6.5:	Special Cases of the Böckenholt (1993) Mixture Model Family	98
Table	7.1:	Aggregate and Segment-Level Results of the Trade Show Performance Study	103
Table	7.2:	Average Price Elasticities within Three Segments	105
Table	7.3:	Application of the EM Algorithm to Synthetic Mixture Regression Data	111
Table	7.4:	Results of the Monté Carlo Study on GLIMMIX Performance	114
Table	7.5:	GLIMMIX Applications in Marketing	118
Table	8.1:	Numbers of Parameters for STUNMIX Models	135
Table	8.2:	Results of STUNMIX Synthetic Data Analyses	137
Table	8.3:	STUNMIX Results for Normal and Log-Normal Synthetic Data	138
Table	8.4:	STUNMIX Marketing Applications	141
Table	9.1:	Concomitant Variable Mixture Model results for Food Concern Data	148
Table	9.2:	Concomitant Variable Mixture Model Results for Ketchup Choice Data	150
Table	9.3:	Estimates from Synthetic Datasets Under Two Conditions	157
Table	9.4:	Marketing Applications of Concomitant Variable Models	158
Table	10.1:	Mixed Markov Results for MRCA Data	162
Table	10.2:	Segment Level S=2 Nonproportional Model Estimates	165
Table	10.3:	Latent Markov Results for MRCA Data	171
Table	10.4:	Conditional Segment Transition Matrix Between Periods 1 and 2	174

List of tables and figures

Table	10.5:	Preference Structure in Three and Four Switching Segments in Two Periods	174
Table	10.6:	Marketing Applications of Dynamic Segmentation	176
Table	11.1:	True and Estimated Joint Segment Probabilities	191
Table	11.2:	Segmentation Structure of Binary Joint Segmentation Model for Banking Services	193
Table	11.3:	Estimated Joint Segment Probabilities for Banking Services Application	194
Table	12.1:	Demographics and Activities by Segment	205
Table	12.2:	Activities, Interests and Opinions Toward Fashion by Segment	207
Table	12.3:	Percentage of Cases Correctly Classified	213
Table	13.1:	Latent Satisfaction Variables and Their Indicators	218
Table	13.2:	Mean Factor Scores in Three Segments	224
Table	13.3:	Estimates of Structural Parameters in Three Segments	225
Table	13.4:	Performance Measures in First Monté Carlo Study	228
Table	13.5:	Performance Measures in Second Monté Carlo Study	229
Table	15.1:	PRIZM™ Classification by Broad Social Groups	246
Table	15.2:	ACORN™ Cluster Classification	250
Table	15.3:	Description of the Dimensions of the GMP System	253
Table	15.4:	Media and Marketing Databases Linked to Geo-demographic Systems	258
Table	16.1:	Rokeach's Terminal and Instrumental Values	263
Table	16.2:	Motivational Domains of Rekeach's Values Scale	264
Table	16.3:	Double-Centered Values	271
Table	17.1:	Estimated Price Elasticities	289
Table	18.1:	Estimated Allocations for Three Profiles in Two Segments	305
Table	18.2:	Comparison of Conjoint Segmentation Procedures	309
Table	18.3:	Mean Performance Measures of the Nine Conjoint Segmentation Methods	313
Table	18.4:	Mean Performance Measures for Each of the Factors	314
Table	18.5:	Parameter Estimates of the Rank-Order Conjoint Segmentation Model	320
Table	19.1	Comparison of Discrete and Continuous Representations of Heterogeneity	328
Table	19.2	Comparison of Segment Level and Individual-Level Approaches	332

Figure	2.1:	Classification of Segmentation Bases	7
Figure	3.1:	Classification of Methods Used for Segmentation	17
Figure	5.1:	Classification of Clustering Methods	42
Figure	5.2:	Hypothetical Example of Hierarchical Cluster Analysis	44
Figure	5.3:	Schematic Representation of Some Linkage Criteria	48
Figure	5.4:	An Example of Cluster Structures Recovered with the FCV	

XVI

		Family	67
Figure	6.1:	Empirical and Fitted Distributions of Candy Purchases	76
Figure	6.2:	Local and Global Maxima	89
Figure	8.1:	TV-Viewing Solution	128
Figure	8.2:	Spatial Map of the Telephone Brands	130
Figure	8.3:	Synthetic Data and Results of Their Analyses with STUNMIX	139
Figure	9.1:	Directed Graph for the Standard Mixture	154
Figure	9.2:	Directed Graph for the Concomitant Variable Mixture	154
Figure	10.1:	Observed and Predicted Shares for the Mixed Markov Model	163
Figure	10.2:	Stepwise Hazard Functions in Two Segments	166
Figure	10.3:	Observed and Predicted Shares for the Latent Markov Model	172
Figure	12.1:	Illustration of the Tailored Interview	197
Figure	12.2:	Cost Versus Accuracy Tradeoff: Discontent and Alienation Scales	198
Figure	12.3:	Number of Items Selected in the Tailored Interview until $p>0.99$	210
Figure	12.4:	Number of Items Selected until $p>0.99$ with Random Item Selection	210
Figure	12.5:	Entropy of Classification at Each Stage of the Interview	212
Figure	12.6:	Number of Respondents Correctly Classified at Each Stage	213
Figure	13.1:	Path Diagram for latent Variables in the Satisfaction Study	219
Figure	14.1:	T=2 dimensional Space for the Schiffman et al. Cola Data	233
Figure	14.2:	Ultrametric Tree for the Schiffman et al. Cola Data	235
Figure	14.3:	Mixture of Ultrametric Trees for the Schiffman et al. Cola Data	237
Figure	16.1:	Values Map: Dimension 2 vs. Dimension 1	272
Figure	16.2:	Values Map: Dimension 3 vs. Dimension 1	273
Figure	16.3:	Regional Positions: Dimension 2 vs. Dimension 1	274
Figure	16.4:	Regional Positions: Dimension 3 vs. Dimension 1	275

Preface

Market segmentation is an area in which we have been working for nearly a decade. Although the basic problem of segmentation may appear to be quite simple -the classification of customers into groups - market segmentation research may be one of the richest areas in marketing science in terms scientific advancement and development of methodology. Since the concept emerged in the late 1950s, segmentation has been one of the most researched topics in the marketing literature. Recently, much of that literature has evolved around the technology of identifying segments from marketing data through the development and application of finite mixture models. Although mixture models have been around for quite some time in the statistical and related literatures, the substantive problems of market segmentation have motivated the development of new methods, which in turn have diffused into psychometrics, sociometrics and econometrics. Moreover, because of their problem-solving potential, the new mixture methodologies have attracted great interest from applied marketing researchers and consultants. In fact, we conjecture that in terms of impact on academics and practitioners, next to conjoint analysis, mixture models will prove to be the most influential methodological development spawned by marketing problems to date.

However, at this point, further diffusion of the mixture methodology is hampered by the lack of a comprehensive introduction to the field, integrative literature reviews and computer software. This book aims at filling this gap, but we have attempted to provide a much broader perspective than merely a methodological one. In addition to presenting an up-to-date survey of segmentation methodology in parts 2 and 3, we review the foundations of the market segmentation concept in part 1, providing an integrative review of the contemporary segmentation literature as well as a historical overview of its development, reviewing commercial applications in part 4 and providing a critical perspective and agenda for future research in part 5. In the second edition, the discussion has been extended and updated.

In part 1, we provide a definition of market segmentation and describe six criteria for effective segmentation. We then classify and discuss the two dimensions of segmentation research: segmentation bases and methods. We attempt to arrive at a set of general conclusions both with respect to bases and methods. The discussion of segmentation methods sets the scene for the remainder of the book.

We start part 2 with a discussion of cluster analysis, historically the most well-known technique for market segmentation. Although we have decided to discuss cluster analysis in some detail, our main methodological focus throughout this book is on finite mixture models. Those models have the main advantage over cluster analysis that they are model-based and allow for segmentation in a framework in which customer behavior is described by an appropriate statistical model that includes a mixture component that allows for market segmentation. In addition, those mixture models have the advantage of enabling statistical inference.

Preface

We start our treatment of mixture models with classical finite mixtures, as described in the statistical literature, and applied to market segmentation a number of decades ago. Those mixtures involve a simple segment-level model of customer behavior, describing it only by segment-level means of variables. We subsequently move towards more complicated descriptions of segment level behavior, such as (generalized linear) response models that explain a dependent variable from a set of independent variables at the segment level, and scaling models that display the structure of market stimuli in a spatial manner. In an appendix to part 2 we describe some of the available computer software for fitting finite mixture models to data, and we include a demonstration of a flexible WINDOWS program, GLIMMIX, developed by I.E.C. PROGAMMA (Groningen, Netherlands).

In part 3, we also discuss the simultaneous profiling of segments and models for dynamic segmentation on the basis of several sets of variables. In the second edition the discussion on the so called concomitant variable models has been extended to show some relationships among them. Part 3 contains a descriptions of a variety of special topics in mixture models for market segmentation, i.e., joint segmentation for several sets of segmentation bases, tailored interviewing for market segmentation, mixtures of structural equation models (LISREL) and mixture models for paired comparison data, providing spatial or tree representations of consumer perceptions. Selected topics in applied market segmentation are contained in part 4, where we discuss selected topics in four application areas of market segmentation that recently have received quite some interest: geo-demographic segmentation, life-style segmentation, response based segmentation, and conjoint analysis. In the final part (5) we step back and try to provide critical perspective on mixture models, in the light of some of the criticisms it has received in the literature, recently. We finish our book with an outlook on future research.

In addition to presenting an up-to-date survey of segmentation methodology in parts 2 and 3, we review the foundations of the market segmentation concept in part 1, providing an integrative review of the contemporary segmentation literature as well as a historical overview of its development, and report academic and commercial applications in part 4. Methodologically, our main focus is on finite mixture models for segmentation problems, including specific topics such as response based segmentation, joint segmentation, dynamic segmentation and the profiling of response-based segments discussed in part 3, but we also provide an overview of the more classical clustering approach to segmentation in part 2.

When planning this monograph, we decided to target our work to two particular segments of readers in the hope that, on the one hand, it will enable academic researchers to advance further in the field and gain a better understanding of the segmentation concept and, on the other hand, it will help applied marketing researchers in making better use of available data to support managerial decisions for segmented markets. Our intention was to produce a review of classic and contemporary research on market segmentation, oriented toward the graduate student in marketing and marketing research as well as the marketing researcher and consultant. To graduate students, this monograph can serve as a reference to the vast literature on market segmentation *concepts* (part 1), *methodology* (parts 2 and 3) and *applications* (part 4).

Preface

To practitioners, it can serve as a guide for the implementation of market segmentation, from the selection of segmentation bases through the choice of data analysis methodology to the interpretation of results. In addition, we hope the book may prove to be a useful source for academics in marketing and related fields who have an interest in classification problems.

We have attempted to provide a maximum of detail using a minimum of necessary mathematical formulation, in particular emphasizing marketing applications. Our standard format for presenting methodology is to start with one or two (previously published) applications of a specific approach, to give the reader a feel of the problem and its solution before discussing the specific method in detail. The reason is that we would like a reader without any prior knowledge of the specific concepts and tools to be able to gain a fairly complete understanding from this monograph. We have tried to explain most statistical and mathematical concepts required for a proper understanding of the core material. Rather than including mathematical and statistical appendices, we have chosen to introduce those concepts loosely in the text where needed. As the text is not intended to be a statistical text, the more statistically oriented reader may find some of the explanations short of detail or too course, for which we apologize. Readers with more background in statistics and/or mixture modeling may want to skip the descriptions of the applications and the introduction of the basic statistical concepts.

Interestingly, when reviewing the literature, we observed that at several stages the further development of the segmentation concept appeared to have been hampered by the current analytical limitations. In other words, the development of segmentation theory has been partly contingent on the availability of marketing data and tools to identify segments on the basis of such data. New methodology has often opened new ways of using available data and new ways of thinking about the segmentation problems involved. We therefore do not see this book as an endpoint, but as a beginning of a new period in segmentation research. Until fairly recently, the application of the mixture technology has been monopolized by the developers of the techniques. We hope this book contributes to the further diffusion of the methodology among academics, students of marketing and practitioners, thus leading to new theoretical insights about market segmentation and questions relating to it.

We use the term "mixture models" throughout this book. A brief explanation is in order. In part of the marketing literature, the models that we are using are called "latent class models". This situation has apparently been caused by the first applications of mixture models in marketing, by Green, Carmone and Wachspress (1976), and Grover and Srinivasan (1987) being applications of a special cases of mixture models formally called "latent class models". The term mixture models in general refers to procedures for dealing with heterogeneity in the parameters of a certain model across the population by imposing a "mixing distribution" on (part of) the parameters of that model. For example, in modeling choice behavior, one may assume that the choice probabilities for the brands in question are heterogeneous across consumers and follow a certain distribution across the population. This distribution can be assumed to be either continuous or discrete, an assumption that has important consequences. In the former case, continuous mixtures arise that are at present regaining popularity in several substantive fields in marketing. If, on the other hand, a discrete mixing distribution is chosen, "finite mixture models" arise, that enable the identification of relatively homogeneous market segments. In contrast, a continuous mixing distribution implies that such groupings of consumers cannot be found in the population. We contrast the continuous and finite mixtures in the last part of this book and extend that discussion in the second edition.

In addition, not only the mixing distribution may be continuous or discrete, but also the distribution describing the variables themselves. For example, choices of brands are typically discrete 0/1 variables, but preference ratings are considered to be measured on interval scales and often described by a continuous distribution. If all measured variables are discrete, such as occurs for example in the analysis of contingency tables, the finite mixture model is called a latent class model. Those latent class models have been developed and applied in particular in psychology and sociology. In marketing, however, more often both continuous and discrete variables are measured. Therefore, we will use the term "finite mixture models", or loosely "mixture models", throughout the book, noting here that latent class models are in fact special cases of the finite mixture model approach.

Obviously, the mixture approach has not been without its critics. Whereas initially the advantages created by the methodology were emphasized, more recently critical questions have been raised and limitations of the techniques have been pointed out. In our monograph, we attempt not to be uncritical of the mixture approach. We discuss the potential limitations because we think the methodology can be properly used only if all of its advantages and disadvantages are fully disclosed. We postpone discussion of the limitations to the last part (5) of the book in order not to confuse the reader about the worth of the tools at an early stage. However, writing this book made even more clear to us the major advantages of the mixture approach and the progress that has been made by the developments in that field. We believe the current critique may lead to adaptation and further refinements of the method, but certainly not to its being discarded in the near future.

Several colleagues provided useful feedback and discussion of drafts of chapters of this book. We end with a word of thanks to them. First, we thank Julie Kaczynski for motivating us to write the book. We also benefited greatly from the very useful comments from Paul E. Green, Frenkel ter Hofstede, Wim Krijnen and Gary Lilien, and the thoughtful reviews by Ulf Böckenholt, Venkatramam Ramaswamy and Gary J. Russell. We are grateful to all those readers that provided us with constructive feedback on the first edition and we are greatly indebted to Bernard van Diepenbeek and Bertine Markvoort, who worked with great perseverance and accuracy on the layout of the second edition.

Michel Wedel   Wagner A. Kamakura

PART 1
INTRODUCTION

This introductory part of our book provides a broad review of the past literature on market segmentation, focusing on a discussion of proposed bases and methods. Chapter 1 looks at the development of market segmentation as a core concept in marketing theory and practice, starting from its foundations in the economic theory of imperfect competition. This first chapter also discusses criteria that must be satisfied for effective market segmentation. Chapter 2 classifies available segmentation bases and evaluates them according to those criteria for effective segmentation. In chapter 3, we present a classification and overview of techniques available for segmentation. Chapter 4 completes our introduction to market segmentation, with an overview of the remaining chapters of this book.

1
THE HISTORICAL DEVELOPMENT OF THE MARKET SEGMENTATION CONCEPT

After briefly introducing the concept and history of market segmentation, we review the criteria for effective segmentation and introduce the topics to be discussed in this book.

Market segmentation is an essential element of marketing in industrialized countries. Goods can no longer be produced and sold without considering customer needs and recognizing the heterogeneity of those needs. Earlier in this century, industrial development in various sectors of the economy induced strategies of mass production and marketing. Those strategies were manufacturing oriented, focusing on reduction of production costs rather than satisfaction of consumers. But as production processes became more flexible, and consumer affluence led to the diversification of demand, firms that identified the specific needs of groups of customers were able to develop the right offer for one or more sub-markets and thus obtained a competitive advantage. As market-oriented thought evolved within firms, the concept of market segmentation emerged. Since its introduction by Smith (1956), market segmentation has become a central concept in both marketing theory and practice. Smith recognized the existence of heterogeneity in the demand of goods and services, based on the economic theory of imperfect competition (Robinson 1938). He stated: "Market segmentation involves viewing a heterogeneous market as a number of smaller homogeneous markets, in response to differing preferences, attributable to the desires of consumers for more precise satisfaction of their varying wants."

Many definitions of market segmentation have been proposed since, but in our view the original definition proposed by Smith has retained its value. Smith recognizes that segments are directly derived from the heterogeneity of customer wants. He also emphasizes that market segments arise from managers' conceptualization of a structured and partitioned market, rather than the empirical partitioning of the market on the basis of collected data on consumer characteristics. Smith's concepts led to segmentation research that partitioned markets into homogeneous sub-markets in terms of customer demand (Dickson and Ginter 1987), resulting in the identification of groups of consumers that respond similarly to the marketing mix. That view of segmentation reflects a market orientation rather than a product orientation (where markets are partitioned on the bases of the products being produced, regardless of consumer needs). Such customer orientation is essential if segmentation is to be used as one of the building blocks of effective marketing planning.

The question of whether groups of customers can be identified as segments in real markets is an empirical one. However, even if a market can be partitioned into homogeneous segments, market segmentation will be useful only if the effectiveness, efficiency and manageability of marketing activity are influenced substantially by discerning separate homogeneous groups of customers.

Recent changes in the market environment present new challenges and opportunities for market segmentation. For example, new developments in information technology provide marketers with much richer information on customers' actual behavior, and with more direct access to individual customers via database marketing and geo-demographic segmentation. Consequently, marketers are now sharpening their focus on smaller segments with micro marketing and direct marketing approaches. On the other hand, the increasing globalization of most product markets (unification of the EC, WTO, and regional accords such as the MERCOSUR) is leading many multi-product manufacturers to look at global markets that cut across geographic boundaries. Those developments have lead to rethinking of the segmentation concept.

Six criteria – *identifiability, substantiality, accessibility, stability, responsiveness and actionability* – have been frequently put forward as determining the effectiveness and profitability of marketing strategies (e.g., Frank, Massy and Wind 1972; Loudon and Della Bitta 1984; Baker 1988; Kotler 1988).

Identifiability is the extent to which managers can recognize distinct groups of customers in the marketplace by using specific segmentation bases. They should be able to identify the customers in each segment on the basis of variables that are easily measured.

The *substantiality* criterion is satisfied if the targeted segments represent a large enough portion of the market to ensure the profitability of targeted marketing programs. Obviously, substantiality is closely connected to the marketing goals and cost structure of the firm in question. As modern concepts such as micro markets and mass customization become more prevalent, profitable segments become smaller because of the lower marginal marketing costs. In the limit, the criterion of substantiality may be applied to each individual customer; that is the basic philosophy of direct marketing, where the purpose is to target each individual customer who produces marginal revenues that are greater than marginal costs for the firm.

Accessibility is the degree to which managers are able to reach the targeted segments through promotional or distributional efforts. Accessibility depends largely on the availability and accuracy of secondary data on media profiles and distributional coverage according to specific variables such as gender, region, socioeconomic status and so on. Again, with the emergence and increasing sophistication of direct marketing techniques, individual customers can be targeted in many markets.

If segments respond uniquely to marketing efforts targeted at them, they satisfy the *responsiveness* criterion. Responsiveness is critical for the effectiveness of any market segmentation strategy because differentiated marketing mixes will be effective only if each segment is homogeneous and unique in its response to them. It is not sufficient for segments to respond to price changes and advertising campaigns; they should do so differently from each other, for purposes of price discrimination.

Only segments that are stable in time can provide the underlying basis for the development of a successful marketing strategy. If the segments to which a certain marketing effort is targeted change their composition or behavior during its implementation, the effort is very likely not to succeed. Therefore, *stability* is necessary, at least for a period long enough for identification of the segments, implementation of the segmented marketing strategy, and the strategy to produce results.

Segments are *actionable* if their identification provides guidance for decisions on the effective specification of marketing instruments. This criterion differs from the responsiveness criterion, which states only that segments should react uniquely. Here the focus is on whether the customers in the segment and the marketing mix necessary to satisfy their needs are consistent with the goals and core competencies of the firm.

Procedures that can be used to evaluate the attractiveness of segments to the managers of a specific firm involve such methods as standard portfolio analysis, which basically contrasts summary measures of segment attractiveness with company competitiveness for each of the segments of potential interest (McDonald and Dunbar 1995).

The development of segmented marketing strategies depends on current market structure as perceived by the firm's managers (Reynolds 1965; Kotrba 1966). The perception of market structure is formed on the basis of segmentation research (Johnson 1971). It is important to note that segments need not be physical entities that naturally occur in the marketplace, but are defined by researchers and managers to improve their ability to best serve their customers. In other words, market segmentation is a theoretical marketing concept involving artificial groupings of consumers constructed to help managers design and target their strategies. Therefore, the identification of market segments and their elements is highly dependent on the *bases* (variables or criteria) and *methods* used to define them. The selection of appropriate segmentation bases and methods is crucial with respect to the number and type of segments that are identified in segmentation research, as well as to their usefulness to the firm. The choice of a segmentation base follows directly from the purpose of the study (e.g., new product development, media selection, price setting) and the market in question (retail, business-to-business or consumer markets). The choice of different bases may lead to different segments being revealed; much the same holds also for the application of different segmentation methods. Furthermore, the choices of methods and bases are not independent. The segmentation method will need to be chosen on the basis of (1) the specific purposes of the segmentation study and (2) the properties of the segmentation bases selected.

The leading reference on market segmentation is the book by Frank, Massy and Wind (1972). They first distinguished the two major dimensions in segmentation research, bases and methods, and provided a comprehensive description of the state of the art at that time. However, the book dates back 25 years. Despite substantial developments in information technology and data analysis technology, no monograph encompassing the scientific developments in market segmentation research has appeared since its publication (we note the recent book by McDonald and Dunbar, 1995, which provides a guide to the implementation of segmentation strategy). Although Frank, Massy and Wind provide a thorough review of segmentation at that time, important developments have occurred since then in areas such as sample design for consumer and industrial segmentation studies, measurement of segmentation criteria, new tools for market segmentation and new substantive fields in which market segmentation has been or can be applied. Our book attempts to fill that gap.

Frank, Massy and Wind (1972) classified research on market segmentation into two schools that differ in their theoretical orientation. The first school has its foundation in microeconomic theory, whereas the second is grounded in the behavioral sciences. The differences between the two research traditions pertain both to the theoretical underpinnings and to the bases and methods used to identify segments (Wilkie and Cohen 1977). We use the Frank, Massy and Wind classification to discuss segmentation bases. However, new segmentation bases have been identified since, and new insights into the relative effectiveness of the different bases have led to a fairly clear picture of the adequacy of each specific base in different situations. From the published literature, it is now possible to evaluate the segmentation bases according to the six criteria discussed above generally considered to be essential for effective segmentation.

The recent research in the development of new techniques for segmentation, especially in the area of latent class regression procedures, starting with the work of DeSarbo and Cron (1988) and Kamakura and Russell (1989), suggests the onset of a major change in segmentation theory and practice. The full potential of the newly developed techniques in a large number of areas of segmentation has only begun to be exploited, and will become known with new user-friendly computer software. In 1978, Wind called for analytic methods that provide a new conceptualization of the segmentation problem. The methods that have been developed recently are believed to meet Wind's requirement, and should become a valuable adjunct to current market segmentation approaches in practice. Moreover, the emergence of those new techniques has led to a reappraisal of the current and more traditional approaches in terms of their ability to group customers and predict behavioral measures of interest.

In addition to the above developments, segmentation research has recently expanded to encompass a variety of new application areas in marketing. Examples of such new applications are segmentation of business markets, segmentation for optimizing service quality, segmentation on the basis of brand equity, retail segmentation, segmentation of the arts market, price- and promotion-sensitivity segmentation, value and lifestyle segmentation, conjoint segmentation, segmentation for new product development, global market segmentation, segmentation for customer satisfaction evaluation, micro marketing, direct marketing, geodemographic segmentation, segmentation using single-source data and so on. A search of the current marketing literature has revealed more than 1600 references to segmentation, demonstrating the persistent academic interest in the topic.

This first part of our book reviews segmentation research, focusing on a discussion of proposed bases and methods. Chapter 2 classifies available segmentation bases and evaluates them according to the six criteria for effective segmentation mentioned previously. In chapter 3, we present a classification and overview of techniques available for segmentation. Chapter 4 is an overview of the subsequent chapters.

2
SEGMENTATION BASES

This chapter classifies segmentation bases into four categories. A literature review of bases is provided and the available bases are evaluated according to the six criteria for effective segmentation described in chapter 1. The discussion provides an introduction to parts 3 and 4 of the book, where some special topics and new application areas are examined.

A segmentation basis is defined as a set of variables or characteristics used to assign potential customers to homogeneous groups. Following Frank, Massy and Wind (1972), we classify segmentation bases into *general* (independent of products, services or circumstances) and *product-specific* (related to both the customer and the product, service and/or particular circumstances) bases (Frank, Massy and Wind 1972; see also Baker 1988; Wilkie and Cohen 1977). Furthermore, we classify bases into whether they are *observable* (i.e., measured directly) or *unobservable* (i.e., inferred). That typology holds for the bases used for segmentation of both consumer and industrial markets, although the intensity with which various bases are used differs across the two types of markets. Those distinctions lead to the classification of segmentation bases first proposed by Frank, Massy and Wind (1972), shown in Figure 2.1.

Figure 2.1: Classification of Segmentation Bases

	General	Product-specific
Observable	Cultural, geographic, demographic and socio-economic variables	User status, usage frequency, store loyalty and patronage, situations
Unobservable	Psycographics, values, personality and life-style	Psychographics, benefits, perceptions, elasticities, attributes, preferences, intention

This chapter provides an overview of the bases in each of the four resulting classes (see also Frank, Massy and Wind 1972; Loudon and Della Bitta 1984; Kotler 1988). Each of the bases is discussed in terms of the six criteria for effective segmentation (i.e., identifiability, substantiality, accessibility, stability, responsiveness and actionability).

Part 1 Introduction

Observable General Bases

A number of bases widely used, especially in early applications of market segmentation research are in this category: cultural variables, geographic variables, neighborhood, geographic mobility, demographic and socio-economic variables, postal code classifications, household life cycle, household and firm size, standard industrial classifications and socioeconomic variables. Media usage also is in this class of segmentation bases. Socioeconomic status is usually derived from sets of measures involving household income and education of household members. An example of a social classification is the following definition of social classes by Monk (1978, c.f. McDonald and Dunbar 1995).

> A – Upper middle class (head of the household successful business or professional person, senior civil servant, or with considerable private means)
>
> B – Middle class (head of the household senior, well off, respectable rather than rich or luxurious)
>
> C1 – Lower middle class (small trades people and white collar workers with administrative, supervisory and clerical jobs)
>
> C2 – Skilled working class (head of the household mostly has served an apprenticeship)
>
> D – Semiskilled and unskilled working class (entirely manual workers)
>
> E – Persons at the lowest level of subsistence (old age pensioners, widows, casual workers, those dependent on social security schemes)

Geodemographic systems aim at neighborhood classifications based on geography, demography and socioeconomics. There is a wide range of commercially available geodemographic segmentation systems (examples are PRIZM™ and ACORN™). ACORN™ (A Classification Of Residential Neighborhoods) is among the most widely accepted systems, comprising the following geodemographic segments (Gunter and Furnham 1992).

- agricultural area's (villages, farms and small holdings),
- modern family housing, higher incomes (recent and modern private housing, new detached houses, young families),
- older housing of intermediate status (small town centers and flats above shops, non-farming villages, older private houses),
- poor quality older terraced housing (tenement flats lacking amenities, unimproved older houses and terraces low income, families),
- better-off council estates, (council estates well-off workers),

Chapter 2 Segmentation bases

- less well-off council estates (low raise estates in industrial towns, council houses for the elderly),
- poorest council estates (council estates with overcrowding and worst poverty, high unemployment),
- multi-racial area's (multi-let housing with Afro-Caribbean's, multi-occupied terraces poor Asians),
- high status non-family area's (multi-let big old houses and flats, flats with single people and/or few children),
- affluent suburban housing (spacious semis big gardens, villages with wealthy commuters, detached houses in exclusive suburbs),
- better off retirement area's (private houses with well of elderly or single pensioners).

Household life cycle is commonly defined on the basis of age and employment status of head(s) of the household and the age of their children, leading to nine classes (cf. McDonald and Dunbar 1995).

- bachelor (young, single, not living at home)
- newly married couples (young, no children)
- full nest I (youngest child under six)
- full nest II (youngest child six or over six)
- full nest III (older married couples with dependent children)
- empty nest I (older married couples, no children living with them, head in labor force)
- empty nest II (older married couples, no children living with them, head retired)
- solitary survivor in labor force
- solitary survivor retired

In different countries, slightly different standard industrial classification schemes are used, but in the United States the following main classification applies (each code is further partitioned into smaller classes).

- 000 – agriculture, forestry and fishing
- 100 – mining
- 150 – construction
- 200 – manufacturing
- 400 – transportation and communication
- 500 – wholesale
- 520 – retail
- 600 – finance, insurance and real estate
- 700 – services
- 900 – public administration

Part 1 Introduction

These segmentation bases are relatively easy to collect, reliable and generally stable. Often the variables are available from sampling frames such as municipal and business registers, so they can be used as stratification variables in stratified and quota samples. Segments derived from the bases are easy to communicate and resulting strategies are easy to implement. The corresponding consumer segments are often readily accessible because of the wide availability of media profiles for most of the bases mentioned. Some differences in purchase behavior and elasticities of marketing variables have been found among these types of segments, supporting the responsiveness criterion for effective segmentation, but in many studies the differences were in general too small to be relevant for practical purposes (e.g., Frank, Massy and Wind 1972; Frank 1968, 1972; McCann 1974). Although the lack of significant findings in those studies supports the conclusion that the general observable bases are not particularly effective, it has not resulted in the demise of such bases. On the contrary, they continue to play a role in simple segmentation studies, as well as in more complex segmentation approaches in which a broad range of segmentation bases are used. Moreover, they are used to enhance the accessibility of segments derived by other bases, as media usage profiles are often available according to observable bases.

Observable Product-Specific Base

The bases in this class comprise variables related to buying and consumption behavior: user status (Boyd and Massy 1972; Frank, Massy and Wind 1972), usage frequency (Twedt 1967), brand loyalty (Boyd and Massy 1972), store loyalty (Frank, Massy and Wind 1972; Loudon and Della Bitta 1984), store patronage (Frank, Massy and Wind 1972), stage of adoption (Frank, Massy and Wind 1972 and usage situation (Belk 1975; Dickson 1982; Loudon and Della Bitta 1984). Although most of these variables have been used for consumer markets, they may also apply to business markets. Many such bases are derived from customer surveys (e.g., Simmons Media Markets), although nowadays household and store scanner panels and direct mail lists are a particularly useful sources.

Twedt's (1967) classification of usage frequency involves two classes: heavy users and light users. Several operationalizations of loyalty have been proposed. Brand loyalty may or may not be directly observable; for example, it has been measured through observable variables such as the last brand purchased (last-purchase loyal) or as an exponentially weighted average of past purchase history (Guadagni and Little 1983). New operationalizations of brand loyalty are based on event history analysis. With respect to stage of adoption, commonly the following stages are distinguished (Rogers 1962).

- innovators (are venturesome and willing to try new ideas at some risk)
- early adopters (are opinion leaders and adopt new ideas early but carefully)
- early majority (are deliberate, adopt before the average person but are rarely leaders)
- late majority (are skeptical, adopt only after the majority)

- laggards (are tradition bound and suspicious of changes, adopt because the adoption is rooted in tradition)

Accessibility of the segments identified from these bases appears to be somewhat limited in view of the weak associations with general consumer descriptors (Frank 1972; Frank, Massy and Wind 1972). Several researchers (Massy and Frank 1965; Frank 1967, 1972; Sexton 1974; McCann 1974) found that consumers who belong to segments with different degrees of brand loyalty do not respond differently to marketing variables. Loyalty was found to be a stable concept. Weak to moderate differences in elasticities of marketing mix variables were found among segments identified by product-specific bases such as usage frequency, store loyalty, etc., indicating that those bases are identifiable and substantial. To a lesser extent they are also stable, accessible and responsive bases for segmentation.

Dickson (1982) provided a general theoretical framework for usage situation as a segmentation basis. When demand is substantially heterogeneous in different situations, situational segmentation is theoretically viable. A classification of situational variables (including purchase and usage situations) has been provided by Belk (1975). He distinguishes the following characteristics of the situation.

- physical surrounding (place of choice decision, place of consumption)
- social surrounding (other persons present)
- temporal perspective (time of day or week)
- task definition (buy or use)
- antecedent states (moods, etc.)

That classification provides a useful framework for the operationalization of usage situational bases. Situation-based variables are often directly measurable, and the segments are stable and accessible (Stout et al. 1977; Dickson 1982). The responsiveness of usage situational segments was investigated by Belk (1975), Hustad, Mayer and Whipple (1975), Stout et al. (1977) and Miller and Ginter (1979). Perceptions of product attributes, their rated importance, buying intentions, purchase frequency and purchase volume were all found to differ significantly across usage situations (Srivastava, Alpert and Shocker 1984). Consequently, the explicit consideration of situational contexts appears to be an effective approach to segmentation.

Unobservable General Bases

The segmentation variables within this class fit into three groups: personality traits, personal values and lifestyle. These bases are used almost exclusively for consumer markets. A comprehensive introduction to this class of segmentation bases is provided by Gunter and Furnham (1992). Such psychographic segmentation bases were developed extensively by marketers in the 1960s in response to the need for a more lifelike picture of consumers and a better understanding of their motivations. They have spawned from the fields of personality and motivation research, which have blended in later years into the psychographics and lifestyle areas. Early landmark papers on psychographics are those by Lazarfeld (1935) and Dichter (1958).

Edward's personal preference schedule is probably the most frequently used scale for measuring general aspects of personality in marketing. Early applications include those of Evans (1959) and Koponen (1960); later studies include those by Massy, Frank and Lodahl (1968), Claycamp (1965) and Brody and Cunningham (1968). Other personality traits used for segmentation include dogmatism, consumerism, locus of control, religion and cognitive style (cf. Gunter and Furnham 1992).

Values and value systems have been used as a basis for market segmentation by Kahle, Beatty and Holmer (1986), Novak and MacEvoy (1990) and Kamakura and Mazzon (1991). The most important instrument for the measurement of human values and identification of value systems is the Rokeach value survey (Rokeach 1973). Rokeach postulates that values represent beliefs that certain goals in life (i.e., terminal values) and modes of conduct (instrumental values) are preferable to others. Those values are prioritized into values systems and used by individuals as guidance when making decisions in life.

Rokeach's instrument includes 18 terminal and 18 instrumental values, to be ranked separately by respondents in order of importance. A shorter and more easily implemented instrument is the list of values (LOV), suggested by Kahle (1983) on the basis of Maslow's (1954) hierarchy. The LOV was especially designed for consumer research. It consists of the following nine items assessing terminal values that are rated on nine-point important/unimportant scales.

- sense of belonging
- excitement
- warm relationship with others
- self-fulfillment
- being well-respected
- fun and enjoyment of life
- security
- self-respect
- sense of accomplishment

Another important scale for assessing value systems was developed by Schwartz and Bilsky (1990) and later modified by Schwartz (1992). It includes 56 values, representing the following 11 motivational types of values.

- self-direction - independent thought and action
- stimulation - need for variety and stimulation
- hedonism - organismic and pleasure-oriented needs
- achievement - personal success
- power - need for dominance and control
- security - safety, stability of society, relationships and self
- conformity - self-restraint
- tradition - respect, commitment and acceptance of customs

- spirituality - inner harmony, meaning in life
- benevolence - prosocial behaviors and goals
- universalism - concern for the welfare of all others and nature

The concept of lifestyle was introduced into marketing research by Lazer (1963) and is based on three components: *activities* (work, hobbies, social events, vacation, entertainment, clubs, community, shopping, sports), *interests* (family, home, job, community, recreation, fashion, food, media, achievements) and *opinions* (of oneself, social issues, politics, business, economics, education, products, future, culture) (Plummer 1974).

Mitchell (1983) proposed a values and lifestyle typology, which is widely known as the VALS™ system. The VALS™ typology is hierarchical and defines a basic system of four categories with nine more detailed segments within them. Drawn originally from Maslow, the following nine segments were in the original VALS™ double hierarchy.

- need driven (survivors, sustainers)
- outer directed (belongers, emulators, achievers)
- inner directed (I-am-me, experientials, socially conscious)
- integrated

A new version of the VALS™ system has been developed recently. The new structure (VALS2™) is not hierarchical like the previous one, but defined by two main dimensions: *resources* defined by income, education, self-confidence, health, eagerness to buy, intelligence, etc. and *self-orientation* (principle-oriented, self-oriented and status-oriented).

Surveys of the literature reveal that general personality measures have a stronger relationship with patterns of behavior across product classes than with behaviors related to a single brand (Frank 1972; Frank, Massy and Wind 1972; Wells 1975; Wilkie and Cohen 1977). General lifestyle segmentation has the same limitations as general personality scales: it is likely to identify factors that influence general behavioral patterns rather than any specific behavior toward a single brand (Ziff 1971; Wells 1975; Dickson 1982). In reviews of the literature, Frank, Massy and Wind (1972) and Wells (1975) concluded that the predictive validity of lifestyle with respect to purchase behavior can be substantially better than that of general observable segmentation bases.

As for the validity of the psychographic segmentation bases, a study by Kahle, Beatty and Holmer (1986) showed that the LOV system of Kahle (1983) better predicted consumer behavior than Mitchell's VALS™ system. However, a subsequent study by Novak and MacEvoy (1990), based on a larger and more representative sample, found that demographic variables had greater explanatory power than the LOV variables. Studies by Lastovicka (1982) and Lastovicka, Murray and Joachimsthaler (1990) also revealed negative evidence for the predictive validity of the original VALS™ system, but demonstrated the potential validity of lifestyle traits in general. Earlier studies (Tigert 1969; Bruno and Pessemeier 1972) showed adequate test-retest reliability for psychographic dimensions.

Personality, values and lifestyle provide a richer perspective of the market based on a more lifelike portrait of the consumer and therefore provide actionable bases that are especially useful for the development of advertising copy. Accessibility, responsiveness and stability are not unequivocally supported in the literature.

Unobservable Product-Specific Bases

This class of bases comprises product-specific psychographics, product-benefit perceptions and importances, brand attitudes, preferences and behavioral intentions. In that order, those variables form a hierarchy of effects, as each variable is influenced by those preceding it (Wilkie and Cohen 1977). Although many of the variables are used more often for consumer markets, most can and have been used for segmenting business markets.

Dhalla and Mahatoo (1976) identified three key areas of product-specific psychographics: value orientations, role perceptions and buying style. Psychographic measures that assess personality traits and lifestyle more immediately related to the choice behavior for the product show a much stronger relationship with that choice behavior than do general psychographic measures (Wells 1975). Unfortunately, little is known of the stability of the segments, and few studies have investigated the reliability and validity of the measures (Lastovicka, Murray and Joachimsthaler 1990).

Consumers' perceptions of brand attributes have been used as a basis for market segmentation since Yankelovich (1964). Segments identified from perceived product attributes are identifiable and substantial (Frank, Massy and Wind 1972). In general, however, purchase behavior toward a product will depend also on the importance consumers attach to each of those attributes. (Fishbein and Ajzen 1975). Perceptual segments are actionable in that they provide information on how to communicate to them. Little research has been done on the stability of perceptual segments, but cognitive states are found to be much less stable than values (Wilkie and Cohen 1977). Consequently, perceptual dimensions have low probity as segmentation bases from both theoretical and strategic points of view (Dhalla and Mahatoo 1976; Wilkie and Cohen 1977; Howard 1985).

Elasticities are often seen as normative ideal bases for segmentation (Massy and Frank 1965; Sexton 1974; Dhalla and Mahatoo 1976). Elasticity is defined here as the relative change in demand in response to a relative unit change in a marketing instrument (mostly price). Claycamp and Massy (1968), however, suggested the individual responses to marketing variables (i.e., change in demand in response to a unit change in the marketing variable) as the normative ideal. Elasticities cannot be observed directly, but must be estimated. Most often, linear or logistic regressions are used for that purpose. A problem in the derivation of individual-level elasticities (required for segmentation purposes) is that the estimates may be unstable, because of the often limited amount of data available at the individual level (cf. Wedel and Kistemaker 1989).

Tollefson and Lessig (1978), Elrod and Winer (1982) and Blozan and Prabhaker (1984) examined the performance of alternative ideal bases with respect to profit maximization, both theoretically and empirically. Unfortunately, the evidence revealed by their research is not conclusive in the sense that arguments supporting the use of one specific measure of demand as a normative ideal basis for the aggregation of consumers into segments have not yet emerged. However, elasticities have over the years become the most commonly cited normative ideal bases for segmentation. Most studies evaluating the effectiveness of other segmentation bases have used elasticities as a benchmark. Note that elasticities capture one of the criteria for effective market segmentation, the *responsiveness* criterion. Market segments that differ in elasticities differ in their response to marketing effort. In other words, the normative ideal has focused on the responsiveness criterion for segmentation, largely neglecting other criteria for effective segmentation.

The concept of benefit segmentation was introduced to the marketing literature by Haley (1968), and has since been applied in both consumer and business markets. Haley argued that the benefits people seek in products are the basic reasons for the heterogeneity in their choice behavior, and benefits are thus the most relevant bases for segmentation. Previously, Yankelovich (1964) and Beldo (1966) had mentioned the identification of differences in consumer benefits as the crucial issue in segmentation research. Wind (1978) listed benefits as a preferred segmentation basis for general understanding of a market and for making decisions about positioning, new product concepts, advertising and distribution because of their actionability. Whereas most authors have used importance weights (derived either directly or indirectly, using preference models) as a basis in benefit segmentation (e.g., Hauser and Urban 1977; Moriarty and Venkatesan 1978; Currim 1981), a few (Frank, Massy and Wind 1972; Dickson and Ginter 1987) suggest the use of ideal points obtained directly or inferred from multidimensional scaling. Benefit segments are potentially identifiable and substantial as demonstrated in several studies (Beldo 1966; Haley 1968, 1984; Myers 1976; Calantone and Sawyer 1978). Moderate differences among benefit segments in brand purchase behavior and strong differences in attitudes and buying intentions have also been demonstrated (Wilkie 1970). In their study on the stability of benefit segments, Calantone and Sawyer (1978) demonstrated that benefit segments are consistent across samples within a given time period (of two years). The segments' benefit importances were stable in time, but segment size, demographics and segment memberships were not. An effective method for assessing benefits in segmentation studies is conjoint analysis (Green and Srinivasan 1978; Cattin and Wittink 1982; Green and Krieger 1991), to be reviewed in a subsequent chapter.

From a theoretical point of view, buying intentions are the strongest correlates of buying behavior discussed so far (Fishbein and Ajzen 1975). Preferences were used as a basis for segmentation by Ginter and Pessemeier (1978) and Sewall (1978), among others. Frank, Massy and Wind (1972) report that preferences and intentions may result in identifiable, substantial segments. Ginter and Pessemeier (1978) related brand preference segments to attribute perceptions and AIO measures to make them accessible and actionable. Responsiveness of the segments is indicated by evidence of the intention-behavior relationship. Little is known about the stability of segments based on preferences or intentions but, from a theoretical point of view, they are expected to be more stable than segments based on purchase behavior itself, in that they are less influenced by the scrambling effects of the purchase environment (Wilkie and Cohen 1977). In comparison with psychographics and benefits, preferences and intentions are less appealing in terms of actionability.

Conclusion

Table 2.1 summarizes our discussion on the various segmentation bases according to the criteria for effective segmentation reviewed previously. Note, however, that virtually all of the evidence on the effectiveness of alternative bases is derived from applications to consumer markets, and that a certain segmentation basis may be preferred depending on the specific requirements of the study at hand. In general, the most effective bases are found in the class of product-specific unobservable bases. Moreover, in modern market segmentation studies, a variety of bases may be combined, each according to its own strengths, as depicted in Table 2.1.

Table 2.1: Evaluation of Segmentation Bases

Bases/ Criteria	Identif- iability	Sustan -tiality	Acces- sibility	Stabi- lity	Actio- nabi- lity	Res- ponsiv- eness
1. General, observable	++	++	++	++	--	--
2. Specific, observable						
Purchase	+	++	--	+	--	+
Usage	+	++	+	+	--	+
3. General, unobservable						
Personality	±	--	±	±	--	--
Life style	±	--	±	±	--	--
Psychographics		--	±	±	--	--
4. Specific, unobservable						
Psychographics	±	+	--	--	++	±
Perceptions	±	+	--	--	+	--
Benefits	+	+	--	+	++	++
Intentions	+	+	--	±	--	++

++ very good, + good, ± moderate, - poor, -- very poor

3
SEGMENTATION METHODS

This chapter classifies current segmentation methods and techniques into four categories. It provides an overview and evaluation of the methods in each of the four categories. Also, it provides an introduction to part 2 of this book, which describes cluster analysis, mixture, mixture regression and mixture scaling methods, currently the most powerful approaches to market segmentation.

Segmentation is essentially a grouping task, for which a large variety of methods are available and have been used. The methods employed in segmentation research can be classified in two ways. First, they can be classified into *a-priori* and *post-hoc* approaches (Green 1977; Wind 1978). A segmentation approach is called *a priori* when the type and number of segments are determined in advance by the researcher and *post hoc* when the type and number of segments are determined on the basis of the results of data analyses.

The second way of classifying segmentation approaches is according to whether *descriptive* or *predictive* statistical methods are used. *Descriptive* methods analyze the associations across a single set of segmentation bases, with no distinction between dependent or independent variables *Predictive* methods analyze the association between two sets of variables, where one set consists of dependent variables to be explained/predicted by the set of independent variables.

A classification of the methods that have been used for segmentation according to those two criteria results in the four categories listed in Figure 3.1. Hybrid forms of segmentation have also been applied, combining both *a-priori* and *post-hoc* types of procedures. They are not included separately in Figure 3.1, but are included in our discussion.

Figure 3.1: Classification of Methods Used for Segmentation

	A priori	Post hoc
Descriptive	Contingency tables, Log-linear models	Clustering methods: Nonoverlapping, overlapping, Fuzzy techniques, ANN, mixture models
Predictive	Cross-tabulation, Regression, logit and Discriminant analysis	AID, CART, Clusterwise regression, ANN, mixture models

In this chapter we briefly describe segmentation methods in each of the four classes. Our purpose is to provide an overview, rather than the details of each method. In part 2 of this book we explain in more detail the most important segmentation methods:

Part 1 Introduction

cluster analysis, mixture regression and scaling. Several methods, such as Q-type factor analysis (Darden and Perreault 1977), have been used in connection with segmentation, but were not designed for and are not specifically suited to identifying or testing groupings of consumers. As we do not consider those methods appropriate for segmentation purposes, they are not discussed here.

A-Priori Descriptive Methods

In a-priori descriptive segmentation, the type and number of segments are determined before data collection. For example, a manager may decide to segment the market for her fast-food chain by usage situation, into the breakfast, lunch and dinner sub-markets. Often, multiple segmentation bases are used to form the segments, and the segments obtained from each of those criteria are assessed by looking at the associations between groupings arising from the alternative bases. In the example, the manager might in addition segment the market by usage situation and location (urban vs. suburban).

Cross-tabulation appears to have been a popular technique for the evaluation of bases in the earlier years of segmentation research. A problem in using contingency tables to measure the association among multiple segmentation bases is that higher order interactions are difficult to detect and interpret in the tables. Green, Carmone and Wachspress (1976) suggested the use of log-linear models for that purpose. The main objective of cross-tabulation and log-linear analysis in such cases is to test segments arising from alternative bases, and to predict one segmentation base from other bases. The methods in this class are suited to quickly obtaining insights about segments and about the associations among segmentation bases, for example, to compare heavy and regular users of a brand by lifestyle (e.g., VALS^{TM}) segments. Although they are not very effective, they continue to be used, especially in hybrid segmentation procedures that combine a-priori and post-hoc methods. Often it is desirable to obtain segments for two separate strata in a population defined a-priori, such as business and private users, users and nonusers, national versus international customers, new customers versus current customers or customer by brand used. In this case a two-step approach is taken. First, a sample is partitioned a priori on the basis of the variable(s) in question. Second, within each of the strata that arise, a post-hoc, mostly clustering-based procedure is used. The hybrid procedure can be seen as combining the strengths of the a-priori and post-hoc (to be discussed below) approaches, but generally its effectiveness depends on the post-hoc procedure used in the second step. The a-priori segmentation in the first step greatly enhances the usefulness of the outcomes for management. For example, Green (1977) proposed such a hybrid approach to segmentation. Shapiro and Bonoma (1984) proposed a hybrid or nested procedure to segment industrial markets. It has been applied by Maier and Saunders (1990), among others.

Chapter 3 Segmentation methods

Post-Hoc Descriptive Methods

In the post-hoc descriptive approach, segments are identified by forming groups of consumers that are homogeneous along a set of measured characteristics. In lifestyle segmentation, for example, consumers are first measured along several demographic and psychographic characteristics; a clustering procedure is then applied to the data, to identify groups of consumers that are similar in terms of their values, activities, interests and opinions. The number of segments and characteristics of each segment are determined by the data and methodology used. Clustering methods are the most popular tools for post-hoc descriptive segmentation. Applications of cluster analysis in marketing research were first reviewed by Punj and Stewart (1983); a more recent comprehensive review is provided by Arabie and Hubert (1994). Next we briefly overview the clustering methods to be discussed in more detail in part 2 of this book.

A major distinction between clustering methods is in the nature of the clusters formed: *nonoverlapping*, *overlapping* or *fuzzy* (Hruschka 1986). In *nonoverlapping* clustering each subject belongs to a single segment only. For example, a consumer belongs to segment A (1) but not to segment B (0) or C (0). In *overlapping* clustering a subject may belong to multiple segments; for example, a subject may belong to segments A (1) and B (1), but not to segment C (0). In *fuzzy* clustering the hard membership or nonmembership of a subject in one (nonoverlapping) or multiple (overlapping) clusters is replaced by the degree of membership in each segment. For example, a subject may belong partly to segment A (0.6), segment B (0.3) and segment C (0.1). Overlapping and fuzzy clustering approaches are consistent with the fact that consumers may belong to more than one segment, possibly in relation to different buying and consumption situations (Arabie 1977; Shepard and Arabie 1979).

Nonoverlapping clustering methods have been the most used in segmentation research. Two major types of non-overlapping cluster techniques can be distinguished: the *hierarchical* (Frank and Green 1968) and the *nonhierarchical* methods. *Hierarchical* clustering methods start with single-subject clusters, and link clusters in successive stages. Two consumers who are placed in the same group at an early stage of the process will remain in the same segment up to the final clustering solution.
Several different hierarchical methods can be distinguished, such as single linkage, complete linkage and minimum variance linkage (or Ward's method). *Nonhierarchical* methods start from a (random) initial division of the subjects into a predetermined number of clusters, and reassign subjects to clusters until a certain criterion is optimized. Two consumers who are placed in the same group at an early stage may end up in different segments. A large number of non-hierarchical methods are available; K-means is the best known and most widely used of those procedures (see chapter 3).

Based on a review of the literature, Punj and Stewart's (1983) main conclusion about the various cluster analysis techniques was that nonhierarchical methods are superior to hierarchical methods. They are more robust to outliers and the presence of irrelevant attributes. A general problem of the nonhierarchical methods is the determination of the number of clusters present in the data (Milligan and Cooper 1985). Among the nonhierarchical methods, minimum variance linkage is superior.

Punj and Stewart (1983) also review applications of clustering methods to segmentation problems in marketing. Hierarchical methods have been used by Greeno, Summers and Kernan (1973, cf. Punj and Stewart 1983) to identify segments on the basis of personality characteristics and behavioral patterns, and by Kiel and Layton (1981, cf. Punj and Stewart 1983) to identify segments of consumers with similar information-seeking behavior in buying new cars. Other and more recent applications of hierarchical methods are abundant. Nonhierarchical methods have been used by Moriarty and Venkatesan (1978, cf. Punj and Stewart 1983) in a benefit segmentation study of buying behavior toward marketing information systems.

A major conceptual problem of the application of hierarchical clustering methods to market segmentation problems is that there are hardly any theoretical arguments to justify the hierarchical relational structure among consumers or firms. On the other hand, such hierarchical relations may be assumed present among brands in market structure studies (cf. DeSarbo, Manrai and Manrai 1994). We therefore conjecture that the benefits of hierarchical methods primarily arise in market structure analysis, and much less in market segmentation research.

Several extensions have been proposed for both the hierarchical and nonhierarchical clustering methods. De Soete, DeSarbo and Carroll (1985) proposed a hierarchical method that clusters objects and simultaneously derives weights for the variables. The weights indicate the importance of the variables in clustering. DeSarbo, Carroll and Clark (1984) extended the K-means algorithm in a similar way. The method also allows for the analysis of several groups of a-priori weighted variables. The usefulness of the algorithms to market segmentation lies in exploring the robustness of cluster structure with respect to alternative sets of variables (such as psychographics and demographics). Surprisingly, the methods have not yet been used in segmentation research.

The overlapping clustering approach was first suggested by Shepard and Arabie (1979). Several generalizations have been proposed (INDCLUS by Carroll and Arabie 1983; GENNCLUS by DeSarbo 1982; CONCLUS by DeSarbo and Mahajan 1984). Those models provide useful approaches to overlapping clustering on the basis of pairwise similarities. Therefore, for segmentation purposes they must be applied to (derived) similarities between all pairs of consumers in a sample. Rather than deriving such similarity measurements from the raw data and applying an overlapping clustering procedure, one may prefer to cluster the raw data directly (using, e.g., K-means). The computational requirements for some of the iterative algorithms may be excessive for real-life segmentation problems with large sample sizes (Arabie et al. 1981), which may also limit their applicability to segmentation problems. We argue that the potential of those methods in market segmentation research is therefore limited.

Two types of fuzzy clustering methods can be distinguished. One is based on the theory of fuzzy sets (Zadeh 1965), which assigns a *degree* of membership for objects to a class (Hruschka 1986). The concepts of fuzzy sets and partial set membership were introduced by Zadeh (1965) as a way of handling imprecision in mathematical modeling. Dunn (1974) and Bezdek (1974) recognized the applicability of Zadeh's concepts to clustering problems and proposed the fuzzy c-means (FCM) algorithm. The FCM algorithm estimates the cluster centroids of a pre-specified number of clusters, and the degree of membership for objects to those clusters. A generalization of the FCM algorithm was developed by Bezdek et al. (1981a,b), called fuzzy c-varieties

(FCV), which identifies not only round clusters, but also clusters with a linear configuration. Hruschka (1986) empirically compared a number of those methods with a nonoverlapping and an overlapping (ADCLUS) procedure, and demonstrated the superiority of the fuzzy methods. Hruschka also provided applications of fuzzy clustering methods to market segmentation. He identified fuzzy segments on the basis of consumer judgments of the appropriateness of brands to different usage situations.

The second type of clustering algorithm which, yields a seemingly "fuzzy" assignment of subjects to segments, is based on the assumption that the data arise from a mixture of (e.g., normal or multinomial) distributions (McLachlan and Basford 1988), and estimates the *probability* that objects belong to each class. Both the fuzzy and the mixture approaches provide membership values that are between zero and one. However, the substantive interpretation of the numbers is quite different for the two methods. In the fuzzy set procedures they provide a consumer's partial membership values for the segments, so that s\he actually belongs to more than one segment. In the case of a mixture model clustering, the underlying assumption is that the consumer belongs only to one segment, but the information in the data for that subject is insufficient to determine uniquely his\her actual segment, and probabilities of segment membership are estimated. A theoretical advantage of the latter assumption is that the membership values are not actual parameters in the mixture model, which is helpful from both computational and statistical points of view.

The mixture methodology is the core of the segmentation methodologies described in chapters 2 and 3 of this book. Latent class analysis (LCA) is a generic name given to a class of methods for the analysis of multiway contingency tables (Goodman 1974; Vermunt 1993). It attempts to explain the observed associations between the factors that make up the table by introducing unobservable underlying classes (clusters). Green, Carmone and Wachspress (1976) first suggested the application of latent class analysis to market segmentation. Grover and Srinivasan (1987) applied the latent class approach for simultaneous segmentation and market structuring. Their application was based on the brand-by-brand cross-classification table of two purchase occasions, and was also generalized to a model that accounts for trends in the segments' market shares over time (Grover and Srinivasan 1989). As those approaches provide insight to the patterns of competition of brands within post-hoc-defined segments, considerable diagnostic advantages are obtained. In other applications of latent class models in marketing, Lehman, Moore and Elrod (1982) identified segments on the basis of the information acquisition process of consumers and used a mixture model to identify two segments of consumers with limited problem solving and routinized response behavior, respectively; and Kamakura and Mazzon (1991) developed a latent class model for identifying of consumer segments with different value systems on the basis of rankings of the values of Kahle's (1983) LOV instrument. The usefulness of that procedure was demonstrated by Kamakura and Novak (1992). Those and other latent class approaches are described in more detail in subsequent chapters of this book.

The post-hoc descriptive procedures are powerful and frequently used tools for market segmentation. Especially useful are the overlapping and fuzzy methods, which have conceptual advantages over the nonoverlapping hierarchical approaches. Unfortunately, computer programs are not yet widely available for the most recent techniques, and the methods do not allow for predictions of behavior, attitudes or preferences from independent measures such as brand attributes or marketing mix variables.

A-Priori Predictive Methods

A-priori *predictive* approaches require the definition of a-priori descriptive segments based on one set of criteria, and the subsequent use of predictive models to describe the relation between segment membership and a set of independent variables. Two types of approaches can be distinguished (Wilkie and Cohen 1977). In *forward* approaches, background characteristics such as sociodemographics and psychographics (see observable and unobservable general segmentation bases in chapter 2) are first used to form a-priori segments that are then related to product-specific measures of purchase behavior. For example, the developer of a new electronic magazine would first identify demographic segments among Internet users, and then test whether those segments can predict readership of the various sections of the magazine. In *backward* approaches the segments are first defined on the basis of product-specific purchase-related variables (such as brand loyalty). The profiles of those segments are then described along a set of general consumer characteristics. For example, the same electronic magazine developer would identify heavy users of electronic magazines, and then verify whether demographic characteristics can discriminate between heavy and regular/light users.

Tabulation appears to be the most popular method in forward segmentation approaches, where the average values of an at least interval scaled dependent variable are presented according to one or more predictor or segmentation variables. Bass, Tigert and Lonsdale (1968) strongly advocated the use of that method. The advantages are that nonlinear as well as interaction effects can be estimated. A difficulty, however, is the extension of the technique beyond two variables. The difficulty is overcome by linear regression, which can estimate the effect of multiple segmentation variables, plus their partial contributions (Wildt and McCann 1980). However, Bass, Tigert and Lonsdale (1968) opposed the use of regression analysis in segmentation. They argued that regression has the individual and not the group (segment) as the unit of analysis. Initially, that standpoint was confirmed by Morrison's (1973) analyses, but Wildt (1976) and Beckwith and Sasieni (1976) produced evidence that supported the use of regression. Umesh (1987), for example, developed a regression model to explain consumers' preferences for transportation to shopping centers. His model was used to estimate the relation between consumer preference and predictor variables such as waiting time, traveling time and cost. The regression coefficients represented the respective importance of the variables in explaining consumer preference. The regression coefficients were in turn hypothesized to depend on factors such as geographic location according to a second regression model explaining differences in attribute importance levels across locations.

A more elaborate forward approach has been proposed and tested by Currim (1981) combined cluster analysis for post-hoc descriptive segmentation with the subsequent estimation of choice models at the segment-level. Currim tested two segmentation bases for post-hoc descriptive segmentation: benefits (measured through individual-level conjoint analysis of urban transportation modes) and usage situation (purpose of the trip). Once the post-hoc segments were identified from each basis, Currim estimated a multinomial logit model within each segment, explaining choices among transportation alternatives as a function of their characteristics. Currim found that the segment approach provided better fit and predictions than the aggregate one. His estimation results show no clear advantage of one segmentation basis (benefit or usage situation) over another. However, he obtained better predictive performance for the segmentation based on usage situation.

A common method applied in backward approaches for market segmentation is discriminant analysis. Th method is useful for the description of segments rather than identifying segments in a market (see, e.g., Frank, Massy and Wind 1972). Multinomial logit models have also been used to relate segmentation bases to purchase behavior (Rao and Winter 1978; Gensch 1985). With the growing availability of scanner panel data, those methods have become more important in forward approaches to market segmentation. The previous limitation of multinomial probit models, that they could not be feasibly computed for more than three or four alternatives, has now been solved by the application of Monté Carlo integration methods, which enhances their applicability in marketing research (cf. Chintagunta 1993; Elrod and Keane 1995).

In summary, a-priori predictive methods are implemented in two stages. First, a-priori segments are formed by using one set of segmentation bases, and then the identified segments are described by using a set of independent variables. The major disadvantage of the methods is that they are based on often relatively ineffective a-priori segmentation bases in the first stage of the process.

Post-Hoc Predictive Methods

Post-hoc predictive methods identify consumer segments on the basis of the estimated relationship between a dependent variable and a set of predictors. The segments formed by post-hoc predictive methods are homogeneous in the relationship between dependent and independent variables. For example, a producer of dairy food products may want to segment consumers in the European Community on the basis of the importance they attach to various characteristics of yogurt, such as fat content, packaging and price. The traditional method for predictive clustering is automatic interaction detection, AID. That method identifies interactive effects of categorical segmentation bases on a dependent variable, such as a measure of purchase behavior (Assael 1970; Assael and Roscoe 1976). It splits a sample into groups that differ maximally according to a dependent variable, such as purchase behavior, on the basis of a set of independent variables, often socioeconomic and demographic characteristics. AID was generalized to cope with multiple dependent variables (MAID, MacLachlan and Johansson 1981) and with categorical dependent variables (CHAID, Kass 1980).

Martin and Wright (1974) developed an AID-like algorithm, SIMS, to obtain profit-maximizing segments (a related method, classification and regression trees or CART,

Part 1 Introduction

Breiman et al. 1984, is nonparametric.). Doyle and Fenwick (1975) showed, however, that unless the sample size is very large, trees are often unstable so that cross-validation results are not satisfactory. Doyle and Hutchinson (1976) provided empirical evidence that for market segmentation, clustering methods are preferable to AID. At the end of the 1970s AID fell into disrepute because it ignores chance capitalization and thus produces low cross-validity in holdout predictions. CHIAID (cf. Magdison 1994) utilizes a Bonferroni adjustment to significance tests, whereby the problem of chance capitalization is largely eliminated. Magdison (1994) extended the CHIAID algorithm to handle a categorical dependent variable scaled by quantitative scores. Magdison used CHIAID in the context of the selection of addresses in direct mail and provided calculations for the profitability of the derived segments.

A particularly interesting technique that is increasingly applied to market segmentation problems is artificial neural network (ANN) analysis. The rationale behind neural nets is that they attempt to mimic the function of the mammal brain. Artificial neurons receive input that is transferred to output through some flexible nonlinear functions. The artificial neurons are organized in layers of three types: the *input layer*, the *output layer* and the *hidden layer* between the other two. The input layer in segmentation studies inputs the segmentation bases, which are used to predict some dependent variable of interest, or segment membership, in the output layer. The analyst must decide on the number of neurons in each hidden layer and the number of such layers. ANN provides particularly powerful predictive algorithms, especially in situations where underlying relations are nonlinear. However, only heuristic procedures are available to decide on the number of layers and nodes within layers, and the parameter estimates lack substantive interpretation in most cases and do not have statistical properties. Moreover studies have demonstrated that ANN may be outperformed by statistical methods such as partial least squares or K-means cluster analysis (Balakrishnan et al. 1995). Recent developments in the ANN literature partly address those problems by estimating the neural networks in a maximum likelihood framework, which enables the analyst to draw statistical inferences on the parameter estimates.

Conjoint analysis (discussed in more detail in chapter 3) constitutes an important area for post-hoc predictive segmentation research. That methodology is particularly useful for post-hoc predictive segmentation because its more recent developments allow for the grouping of consumers according to how they respond to product features in making choice decisions. Componential segmentation (Green 1977; Green and DeSarbo 1979) involves an extension of conjoint analysis in which both product profiles and respondent profiles are considered. The respondent profiles are generated on the basis of predefined (often demographic) consumer descriptors. The hypothetical product profiles are presented to respondents, selected from an available sampling frame related to the specifications of the respondent profiles. The componential segmentation model makes possible the prediction of new product preferences for every segment defined as a combination of the segmentation variables. However, the strength of the model depends on the particular consumer descriptors (typically observable segmentation bases) used.

A two-stage procedure is often used for benefit segmentation with conjoint analysis (e.g., Hauser and Urban 1977; Moriarty and Venkatesan 1978; Currim 1981). At the first stage, benefit importance values (or part-worths in conjoint analysis) are obtained at the individual level either by direct rating of product attributes or statistical estimation. At the second stage, subjects are clustered on the basis of similarity of those importance values and a prediction equation is determined for each segment. Moore (1980) provided some empirical evidence that the two-stage clustering approach, being a post-hoc approach, is superior to componential segmentation. The validity of segments obtained with the two-stage procedure, however, depends on the validity and reliability of the idiosyncratic importance values (Slovic and Lichtenstein 1971; Urban and Hauser 1980; Brookhouse, Buion and Doherty 1986; Wedel and Kistemaker 1989). Also, clustering subjects on the basis of importance values does not maximize the accuracy with which overall product evaluations are predicted (Kamakura 1988). Several procedures have been developed that alleviate the disadvantages of the two-stage procedure. They are reviewed only briefly here because they are discussed in detail in other parts of the book.

Hagerty (1985) proposed a segmentation method tailored to conjoint analysis that estimates a set of weights for each subject in such a way that the predictive accuracy of a conjoint model is optimized. The optimal scheme for weighting consumer responses is found by using Q-factor analysis. A problem associated with the optimal weighting method is the difficulty of interpreting the solution in terms of segments (Stewart 1981).

Christal (1968), Bottenberg and Christal (1968), Lutz (1977), Ogawa (1987) and Kamakura (1988) have presented hierarchical predictive segmentation approaches (the methods of Ogawa and Kamakura are tailored to conjoint analysis). The procedures aggregate consumers in such a way that the accuracy with which preferences are predicted from product attributes or profiles is maximized. A drawback related to the hierarchical algorithm used in those methods is that misclassifications at earlier stages of the algorithm may carry on to higher levels. Models that are over-parameterized at the individual level (i.e., contain more parameters than the number of available observations per individual) cannot be handled and the methods yield nonoverlapping partitions.

The disadvantages are overcome in clusterwise regression procedures. Clusterwise regression is a method for simultaneous prediction and classification, and was originally proposed by Späth (1979, 1981 and 1982). In a regression context the method clusters subjects nonhierarchically in such a way that the fit of the regression within clusters is optimized. Both DeSarbo, Oliver and Rangaswamy (1989) and Wedel and Kistemaker (1989) recognized the applicability of clusterwise regression to market segmentation and extended the method to handle more than one observation per subject. The method of DeSarbo et al. deals with overlapping clusters and multiple dependent variables. Wedel and Steenkamp (1989) developed a method that also handles partial membership of consumers in segments. Wedel and Steenkamp (1991) extended the procedure to allow for a simultaneous grouping of both consumers and brands into classes, making possible the identification of market segments and market structures at the same time. The methods are in the class of fuzzy methods as they allow consumers (and brands) to have certain degrees of membership in several classes.

Although the fuzzy clusterwise regression procedures are powerful approaches to segmentation in which classification and prediction are combined, the disadvantages is that users must subjectively specify fuzzy weight parameters that influence the degree of separation of the clusters and that the statistical properties of the estimators are not established.

The disadvantages are largely alleviated by latent class (or mixture) regression methods. Such methods fit into three categories: mixture, mixture regression and mixture multidimensional scaling models. They are discussed extensively in part 2 of this book. Mixture regression models simultaneously group subjects into unobserved segments and estimate a regression model within each segment, relating a dependent variable to a set of independent variables (cf. Wedel and DeSarbo 1994). A recent methodological extension is the inclusion of concomitant variables, leading to models in which latent segments are estimated and simultaneously the segments are profiled with consumer descriptor variables (Kamakura, Wedel and Agrawal 1994). That circumvents the necessity of profiling the segments in a second step of the analysis. Mixture multidimensional scaling (MDS) models can be seen as "data unmixing" procedures, as they simultaneously estimate market segments as well as perceptual or preference structures of consumers in each segment (i.e., a brand map depicting the positions of the brands on a set of unobserved dimensions assumed to underlie perceptual or preference judgments). A review of those procedures is provided by DeSarbo, Manrai and Manrai (1994).

Mixture, mixture regression and mixture MDS methods currently provide the most powerful algorithms for market segmentation. They are the core of the methodological portion of this book, and are described in detail in Parts 2 and 3. A particularly useful member of this family of models for purposes of market segmentation is the clusterwise logit model, developed by Kamakura and Russell (1989) for the analysis of consumer choice behavior. Applied to household scanner data, the clusterwise logit model identifies consumer segments that are homogeneous in their brand preferences and in their response to price and sales promotions, thus providing a direct linkage between actual behavior in the market place and managerially actionable variables of the marketing mix. The clusterwise logit model has been extended to include the aforementioned concomitant variables (Gupta and Chintagunta 1994; Kamakura, Wedel and Agrawal 1994) that simultaneously link segment membership to observable characteristics of each household, resulting in a model that produces identifiable, accessible, actionable and responsive segments. Moreover, the methods conform to the normative ideal postulated in the early years of market segmentation, and solve a number of problems that have previously hampered the implementation of that ideal.

Normative Segmentation Methods

In accordance with its orientation in microeconomic theory, the microeconomic school of segmentation research has suggested that segments should be defined on the basis of consumers' demand functions (Massy and Frank 1965; Claycamp and Massy 1968). Dickson and Ginter (1987) argue that segmentation entails a grouping of demand functions rather than a grouping of consumers. Normative segmentation approaches can be classified according to the types of models used. Typically, mathematical models are used to describe and explain consumers' purchase behavior. Naert and Leeflang (1978) classify them as *descriptive* models that provide predictions at the (cross-sectional) behavioral level, *demand response* models that assess consumers reactions to marketing mix variables, and *policy* models that entail a profit function which is maximized to yield an optimal allocation of marketing instruments.

A variety of *descriptive normative* approaches have been proposed in the segmentation literature (e.g., Lessig and Tollefson 1971; Frank and Strain 1972; Sexton 1974; Frank 1972; Assael and Roscoe 1976). A variety of a-priori and post-hoc methods (automatic interaction detector, clustering and canonical correlation) have been used to identify segments on the basis of purchase behavior (Blattberg and Sen 1974, 1976; Blattberg et al. 1978; Starr and Rubinson 1978; Blattberg, Buesing and Sen 1980). Essentially those descriptive normative ideal approaches are standard approaches of segmentation applied to bases that can be considered ideal from a normative point of view.

Demand response models relate changes in the number of purchases of a brand (Wildt and McCann 1980), the brand's market share (Massy and Frank 1965; Sexton 1974; McCann 1974) or competitive reactions across segments (Plat and Leeflang 1988) to changes in marketing instruments. Elasticities are often seen to be the normative ideal basis for segmentation (Massy and Frank 1965; Sexton 1974; Dhalla and Mahatoo 1976; Russell and Kamakura 1994). Elasticity is defined here as the relative change in demand for a one-unit relative change in a marketing instrument (mostly price). Claycamp and Massy (1968), however, suggested that individual marginal responses to marketing variables (i.e., the change in demand for a one-unit change in the variable) are more appropriate for normative segmentation. Because they are defined in relation to regular demand level, elasticities tend to hide the true magnitude of the effect. On the other hand, marginal responses make comparisons more difficult because they depend on the measurement units of the output and inputs.

A problem in the derivation of elasticities at the consumer level, which are required for demand response segmentation purposes, is that estimates may be unstable (cf. Wedel and Kistemaker 1989) because of limited amounts of data at the individual level. That holds especially for elasticities estimated with individual-level logit models, as the estimates are only asymptotically consistent (which means they converge in probability to their true value only for large numbers of observations). Because of that limitation, econometric regression-type methods have been used to estimate the parameters of the models mostly within segments defined a priori. The potentially powerful approach to segmentation on the basis of demand response is therefore used mostly in combination with such rather weak a-priori segmentation designs based on general consumer descriptors as segmentation bases (Massy and Frank 1965; Frank 1972; McCann 1974; Plat and Leeflang 1988). An example of post-hoc segmentation using demand response models is given by Elrod and Winer (1982), who estimated price elasticities for each of 794 households and applied hierarchical clustering to form homogeneous segments in their response to price. However, in their approach the individual-level estimates of elasticities may be unstable because of limited data, and thus negatively affect the segments identified on the basis of those estimates. A solution to those problems has been proposed by the application of mixture regression models to household scanner panel data (Kamakura and Russell 1989; Bucklin and Gupta 1992; Gupta and Chintagunta 1994; Kamakura, Kim and Lee 1996), which produces segments that are homogeneous in their price and sales promotion elasticities without the direct estimation of those elasticities for each household.

The authors who follow the *policy-oriented* normative approach have the objective of obtaining the optimal allocation of marketing variables to a-priori defined segments, from the maximization of a profit function (Claycamp and Massy 1968; Frank, Massy and Wind 1972; Mahajan and Jain 1978; Tollefson and Lessig 1978; Winter 1979). Unfortunately, empirical applications of the approaches are lacking and estimation procedures are often not provided or are not computationally feasible (Elrod and Winer 1982).

Conclusion

To summarize our evaluation of segmentation methods, Table 3.1 presents an overview of the procedures discussed. We evaluate the methods on their effectiveness for segmentation and prediction, on their statistical properties, on the availability of computer programs and on applicability to segmentation problems. The evaluations are based on our subjective assessment of performance on those criteria, but nevertheless the reader may find such a classification a useful guide to segmentation methods.

Table 3.1:Evaluation of Segmentation Methods

Methods/Criteria	Effectiveness for segmentation	Effectiveness for prediction	Statistical properties	Application known	Availability of programs
1. A-priori, descriptive					
- log linear models	±	--	+	++	++
- cross tabs	±	--	++	++	++
2. A-priori, predictive					
- regression	-	++	++	++	++
- discriminant analysis	-	++	++	++	++
3. Post-hoc, descriptive					
- non overlapping	++	--	-	++	++
- overlapping	++	--	-	--	-
- fuzzy	++	--	-	++	+
4. Post-hoc, predictive					
- AID	±	+	-	++	+
- 2-stage segmentation	+	++	-	+	++
- clusterwise regression	++	++	±	+	+
- mixture regression	++	++	+	+	+
- mixture MDS	++	++	+	±	-

++ very good, + good, ± moderate, - poor, -- very poor

4
TOOLS FOR MARKET SEGMENTATION

In this chapter we reiterate the main conclusions drawn from the two preceding chapters about segmentation bases and methods, and introduce further developments, to be presented in the subsequent chapters.

The selection of both segmentation bases and methods is crucial to the effectiveness of the segmentation study in providing useful input for strategic management decisions. Normative segmentation has been postulated as the ideal approach, but has hardly been applied in practice. Elasticities with respect to marketing variables are generally agreed upon as the normative ideal basis for segmentation, even though their use in practice is not without problems. Normative segmentation has been useful in guiding scientific thought about segmentation and in evaluating the effectiveness of other segmentation bases. However, the application of post-hoc approaches for normative segmentation has traditionally been hampered by difficulties in the estimation of individual demand and by excessive computational requirements. Given elasticities as the normative ideal basis, the question remains: How do we obtain segments that approach the normative ideal from samples of consumers or firms? The mixture logit model proposed by Kamakura and Russell (1989) has provided a solution by its simultaneous segmentation and estimation procedure. The procedure reveals segments that differ in their price and promotion elasticities in a way that maximizes differences in utility among segments, and therefore represents a feasible implementation of the normative ideal concept, the more so in view of the wide availability of scanner panel data. A general modeling framework including that model and several of its extensions is discussed in part 2.

Situational variables have been shown to have potential for segmentation purposes. However, they are preferably used in addition to, rather than instead of, consumer segmentation. Domain- or product-specific lifestyle variables and benefits appear to be the most effective bases currently, as they satisfy several requirements such as stability, responsiveness, accessibility and actionability. Several chapters in this book pertain to such approaches to segmentation. Chapter 16 specifically addresses lifestyle segmentation, whereas chapter 18 covers benefit segmentation using conjoint analysis. Furthermore, those topics are elaborated upon in several other chapters; for example, chapter 12 explicitly examines tailored interviewing for lifestyle segmentation.

However, other segmentation bases should not be discarded. In modern segmentation research a variety of different types of bases should be used, each according to its own strengths. General and observable bases, for example, stand out in stability and accessibility, and can therefore be used for segment description and in media selection for consumer markets. In a similar vein, such variables are typically used as "concomitant variables" in the concomitant variable mixture models for segment description discussed in chapter 9. Personality and lifestyle variables provide real-life impressions of consumers within each segment, and hence are well suited for the development of effective advertising messages. Value segmentation, discussed in chapter 16 is useful for identifying segments that differ in broad patterns of buying and consumption behavior, such as food products, and for international market segmentation. The demands placed on the market research industry for the collection of data from consumers or firms for segmentation purposes are constantly increasing. Managers nowadays need insights not only on the buying behavior of segments, but also on their demographic and socioeconomic profiles, their lifestyle, attitudes and media exposure. They also want to know how to find and reach those consumers. In chapter 15 we discuss geodemographic segmentation, which has been widely used to link various databases at a certain level of aggregation (e.g., ZIP+4 codes).

A topic that appears to be largely overlooked in the segmentation literature is the optimal sample design for segmentation studies or, more generally, the effects of sample design on derived segments. For example, in what way should the sample design be accommodated in segmentation methodology? Do the segments derived from disproportionally stratified samples reflect the segments in the population from which the samples were drawn? In what way should surveys or panel studies be conducted to optimize the ability to draw inferences on segments in the population? We address those questions in chapters 5 and 6.

Among to segmentation methods, a-priori segmentation procedures are probably the simplest but least effective. Their most effective use is probably in hybrid procedures, where they are combined with post-hoc methods. Post-hoc methods, and especially clustering methods, are relatively powerful and frequently used in practice. Therefore we discuss clustering methods in more detail in chapter 5. The use of nonoverlapping clustering seems to be unnecessarily restrictive in many cases, however. Fuzzy clustering methods allow customers to belong to a number of segments, or attach probabilities of membership to the subjects or firms being segmented. The development of such clustering procedures thus affords a better theoretically grounded representation of markets. From a managerial perspective, directing marketing efforts to fuzzy segments is not conceptually more difficult than directing them to the traditional nonoverlapping segments; fuzzy segments can be profiled along general consumer descriptors, which enables the marketing manager to develop actionable strategies to target them. In chapter 5, we therefore pay attention to this important class of clustering methods. A class of clustering methods that has become important in segmentation research is the clusterwise regression methods, also treated in that chapter. Such methods simultaneously group customers into segments, and estimate a response relation within each segment. The discussion of clusterwise regression methods provides an introduction to the mixture regression methods discussed in chapter 6.

One of the major problems in the application of cluster analysis methods to market segmentation research is the plethora of methods available to the marketing researcher and the ad hoc decisions that must be made in their application. The multitude of techniques has confused marketers and has shifted discussion of research from methodological developments to issues related to the selection of appropriate methods. Moreover, it has impeded meta-research directed at integrating findings from market segmentation research (Punj and Stewart 1983). Our findings support the conclusion of Punj and Stewart. Although more than 1600 references on market segmentation were reviewed, we could not form empirical generalizations on the basis of findings for specific markets or problems because of the enormous variety of segmentation procedures used and a lack of concise descriptions of the motivation for their use. Often, cluster analysis has been used as a heuristic technique for data aggregation only, and theoretical considerations underpinning the application of clustering methods in segmentation research are limited. In most published research, we find a lack of specificity about the method of clustering used to arrive at the results and about the decisions on similarity measures, algorithms, data transformations, missing values and so on. It may reflect either a lack of concern or a lack of information on the part of researchers. We believe it is crucial in segmentation research that the selected methods optimally serve the purpose of the study and the structure of the data collected. Whereas segmentation research has traditionally been constrained to the application of available clustering methods to available data, modern research in the area has focused on the development of segmentation bases and methods that are tailored to specific segmentation problems, thus producing an optimal match of purpose, method and data. At the same time, the application of a uniform methodology, the mixture approach presented in this book, will contribute to the possibility of empirical generalization in future segmentation research.

The mixture regression and mixture scaling methodologies discussed in chapters 7 and 8 afford the flexibility currently required by the large number of substantive marketing applications. The methodologies presented in those chapters are general enough to subsume most of the mixture segmentation models presented in the marketing literature to date. They enable the researcher to build models to represent the market phenomena studied and to segment the population on the basis of the parameters (or subsets of parameters) of those models. The statistical basis of the methodology allows for statistical tests that provide answers to questions such as: Are segment structures similar in different countries? Are the effects of price or advertising statistically different across segments? What is the number of segments present in the population? Do segment structures change over time? The statistical approach clearly is a major step forward in segmentation research.

In part 2 we discuss standard mixture models that can be used for identifying segments; these models are analogous to standard clustering algorithms, falling in the category of descriptive segmentation methods, but have the advantages over clustering algorithms of taking a statistical approach. In part 2 we also discuss mixture regression models that allow for simultaneous classification and prediction, as well as mixture unfolding methods that enable one to simultaneously identify segments and latent variables underlying consumers' judgments for each of those segments. Both of these classes of mixture models can be used to translate marketing data into strategic information, as is illustrated in those chapters by several examples.

We then discuss a class of mixture models that can be used to identify segments and simultaneously profile them with consumer descriptors: concomitant variable mixtures. Those models add to the potential to implement segmentation strategies by facilitating access to segments. The approach is extended to models that allow changes in segment structure over time in the same chapter. These models conceptually extend market segmentation to deal with a dynamic marketplace.

In part 3 we describe a number of special topics, in particular, extensions of the models in part 2. This part illustrates the flexibility of the mixture model approach to describe a wide range of market phenomena. We cover four selected topics. The first is a model for simultaneously identifying multiple segment structures on the basis of several segmentation bases. The approach allows one to examine the relationships among the derived segmentation schemes. Then we discuss a tailored interviewing approach targeted to life-style segmentation that is based on a mixture model. This approach yields a substantial reduction in the length of questionnaires where the purpose is to classify respondents into a number of segments identified in prior research. We also describe a recently developed mixture segmentation model for LISREL applications. LISREL is a very flexible model that has gained popularity in the consumer behavior and services marketing literature. The mixture LISREL model allows one to build response-based segmentation models in which the variables in question are described by a LISREL model. Finally, we discuss mixture models for the analysis of paired comparison data, and in particular mixture of trees and mixture of spaces.

Among practitioners, there is both an interest in and a need for the next generation of segmentation models. The mixture model approach enables them to handle a number of well-known segmentation problems more satisfactorily than traditional clustering-based segmentation techniques. This holds in particular for geo-demographic segmentation and for life-style segmentation, application areas that have received continuing interest. Those application areas are discussed in chapters 15 and 16. Whereas in those two chapters the focus is on the description of a sample in terms of underlying segments, the possibilities of mixture models that allow simultaneous classification and prediction may be of even greater interest to applied market researchers. The possibility of simultaneous classification and prediction solves long-standing problems in applied marketing research. Examples arise in direct marketing, where individuals must be selected from customer lists, based on predictions of their response to future mailings, as well as in segmentation on the basis of household level scanner panel data, where purchases of scanned products are to be predicted and segments identified simultaneously. Another example occurs in conjoint analysis, where individuals in a sample are grouped into segments and segment-level part-worths must be simultaneously estimated so that responses to new product design can be predicted and simulations performed at the segment level. The latter two applications are discussed in chapters 17 and 18 of the book. A Monte Carlo study is discussed that shows the advantages of the mixture regression approach to segmentation for conjoint analysis over an array of alternative procedures that has been proposed in the literature.

In part 5 of the book, we draw up conclusions and formulate directions for future research. We discuss how the identification of market segments relates to the formulation of segmentation strategy, and the demands that this linkage places on the further development of mixture model approaches. We address a fundamental assumption of market segmentation in general and mixture models in particular: that a limited number of segments in fact exists in the population. We contrast this to the hierarchical Bayes approach that recently has received considerable attention among marketing researchers. The latter approach in contrast assumes a continuous distribution of heterogeneity; advocates of that approach have criticized finite mixture models for their assumption of discrete underlying heterogeneity. We assess the pro's and con's of both classes of methods in the final chapter. We conclude that much of the applicability of the methods depends on the need for individual level versus segment level analyses, and that the methods have different advantages and disadvantages, depending on the particular substantive field of application. Our conclusion is that in spite of these criticisms the mixture model approach will retain its value for market segmentation, but that there are areas where other approaches, such as the hierarchical Bayes approach, are more appropriate. We end with formulating an agenda for future research.

A wider implementation of mixture models in market segmentation research has been hampered by the lack of user friendly software. The popularity of the clustering approach, for example, can be ascribed, largely to its availability in the most popular statistical packages, such as BMDP, GENSTAT, SAS, S-PLUS, SPSS and so on, as well as in some more specific software packages such as CLUSTAN and PC-MDS. Mixture model algorithms have not met that level of availability so far. However, a WINDOWS software package has been recently developed by one of the authors. The package, called GLIMMIX, includes most of the models described in chapters 6 and 7. We describe the program and a few others in this book.

The recent development and application of new approaches to segmentation (i.e., mixture models) suggest the onset of a dramatic change in segmentation theory and practice. The full potential of the newly developed approaches in applied segmentation research is yet to be exploited and will become known with the new user-friendly computer software GLIMMIX. In 1978, Wind called for analytic methods that provide a new conceptualization of the segmentation problem. We believe the methods presented in this book meet Wind's requirements, and hope they will become a valuable tool in market segmentation approaches in marketing academia as well as in marketing practice. A more uniform and theoretically sound approach to market segmentation will contribute to the formation of empirical generalizations about market segments in the future.

PART 2
SEGMENTATION METHODOLOGY

This second part contains a more detailed technical discussion of the segmentation methods reviewed in chapters 3 and 4. Aside from the technical description of each method, we also review their application in Marketing. We start (in chapter 5) with a description of the non-overlapping clustering algorithms traditionally used in market segmentation. Chapter 5 also includes a discussion of more recent overlapping and fuzzy clustering algorithms.

The subsequent chapters in this second part cover the more recent finite-mixture models, which are the main methodological focus of our book. Chapter 6 introduces the basic finite-mixture model and methods to fit to observed data. Chapters 7 and 8 extend the basic finite-mixture model to estimate segment-level regression coefficients in the regression mixture approach (chapter 7) or map coordinates and segment-level ideal points in the mixture unfolding models (chapter 8).

The last two chapters in this part present further extensions of the finite-mixture approach. In chapter 9, a concomitant-variable sub-model is appended to the finite mixture model, so that segments can be profiled according to other observed characteristics of their members, at the same time these segments are identified and their parameters estimated. Chapter 10 presents extensions of the finite-mixture model that allow for non-stationarity in segment parameters and composition, thus accounting for the dynamic evolution of market segments.

5
CLUSTERING METHODS

In this chapter, we describe the hierarchical and nonhierarchical descriptive clustering methods traditionally used for market segmentation. We also discuss fuzzy clustering and clusterwise regression methods. The latter two sections also provide an introduction to chapters 6 and 7.

Cluster analysis is a convenient method commonly used in many disciplines to categorize entities (objects, animals, individuals, etc.) into groups that are homogeneous along a range of observed characteristics. Once those homogeneous groups are formed, the researcher can focus attention on a small number of groups rather than the large number of original entities. Instead of values of variables for a large number of objects, the researcher works only with the values of those variables for a limited number of homogeneous groups of the same objects. Cluster analysis is therefore often seen as an exploratory method of data analysis, intended more for generating hypotheses than for testing them. In marketing research, however, cluster analysis is used for more than data exploration; clustering methods are commonly used in marketing for the identification and definition of market segments that become the focus of a firm's marketing strategy. We established in chapter 3 that cluster analysis has long been the dominant and preferred method for market segmentation. Part of its popularity undoubtedly stems from its availability in many standard computer packages. Cluster analysis is in fact not a single technique, but encompasses a relatively wide variety of techniques that attempt to form clusters, groups with internal cohesion and external isolation (Gordon 1980).

Cluster analysis originated in the biological sciences, where it was developed to provide taxonomies of animal and plant species. The use and development of the technique spread across a large number of scientific disciplines such as medicine, psychology, sociology, food and sensory research, business economics, marketing research and many others. Cluster analyses has been known in those disciplines under different names, such as numerical taxonomy, Q-analysis, unsupervised pattern recognition, clumping and grouping analysis. The problem addressed by those techniques can be stated as follows (Everitt 1992): Given a collection of N objects, each of which is described by a set of K variables, derive a partition into a number of clusters or segments that are internally homogenous and externally heterogeneous. Both the number of classes and their properties in terms of the K variables must be determined. In market segmentation the objects are consumers or firms and the variables are the segmentation bases. We define the following basic notation to be used throughout this chapter.

Part 2 Segmentation methodology

Notation

Let:

$n, m = 1,..., N$ indicate objects (consumers/firms),
$k = 1,..., K$ indicate segmentation bases, variables or repeated measures on the objects,
$s, t = 1,..., S$ indicate clusters or segments,
$p = 1,..., P$ indicate independent variables,

y_{nk} be the value of the dependent segmentation variable k for consumer/firm n,
x_{nkp} be the value of the pth independent variable for repeated measure k on object n,
p_{ns} be the membership of consumer/firm n in segment s,
S_{nm} be the similarity between objects n and m,
D_{nm} be the distance between objects n and m.
$\mathbf{Y} = ((y_{nk}))$,
$\mathbf{P} = ((p_{ns}))$,
$\mathbf{X} = (((x_{nkp})))$.
$\mathbf{y} = \text{vec}(\mathbf{Y})$.

Two major groups of clustering methods can be distinguished by their use of dependent and independent variables (see chapter 3). Methods that distinguish dependent and independent variables are called *predictive* clustering methods, or clusterwise regression methods. Such methods form clusters that are homogeneous on the estimated relationship between the two sets of variables. *Descriptive* or traditional clustering methods do not make a distinction between dependent and independent variables, forming clusters that are homogeneous along a set of observed variables. Each of those two groups of procedures fits into three major categories of cluster analysis that we distinguish here: *nonoverlapping methods, overlapping methods and fuzzy methods.*

We distinguish the three major forms of clustering according to the form of the partitioning matrix **P** they produce. The matrix **P** has N rows, corresponding to consumers/firms, and S columns, corresponding to segments. The entries in the partitioning matrix **P** indicate the assignments of consumers to segments. In the case of *nonoverlapping clusters* a consumer belongs to one and only one segment, and consequently the matrix **P** has only one element in each row equal to one, the other elements in that row being zero. An example of the matrix **P** for nonoverlapping segments is shown below. Here subject 1 belongs to segment 1, subject 2 to segment 3, and so on.

Matrix **P** for Nonoverlapping Clusters

Subject	Segment	1	2	3
1		1	0	0
2		0	0	1
3		0	1	0
4		1	0	0

Two different types of nonoverlapping clustering methods are commonly distinguished: *hierarchical* and *nonhierarchical* methods. *Hierarchical* methods do not identify a set of clusters directly. Rather, they identify hierarchical relations among the N objects on the basis of some measure of their similarity. Nonhierarchical methods derive a partitioning of the sample into clusters directly from the raw data matrix **Y**.

Methods providing overlapping and fuzzy partitionings relax the assumption of external isolation of the clusters. In the situation of *overlapping clusters,* a consumer may belong to more than one segment. Therefore one row of the matrix **P** may contain several entries equal to one. An example of the partitioning matrix **P** for overlapping clusters is shown below, where subject 1 belongs to both segments 2 and 3, subject 2 belongs to segments 1 and 2, and so on.

Matrix **P** for Overlapping Clusters

Subject	Segment	1	2	3
1		0	1	1
2		1	1	0
3		0	1	0
4		1	0	1

In the case of *fuzzy clusters*, consumers have partial membership in more than one segment and the matrix **P** contains nonnegative real numbers that sum to one for each row. Two types of fuzzy clustering can be distinguished: the procedures based on fuzzy set theory and the mixture procedures. The mixture and fuzzy procedures are conceptually different. *Mixture procedures* assume segments are nonoverlapping but, because of the limited information contained in the data, subjects are assigned to segments with uncertainty, reflected in probabilities of membership contained in the matrix **P**. The *fuzzy set procedures* assume consumers actually have partial memberships in several segments. Here the partitioning matrix **P** contains partial membership values, which also are between zero and one and sum to one for each subject. In the example of the membership matrix, subject 1 has partial membership of 0.1 in segments 1 and 2 and a partial membership of 0.8 in segment 3; subject 2 has partial memberships of 0.6 and 0.4, respectively, for segments 1 and 2.

Part 2 Segmentation methodology

Matrix **P** for Fuzzy Clusters

Subject	Segment	1	2	3
1		0.1	0.1	0.8
2		0.6	0.4	0.0
3		0.2	0.3	0.5

Figure 5.1 summarizes our classification of clustering methods. In the remainder of this chapter, we discuss the non-overlapping, overlapping and fuzzy set methods in detail. The mixture methods of clustering are discussed in the next two chapters. For more extensive introductions to clustering, see Everitt (1992), Gordon (1980), Kaufman and Rousseeuw (1990) or general texts on multivariate analysis such as Dillon and Goldstein's (1990). We draw upon those references in the subsequent sections, but avoid extensive referencing in the text. First, we provide an example of a clustering approach to market segmentation.

Figure 5.1: Classification of Clustering Methods

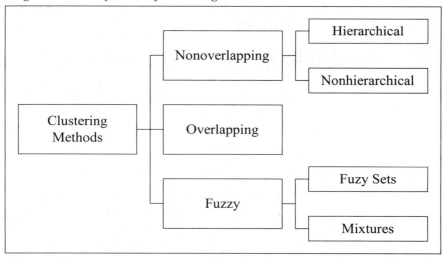

Example of the Clustering Approach to Market Segmentation

An illustrative application of cluster analysis to market segmentation has been provided by Singh (1990). The purpose of his analysis was to identify segments of consumers that exhibit different styles of complaint behavior with respect to services that were hypothesized to evoke differences in dissatisfaction (he used four types: grocery shopping, automotive repair, medical care and banking). From a random sample from a telephone register, usable data were obtained for 465 subjects.

Ten complaint intentions were reduced to three factors, by a confirmatory factor analysis approach. The factors were named Voice (propensity for complaints directed at the seller), Private (propensity for complaining to friends/relatives), and Third party (propensity for complaining to parties not involved in the exchange). The factor scores for each respondent were subsequently used in a cluster analysis (after examination of outliers), across the four services. The sample was split in halves and initial cluster analysis was carried out on one half. A hierarchical method (Ward's) was used, and the hierarchical structure suggested between three and seven segments. A nonhierarchical (K-means) method was then applied for each number of clusters between three and seven. The cluster centroid estimates from the hierarchical solution were used as initial cluster centers. A similar analysis was performed on the holdout sample, and the correspondence of the cluster solutions between the two samples (measured by the coefficient Kappa) was used to verify the validity of the clustering results. The four-cluster solution provided the maximum agreement (a Kappa value of 0.89). *Passives* (14% of the sample) were dissatisfied consumers whose intentions to complain were below average on all three factors, and were the least likely to complain. *Voicers* (37%) were dissatisfied consumers who were below average on private and third party intentions. They tended to complain actively to the service provider and had little desire to engage in negative word of mouth or to go to third parties. *Irates* (21%) had above average-intentions to respond privately. They intended not to complain directly to the seller, but engaged in negative word of mouth, and were not likely to take third-party actions. *Activists* (28%) were characterized by above-average complaint intentions on all three dimensions, but especially for complaints to third parties.

The external validity of the four segments was checked with self-reported complaining behavior. The a-priori hypothesized pattern was significantly supported by the data. For example, Voicers and Activists reported the highest percentages of complaints to sellers (83% and 72%, respectively, compared to 36% for the Passives). The segments were profiled, by discriminant analysis on variables such as expected payoffs from complaining, attitudes toward complaining, prior experience of complaining and demographic characteristics.

Nonoverlapping Hierarchical Methods

Hierarchical classifications typically result in a dendrogram, a tree structure that represents the hierarchical relations among all objects being clustered. The original purpose of hierarchical clustering procedures was to reveal the taxonomy or hierarchical structure among objects such as species of animals. The procedures have been widely applied in marketing to identify the hierarchical structure of product markets. They have also been used to identify market segments, as shown in the preceding example (see also Bass, Pessemeier and Tigert 1969; Montgomery and Silk 1972). Formally speaking, clusters themselves are not derived directly by the hierarchical methods, and a researcher seeking a solution with a certain number of clusters will need to decide how to arrive at those clusters from the tree representation

Part 2 Segmentation methodology

produced. Figure 5.2 is an example of such a hierarchical cluster solution for a market segmentation problem. The figure shows the derived hierarchical structure, starting with the single-subject (denoted by A, B, etc.) clusters on the left side. Two possible segment solutions, for six and two segments at levels of similarity of 10 and 20, respectively, are also indicated.

Such hierarchical structures can be derived by basically two types of algorithms: *agglomerative* and *divisive* methods (cf. Everitt 1992). *Agglomerative* methods start with single-subject clusters, as on the left of Figure 5.2, and proceed by successively merging those clusters at each stage of the algorithm until one single group is obtained. *Divisive* methods start with all subjects in one single group, as on the right of Figure 5.2, and successively separate each group into smaller groups until single-subject groups are obtained. The latter class of methods has been less popular in applied segmentation research.

Figure 5.2: Hypothetical Example of Hierarchical Cluster Analysis

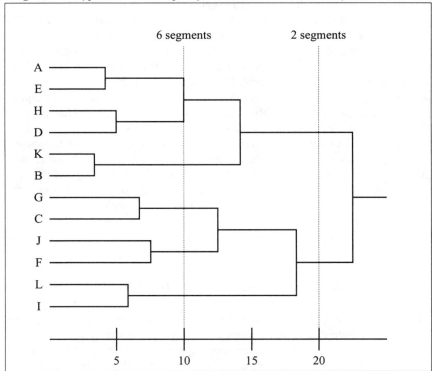

Similarity Measures

Hierarchical cluster algorithms operate on the basis of the relative similarity of the objects being clustered. A variety of similarity, dissimilarity and distance measures can

Chapter 5 Clustering methods

be used in hierarchical cluster analysis. Those similarity measures assess the strength of the relationship between the objects clustered and are derived from the variables measured on the objects. The type of similarity measure used is at the discretion of the researcher, and will depend on whether the scales of the segmentation bases are metric or nonmetric and on the type of similarity desired.

Similarity measures are typically non-negative ($S_{nm} \geq 0$) and must be symmetric ($S_{nm} = S_{mn}$). Those properties are shared by distance measures, which must also satisfy the triangle inequality ($D_{nm} \leq D_{nl} + D_{ml}$), and are sometimes also definite ($D_{nm} = 0$ if n = m). The most important similarity coefficients are stated in Table 5.1.

The similarity/dissimilarity measures in Table 5.1 are symmetric (i.e. $S_{nm} = S_{mn}$), and are between zero and one (the correlation is between -1 and 1). Dissimilarity is defined by $1 - S_{nm}$. The similarity measures for binary (0/1) variables are based on the two-way cross-classification of the two objects, counting the number of times the objects both have an attribute present (1,1; denoted by "a" in Table 5.1), both have an attribute absent (0,0; denoted by "d") or where one has the attribute and the other does not (1,0: "b"; 0,1: "c"). An extensive overview of those measures can be found in the CLUSTAN manual (Wishart 1987). No definitive rules can be given for the selection of one of the similarity measures. The most important distinctions are whether or not (1) negative matches are incorporated into the similarity measure, (2) matched pairs of variables are weighted equally to or carry more weight than unmatched pairs and (3) unmatched pairs carry more weight than matched pairs.

In many market segmentation applications, the absence of a characteristic in consumers or firms might not reflect similarity among them. However, the simple matching coefficient, S1, as well as S4, S7, S8 and S9, equate joint absence of a characteristic to similarity and may therefore be of limited value in segmentation research. Measures such as the Jaccard coefficient (S2), as well as the measures S3, S5, S6, S10 and S11 do not make that assumption. Note, for example, that the Russell and Rao coefficient, S6, includes matched absence d in the denominator. The Czesanowski coefficient (S3) assigns double weight to matched pairs to compensate for the neglect of unmatched pairs. The measure proposed by Sokal and Sneath (S4) weights both presence and absence of attributes twice. The phi coefficient (S12) is based on the well-known chi-squared statistic for two-way tables. It is the analogue of the correlation coefficient for binary variables. For categorical data with more than two levels (say Q levels), an appropriate measure can be obtained by splitting them into Q categories, and assigning $S_{nmq} = 1$ if n and m have the same value of Q and zero otherwise, then calculating the average of S_{nmq} across all Q levels. Alternatively, as S12 is based on the chi-squared statistic, it can be extended for this situation.

For ordinal variables, many authors recommend treating the ranks as metric data and applying some distance metric. When the maximal rank M_k differs among variables, it is useful to convert the rank ordered variables to the zero-one range by dividing each observed rank minus one by $M_k - 1$.

The measures for metric variables reported in Table 5.1 are distance or dissimilarity measures (except for the correlation coefficient D1). All are based on the differences of the objects with respect to a variable, summed across all variables. The most well known measure, the Euclidian distance (D2), is not invariant to scale changes in the

Part 2 Segmentation methodology

variables **Y**. That problem is corrected by the standardized measures D2b and D2c. These measures all assume the variables are uncorrelated. The Mahalanobis distance (D4) accounts for correlations among the variables and is scale invariant. The Minkowski metric, for several values of r, includes several other measures as special cases (e.g., D2 and D3 arise from r = 2 and r = 0, respectively). Dissimilarity measures can be characterized by whether they account for *elevation*, *scatter* and *shape* of the variables. Elevation is equivalent to the intercept in a regression of two variables, shape is equivalent to the direction of the regression line and scatter is equivalent to the spread of the observations around the line. The measure of angular distance (D6) accounts for the elevation, scatter and shape of the original variables. The measures D2, D3 and D4 include scatter and shape, whereas the correlation coefficient (D1) includes scatter only. The choice among those measures will depend on which of the aspects (elevation, scatter or shape) is to be included in a measure of similarity.

Some of the coefficients for binary variables and metric variables are related. For example, using S1 or S2 for binary variables and the standard transformation of similarity to distance, $D_{nm} = (2 - 2S_{nm})^{1/2}$, results in the Euclidian distance. For mixed sets of variables, Gower (1971) has proposed a coefficient that amounts to calculating a measure for metric variables and for binary variables separately. The similarity S_{nm} is obtained by averaging over k (possibly a weighted average can be calculated).

Table 5.1: The most Important Similarity Coefficients

Binary Data

$a = \sum_k y_{nk} y_{mk}$

$b = \sum_k (1 - y_{nk}) y_{mk}$

$c = \sum_k y_{nk} (1 - y_{mk})$

$d = \sum_k (1 - y_{nk})(1 - y_{mk})$

(S1)	Simple matching	(a + d)/(a + b + c + d)
(S2)	Jaccard	a/(a + b + c)
(S3)	Czesanowski	2a/(2a + b + c)
(S4)	Sokal and Sneath 1	2(a + d)/(2a + 2d + b + c)
(S5)	Sokal and Sneath 2	a/(a + 2b + 2c)
(S6)	Russel and Rao	a/(a + b + c + d)
(S7)	Hamann	(a + d − b − c)/(a + b + c + d)
(S8)	Rogers and Tanimoto	(a + d)/(a + b + 2c + 2d)
(S9)	Yule's Q	(ad − bc)/(ad + bc)
(S10)	Kulczynski	a/(b + c)
(S11)	Ochiai	$a/\{(a + b)(a + c)\}^{1/2}$
(S12)	Psi	$\{\chi^2/(a + b + c + d)\}^{1/2}$

Chapter 5 Clustering methods

Table 5.1 – continued

Metric Data

σ_n^2	= variance of measures on object n	
R_k	= range of variable k	
w_k	= a weight	
(D1)	Correlation coefficient	$\sum_k (y_{nk} - y_n)(y_{mk} - y_m) / \sigma_n \sigma_m$
(D2)	Euclidian distance	$\sum_k w_k (y_{nk} - y_{mk})^2)^{1/2}$
	(D2a) Unstandardized	$w_k = 1$
	(D2b) Pearson distance	$w_k = 1/\sigma_k^2$
	(D2c) Range standardized	$w_k = 1/R_k^2$
(D3)	City block distance	$\sum_k |y_{nk} - y_{mk}|$
(D4)	Mahalanobis distance	$(y_n - y_m)' \sum^{-1} (y_n - y_m)$
(D5)	Minkowski distance	$(\sum_k w_k (y_{nk} - y_{mk})^r)^{1/r}$
(D6)	Angular distance	$\sum_k y_{nk} y_{mk} / (\sum_k y_{nk}^2 y_{mk}^2)^{1/2}$
(D7)	Canberra distance	$\sum_k |y_{nk} - y_{mk}| / (y_{nk} + y_{mk})$

Mixed Data

	Gower distance	$1 - \sum_k w_k |y_{nk} - y_{mk}| / K$
		Metric data: $w_k = 1/R_k$;
		nonmetric data $w_k = 1$

In general, the researcher needs to investigate each set of segmentation variables and theoretical arguments–if available–should lead to the selection of the most appropriate similarity measure. Punj and Stewart (1983) reached an important conclusion about the use of (dis)similarity measures. From their review of the literature, they concluded that the choice of such a measure does not appear to be critical to recovering the underlying cluster structure. The choice of the clustering algorithm may be more important. However, for clustering algorithms that are sensitive to outliers in the data (see below), they concluded that Pearson correlations may be a preferred choice. A similar effect can be obtained by standardizing the data before clustering (the issue of standardization is discussed more extensively below). When dependencies among the variables are expected, it is useful to select a measure that corrects for such dependencies, such as the Mahalanobis distance (D4).

Agglomerative Cluster Algorithms

The similarity $S = ((S_{nm}))$ or distance $D = ((D_{nm}))$ matrices among objects constitute the basic input to hierarchical clustering algorithms. Agglomerative clustering algorithms produce a series of partitions of the data that arise by fusing groups of objects at successive stages of the algorithm, starting with single-object clusters. Several algorithms are available for that purpose. They can be distinguished on the basis of the way they define the distance between an individual object and a cluster, or between two clusters, in later stages of the algorithms, as well as in the way the new cluster center is computed. Table 5.2 lists the definitions of cluster distance for the most popular hierarchical algorithms.

At each stage of the *single linkage* cluster algorithm, the distance between two clusters is defined as the shortest distance between any two members in the two clusters. If the distance is defined as the longest distance between any two members of the two clusters, the *complete linkage* algorithm arises. In *average linkage,* distance is defined as the average distance between all pairs of members of the two clusters. The measure accounts for the complete cluster structure at each stage of the algorithm. A variant of that method is the *weighted average linkage* algorithm, where the distance between any pair of members in two clusters is weighted by the sizes of the respective clusters. It alleviates the problem of average linkage, that the larger cluster receives more weight, and the new cluster center is more influenced by the largest of the clusters merged.

In *centroid linkage*, the distance between two clusters is defined as the Euclidian distance between the cluster centers (average of the variables for the members of that cluster). The method has the same disadvantage as the average linkage method, that the new cluster is more influenced by the largest of the two clusters being merged. *Median linkage* alleviates that problem by calculating the new cluster center not as the average of the two centroids, but as the middle of the line connecting the two previous cluster centers. Some of the linkage criteria are represented schematically in Figure 5.3.

Figure 5.3: Schematic Representation of Some Linkage Criteria

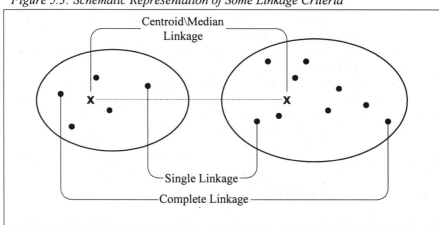

Ward's method or *minimum variance linkage* (A6) selects clusters to be linked at each stage so that their linkage produces the smallest possible increase in the pooled within-cluster variance. The *p-nearest neighbor* clustering method is based on the neighborhood value, which is defined as follows. Suppose subject m is the ath nearest neighbor to subject n, and subject n is the bth nearest neighbor of subject m, defined on the basis of some distance measure. Then their neighborhood value is a + b. Segments are formed on the basis of neighborhood values exceeding a threshold p. More extensive descriptions of these and other algorithms can be found in several textbooks (Gordon 1980; Kaufman and Rousseeuw 1990; Everitt 1992).

It has been argued that on mathematical grounds, the single linkage algorithm should be preferred (cf. Everitt 1992, p.70). However, the single linkage criterion has the often undesirable property of chaining, where the method fails to recover distinct clusters but links each object successively to the previous ones. Other hierarchical methods tend to be biased toward finding round or spherical clusters, especially the centroid (A4) and minimum variance (A6) methods. A disadvantage of the centroid method is that the centroid of a newly formed segment will be closer to the centroid of the largest of the two groups being fused. As stated above, the median method overcomes that problem but appears to be unsuitable for measures such as correlation coefficients. In general, no hierarchical method can be universally designated as superior (see Milligan 1980). Single, centroid and median linkage methods are less - affected by outliers, but are affected by noise in the data. Complete linkage and centroid linkage methods are less likely to produce clusters of extremely unequal sizes. Average linkage performs poorly in the presence of outliers, but is less affected by noise in the data or clusters of unequal sizes. In addition, median and centroid linkage algorithms have serious problems of inversion and non-uniqueness, discussed later in this chapter.

Table 5.2: Definition of Cluster Distance for several Types of Hierarchical Algorithms.

	Algorithm	Recurrence relation of distance
(A1)	Single linkage	Smallest distance between members of two segments
(A2)	Complete linkage	Largest distance between members of two segments
(A3)	Average linkage	Average distance between members of two segments
(A3a)	Weighted average linkage	Weighted average distance between members of two segments
(A4)	Centroid linkage	Distance between segment averages of the variables
(A5)	Median linkage	Distance between segment medians of the variables
(A6)	Minimum variance linkage	Minimum increase in total sum of squares
(A7)	p-Nearest neighbors	Mutual neighborhood value
(A8)	Clusterwise regression	Minimum increase in total residual sum of squares of within cluster regressions

Minimum variance (Ward's) clustering is affected negatively by outliers in the data and is likely to produce clusters of different sizes, but is less affected by random noise in the data. In general, the method performs well in recovering true clusters. Apparently average linkage and minimum variance clustering outperform the other methods over a wide variety of conditions for segmentation purposes (Punj and Stewart 1983), whereas the p-nearest neighbors procedure seems able to uncover types of cluster structure that cannot be identified by many of the other hierarchical procedures (Gowda and Krishna 1978).

Divisive Cluster Algorithms

An important divisive hierarchical clustering method mentioned in most textbooks on clustering is the *splinter average distance method* (Macnaughton-Smith et al. 1964). That *polythetic* (i.e., based on several variables) method successively accumulates "splinter groups" by separating out subjects whose distance to the other individuals in the main group is larger than that to the splinter group. When no such subject can be found, the splinter group is further partitioned. Another well-known method, proposed by Edwards and Cavalli-Sforza (1965), minimizes a least squares criterion as does Ward's agglomerative method (and the K-means procedure to be discussed later). The method imposes severe computational burdens, however, because all possible divisions of a cluster into smaller clusters must be tried. Some proposed extensions alleviate that problem somewhat.

Monothetic methods (i.e., based on a single variable at each stage) are mostly used for binary variables. Here, for each cluster the best division of the cluster into two sub-clusters is found in terms of presence or absence of a binary attribute. One new cluster contains all objects with the attribute and the other cluster contains the objects without the attribute. In each step, the attribute is chosen so that the similarity between the two new clusters, measured over all attributes, is minimized. In principle, any of the above similarity measures for binary variables can be used for that purpose. Each of the new clusters is then examined for a new subdivision, and the cluster that divides into the two most dissimilar sub-clusters is split. The process is continued until a predetermined number of clusters is reached. The process proceeds from right to left in Figure 5.2, where potentially different binary variables are used at each node in the hierarchy. The divisive methods are much less popular that the agglomerative methods, and to our knowledge have not been used in segmentation research.

Ultrametric and Additive Trees

The hierarchical clustering methods discussed above have some important limitations: they are deterministic, do not accommodate the random measurement error in segmentation variables, do not optimize a certain criterion of interest (with the exception of Ward's method) and the choice of the similarity measure and the cluster algorithm is largely subjective. Those limitations are alleviated by procedures that have been proposed to fit so-called ultrametric and additive tree structures to similarity data. The ultrametric and additive trees are similar to that in Figure 5.2, but satisfy certain properties or constraints.

The basic principle of the approaches is to minimize a least-squares function of observed distances relative to the distances fitted by a prespecified tree structure (Δ_{nm}): $\Sigma_{n<m} (D_{nm} - \Delta_{nm})^2$, subject to constraints on the distances, so that they represent the respective trees. The constraints are respectively the ultrametric inequality, $D_{nm} \leq \max(D_{nl}, D_{ml})$, and the additive inequality, $D_{nm} + D_{kl} \leq \max(D_{nk} + D_{ml}, D_{nl} + D_{mk})$. The choice between the two types of tree models depends on the availability of substantive arguments motivating either of the two types of constraints or on the relative fit of the models. We do not discuss these procedures in detail here (see chapter 14); but an introduction is provided by Corter (1996) and an overview by DeSarbo, Manrai and Manrai (1993). Moreover, the procedures are suitable to problems in which a hierarchical structure is assumed to underlie the data, such as in the derivation of competitive market structure. For segmentation problems, theoretical arguments to justify the hierarchical relational structure among consumers or firms derived by these methods seem to be lacking, except perhaps for business-to-business applications involving a smaller sample of customers logically classified in a hierarchical (SIC) structure. We therefore conjecture that the benefits of ultrametric and additive tree models are obtained primarily in market structure analysis and much less in market segmentation research. Note, however, that much the same argument applies to the application of hierarchical methods to market segmentation problems.

Hierarchical Clusterwise Regression

Whereas all of the preceding methods are descriptive and do not distinguish dependent and independent variables, Kamakura (1988) proposed a hierarchical clusterwise regression procedure that does allow for predictions within classes (A8). Originally the method was designed for a specific type of data (conjoint data, see part 3), but it is applicable to a much wider range of problems. The purpose here is to predict some dependent measure of interest, say preferences obtained from K repeated measures on n subjects, from a set of P independent variables, such as P characteristics of the products. At the first stage of the algorithm, a regression equation is estimated for each subject yielding the regression coefficients of the P independent variables for each of the N subjects, where $ß_n$ is a $(P \times 1)$ vector containing the P regression weights for subject n. In the second stage, the algorithm tries all possible fusions of two subjects and re-estimates the regression coefficients for all clusters of two subjects that arise (the coefficients are estimated across n and k within the clusters). The fusion of subjects that yields the minimum increase in the total residual sum of squares of the regression across all clusters is retained, and the two subjects are combined. The agglomerative process is thus similar to that of Ward's method. In each successive stage, segments are linked that provide the smallest possible increase in the pooled within-segment error variance. Kamakura shows that minimizing the residual sum of squared regression residuals is equivalent to maximizing the expected predictive accuracy across segments. Empirical comparisons and applications support the validity of the procedure. However, replications by Green and Helsen (1989) showed that the predictive fit for the segment-

Part 2 Segmentation methodology

level solution is essentially comparable to the results obtained when a regression equation is estimated for each consumer.

Nonoverlapping Nonhierarchical Methods

The nonhierarchical clustering methods differ from the hierarchical clustering ones in that they do not seek a tree structure in the data. Rather, they partition the data into a predetermined number of segments, minimizing or maximizing some criterion of interest. The methods in this class differ according to the criterion that is optimized and the algorithms that are used in the optimization process. The most important criteria that have been proposed are summarized in Table 5.3. They are based on the within- and between-segment sum-of-squares matrices of the K segmentation variables.

Table 5.3: Optimization Criteria in Nonhierarchical Clustering

T = total $K \times K$ dispersion matrix
W = pooled within-segment $K \times K$ dispersion matrix
W_s = within-segment $K \times K$ dispersion matrix for segment s
B = between-segment $K \times K$ dispersion matrix
σ_s = within-segment regression residual sum of squares
P_{sj} = choice probability of brand j in segment s
A_{nj} = 1 if brand j is chosen by subject n, = 0 otherwise.

(O1) $\text{trace}(W)$
(O2) $\det(W)$
(O3) $\text{trace}(BW^{-1})$
(O4) $\det(BW^{-1})$
(O5) $\Pi_s \det(W_s)^{N_s}$
(O6) $\Pi_s (N_s-1) \det(W_s)^{1/K}$
(O7) $N \ln \det(W) - 2\Sigma_s N_s \ln N_s$
(O8) $\Sigma_s (N_s \ln \det(W_s) - 2N_s \ln N_s$
(O9) $\Sigma_{s,k} \{Y_n \text{ EQ INT}(\Sigma_n Y_n / N_s + 0.5)\}$
(O10) $\Sigma_s \sigma_s^2 / S$
(O11) $\Pi_s (\sigma_s)^{2N_s/N}$
(O12) $\Sigma_s L_s, L_s = \Pi_{n,j} P_{sj}^{A_{nj}}$ with $n \in s$

Most of the above criteria, with the exception of O9, apply to variables that are at least interval-scaled. The criterion defined by O1 is equivalent to minimizing the sum of the squared Euclidian distances between individuals and their segment mean. That criterion is in fact the K-means criterion, probably the most frequently used nonhierarchical procedure in segmentation research. A problem related to the criterion is that it is scale dependent; different solutions are obtained when the data are transformed. Further, the criterion tends to yield clusters that are spherical. The O2 criterion does not have the problems of scale dependency or sphericity of clusters. It can be derived from the likelihood under the assumption that the data arise from a mixture of S multivariate normal distributions (Scott and Simons 1971) or from arguments derived from Bayesian statistics (Binder 1978). The criteria O3 and O4 are related to multivariate analysis of variance and canonical variate analysis, respectively; O4 is a generalization of the Mahalanobis distance to several groups. The criteria O1 through O4 assume that all clusters have the same shape, which is overcome by criteria O5 through O8. Criteria O7 and O8 also alleviate the problem with the preceding criteria of tending toward segments of equal size. The criterion defined by O9 gives maximal predictive classification for binary variables (Gower 1974). It is equal to the sum over the segments of the number of agreements among members of each segment and the respective class predictor. The class predictor is defined to be a variate that contains one value (zero or one) for each variable, corresponding to the value of that variable that is more prevalent in the class.

Nonhierarchical Algorithms

The most certain way to find the optimal partitioning of subjects or firms in a predetermined number of segments is to investigate all possible partitionings of the N objects into S segments, which is infeasible for most segmentation problems. That drawback has lead to the development of exchange-type algorithms. Such algorithms move subjects around the segments, and keep new partitions thus obtained only when they improve the criterion of interest. The algorithms start from an initial partition of the data into S segments. The five dominant methods are *Forgy's* method, *Jancey's* method, *MacQueen's* method, the *convergence method* and the *exchange algorithm* of Banfield and Bassil (1977; cf. Milligan 1980). The methods differ in the iterative process and the recalculation of cluster centers. These algorithms try each possible transfer of an object from its current segment to the other segments; the transfer is executed if it improves the criterion. In MacQueens" method, the convergence method and the exchange algorithm, cluster centers are updated each time a subject is successfully transferred. In Forgy's and Jancey's methods each cluster center is recalculated only after a complete pass through the data, that is, after all subjects have been tentatively moved once, which is more computationally efficient. All possible transfers are tried and the algorithms stop if no more improvement in the criterion value is achieved. MacQueen's method, however, stops after one complete pass through the data . The convergence method and the exchange algorithm are the most general procedures. The exchange algorithm adds to the convergence method in that it has a second phase, a swapping phase in which all possible swaps of two subjects between two segments are tested.

When the algorithms are started from a random initial partition of the data, there is not much difference in their performance and they tend to converge to local optima. In other words, the final solutions obtained from different starting solutions may not be the same, a situation that is generally exacerbated by the absence of well-defined clusters in the data, as well as the absence of a good starting partition. The more advanced procedures tend to be less affected by such convergence to local optima, but for good starting partitions (e.g., obtained by a hierarchical clustering procedure) their performance is comparable. Initial partitions can be obtained in several ways: (1) at random, (2) on the basis of a-priori information, (3) by assignment, given S initial cluster centers, of each subject to its nearest cluster center and (4) by hierarchical clustering. Random starting partitions are generally not recommended, unless a large number of random starts are investigated, to prevent local optima. The use of hierarchical clustering solutions appears to be a preferred choice for starting points.

Several approaches other than the "hill-climbing" heuristics described above have been used, including dynamic programming (Jensen 1969), steepest descent numerical optimization (Gordon and Henderson 1977), branch and bound algorithms (Koontz, Narenda and Fukunaga 1975), simulated annealing (Klein and Dubes 1989), and artificial neural networks (Balakrishnan et al. 1995). Especially appealing is the work in dynamic programming and simulated annealing, as those procedures were designed to overcome the problem of local optima found with the exchange-type algorithms and other procedures.

From a review of the literature, Punj and Stewart (1983) concluded that nonhierarchical models are generally superior to hierarchical models. They are more robust to outliers and the presence of irrelevant attributes. Among the nonhierarchical methods, the minimization of det(**W**) appears to be superior. The performance of nonhierarchical methods depends, however on the starting partition chosen. In the absence of any prior information, a hierarchical method such as Ward's minimum variance can used to obtain a rational starting partition.

Determining the Number of Clusters

A major issue in nonhierarchical clustering is the determination of the appropriate number of segments (S). The analyses described above are all conditional on an assumed number of segments. However, the true number of segments in the data is unknown in most cases and must be determined. Several methods have been proposed for determining the number of clusters. The most popular approach in practice is to plot the minimum value of the cluster criterion attained against the number of segments. Large changes in the value of the criterion from one cluster solution to another with one less cluster would be suggestive of the true number of segments. That procedure, like the procedures used to determine the number of segments in hierarchical clustering, is rather subjective. Moreover, in an extensive simulation study it did not appear to perform well; it recovered the true number of clusters in only about 28% of the times across a variety of conditions. A more formal procedure, proposed by Calinski and Habarasz, involves taking the value of S for which $(S - 1)\text{trace}(\mathbf{W})/(N - S)\text{trace}(\mathbf{B})$ is minimal (cf. Everitt 1992). That procedure generally performs best and recovers the true clusters in more than 90% of the cases (Milligan and Cooper 1985).

Nonhierarchical Clusterwise Regression

The non-hierarchical procedures based on maximizing the criteria O1 through O9 are descriptive clustering methods in the sense that they do not distinguish between dependent and independent variables. Several authors have proposed nonhierarchical methods to form clusters that are homogeneous in terms of the relationship between a dependent variable and a set of independent variables (Späth 1979, 1982; Katahira 1987; Wedel and Kistemaker 1989). Those methods optimize the criteria O10, O11 or O12 in Table 5.3. The clusterwise regression method proposed by Wedel and Kistemaker operates as follows:

1. A nonoverlapping starting partition of the sample (random, or derived from the application of, for example, a K-means clustering to the dependent variable) is obtained for a given number of segments S. The partitioning results in a preliminary assignment of the N objects to the sets C_s.

2. Within each of the resulting segments, the dependent variable $y_n = (y_{nk}, n \in C_s, k = 1...K)$, measured on N subjects (possibly with K repeated measurements, e.g. for K brands), is regressed on P independent variables $X_n = ((x_{nkp}, n \in C_s, k = 1... K), p = 1...P)$, according to: $y_n = X_n\beta_s + \varepsilon_n$.

3. An exchange algorithm, based on that of Banfield and Bassil (1977), is used to transfer objects between the segments in such a way that the residual sum of squared errors within the segments is minimized.

That procedure has an advantage over the hierarchical procedure of Kamakura (1988), as it can also be applied in situations where the number of observations for each subject is smaller than the number of explanatory variables (K < P), when the individual-level models in the hierarchical procedure cannot be estimated. The procedure shares the general advantages of nonhierarchical over hierarchical procedures outlined above. Wedel and Kistemaker (1989) show that maximizing the criterion O10 is identical to maximizing the likelihood under the assumption that the dependent variable is normally distributed with equal variances within segments, whereas O11 is obtained if the variances are assumed to differ across segments. Note that these algorithms are equivalent to nonhierarchical algorithms O1 or O2 minimized with the exchange algorithm. The authors suggest determining the number of segments on the basis of the plot of the criterion minimized against the number of clusters, but the Calinski-Habarasz measure calculated on the regression residuals may be better. After applying the clusterwise regression procedure to data, the researcher obtains a set of regression coefficients for each segment, relating the dependent variable to independent variables, as well as the zero-one segment membership of the objects in the sample.

The procedure proposed by Katahira (1987) is a clusterwise logistic regression model maximizing O12. Here, the dependent variable is multinomial, for example, the choices among a set of brands. The model is tailored to the analysis of choice behavior, and the choice probability for brand k in segment s is formulated as:

$P_{sk} = \exp(\mathbf{X_{nk}\beta_s})/\Sigma_k \exp(\mathbf{X_{nk}\beta_s})$. That is the ordinary multinomial logit model for explaining choice probabilities from explanatory variables in segment s. Katahira also uses an exchange algorithm to maximize the likelihood of the multinomial logit model (O12). His procedure estimates the segments and at the same time estimates the multinomial logit model within each of those segments.

A problem with the clusterwise regression procedures is that the usual significance tests for the regression coefficients are not valid. Wedel and Kistemaker therefore propose the use of Monté Carlo significance tests. In those tests, the test statistic (e.g., a t-value for a regression coefficient) is calculated from the data and the model in question, and compared with the distribution of the t-statistic calculated from a number of data sets, M (often M = 20), in which the dependent variable is randomized. If the test statistic from the data exceeds $M(1 - \alpha/2)$ values of the statistic calculated from the random data, the coefficient is significant at a significance level of α.

Wedel and Kistemaker use a scree plot of the log-likelihood, whereas Katahira plots the Akaike information criterion to determine the optimal number of segments. Both of those procedures are heuristic, however, and the selection of the appropriate number of segments remains a topic for future research. Monté Carlo studies of the performance of the clusterwise regression models on synthetic data, performed by Katahira and by Wedel and Kistemaker, supported the validity of the methods.

Miscellaneous Issues in Nonoverlapping Clustering

Variable Weighting, Standardization and Selection

Here, we discuss miscellaneous issues in clustering, such as the selection, weighting and standardization of variables, missing values, outliers and cluster validation. This section is based largely on the book by Everitt (1992) and the series of articles by Milligan (1995) in the newsletter of the Classification Society of North America.

In market segmentation research, the choice of variables, or segmentation bases, should be guided by theory on the substantive domain of application. The choice of bases is based on the researcher's judgement of their relevance for the type of segments being sought and the substantive segmentation problem at hand. Only variables deemed relevant for the segmentation should be included, as the addition of irrelevant variables can dramatically interfere with the recovery of the underlying segments (Milligan 1980). Variable selection and variable weighting are two sides of the same coin; assigning a variable weight of zero is equivalent to not selecting it. However, there are important pragmatic differences: careful selection of variables results in a reduced data-collection effort and in clustering models that are more parsimonious and easier to interpret (Gnanadesikan, Kettenring and Tsao 1995). Whereas techniques for variable selection are abundant in classical linear models and discriminant analysis, those in cluster analysis are limited to block clustering algorithms (Duffy and Quiroz 1991) and the approach of Fowlkes, Gnanadesikan and Kettenring (1988). The latter is a computer-intensive procedure that selects variables for complete linkage in a forward-selection fashion. At the first stage of the procedure, a separate dendrogram is produced on the basis of Euclidean distances between the objects calculated for each of the K variables. For each tree, the observations are then partitioned into S clusters. The results are compared for each variable by using a criterion that assesses the strength of the clusters. The variable that results in the best separation is selected. In the second stage, the variable selected is combined with each of the remaining variables, and (K - 1) trees are produced. Partitions are made and the next variable is selected that exhibits the most separation between clusters. The procedure continues in that way and stops when meaningful additions of variables can no longer be found.

If the researcher has prior evidence that some variables may be less relevant to the clustering than others, a-priori weights can be assigned to those variables, for example, in calculating the similarity measures. The problem in practice is that it is very difficult to arrive at such weights (Everitt 1992). Procedures have been developed to estimate optimal variable weights, which may alleviate the problem of irrelevant variables. De Soete, DeSarbo and Carroll. (1985) and De Soete (1988) designed a hierarchical procedure that simultaneously derives variable weights to optimally satisfy the ultrametric inequality. Milligan (1989) showed that DeSoete's procedure was effective in coping with irrelevant variables, and that the recovery of the underlying cluster structure was greatly enhanced when three masking variables were present. DeSarbo, Carroll and Clark (1984) extended the K-means algorithm in a similar way. Their method also allows for the analysis of several groups of a-priori weighted variables. Variable weights are estimated to optimize the clustering results with respect to a measure of fit.

If variables are measured on widely different scales, and the similarity measure applied is not scale invariant, standardization of variables may be appropriate in segmentation. Standardization in fact is a specific form of variable weighting. An important question is how to standardize. The use of the traditional z-score is often not optimal. Milligan and Cooper (1988) studied eight forms of standardization, and concluded that standardization by range performed consistently well across a wide variety of conditions. Most commonly, variables are standardized prior to the analysis. That procedure may not be appropriate, however: if a variable has large variance because it separates the segments in the data, standardization may reduce differences between the segments on that variable. A solution may be to apply a weighted clustering procedure, as described above, to the standardized data. Another solution could be to use the within-group standard deviations for standardization. However, those are not known until after the clustering. A multistage clustering approach could be used, where in the first stage raw data are subjected to clustering and after the first stage the data are standardized within segments. Subsequent stages would continue to refine the within-segment standard error estimates and the segment memberships. Preliminary work on such an algorithm seems promising (Milligan 1995), but more work needs to be done. An alternative approach, suggested by Gnanadesikan, Harvey and Kettenring (1994), is to use the data points with the smallest interpoint distances to estimate the standard errors of the variables. That approach assumes homogeneity of within-cluster variances for each variable.

Gnanadesikan, Kettenring and Tsao (1995) investigated various forms and procedures for variable weighting and scaling in cluster analysis. The main conclusions that can be drawn about variable weighting, scaling and selection are that: (1) scaling of variables is generally ineffective, but range scaling is superior to variable standardization, (2) weighting based on estimates of within-cluster variability works well and dominates various forms of variable scaling, (3) simultaneous estimation of weights to optimize a hierarchical tree is ineffective (the procedure of De Soete, DeSarbo and Carroll 1985), (4) optimizing a K-means procedure is more effective (DeSarbo, Carroll and Clark 1984) but never a top performer, and (5) forward variable selection is often, but not always, among the best performers (Fowlkes, Gnanadesikan and Kettering 1988).

At this stage, no definitive conclusions about which procedure performs best can be drawn. However, the procedures of using many variables in standard distance-based clustering methods and/or scaling the variables before clustering should be discouraged. Also, the common procedure of using principal components analysis to reduce the number of variables before clustering cannot be recommended, because PCA discards relevant distance information and several arbitrary decisions are involved. The tandem combination of principal components analysis and clustering is not justifiable on theoretical or empirical grounds (cf. Arabie and Hubert 1994).

Outliers and Missing Values

Further potential problems with clustering methods are posed by outliers and missing values. In cluster analysis, it is not clear a priori what an outlier is. It may be an entity that is outside the measurement space of its segment. On the other hand, one or more outliers may represent a unique segment, or a very small segment that is insufficiently covered by the sampling procedure used. Outliers can be identified by using a clustering procedure such as single linkage, which tends to isolate anomalous values. Strategies to cope with outliers are to remove them from the data once they are identified or to use clustering procedures that are less sensitive to the presence of outliers, such as Ward's hierarchical method or many of the nonhierarchical procedures. Cheng and Milligan (1996) proposed a way to measure the impact of individual data points in clustering. The influence of a point on the resulting cluster partition is related to the extent to which different partitions result with or without that point in the data set. The authors propose that a modification of the Hubert and Arabie (1985) Rand index be used for that purpose.

Missing values can likewise be addressed in various ways. First, in hierarchical cluster analysis, the similarity of each pair of objects can be calculated on the basis of variables for which values are measured. Second, all entities with one or more missing values can be eliminated from the data. Third, missing values can be estimated by the mean of the variable in question, or from regressions on the remaining variables. Here, segment-specific calculations would be preferable, but again the segments are unknown a priori. A problem with such imputation procedures is that the statistical properties of the completed data set are unknown. A solution is to use multiple imputation procedures (Rubin 1987). However, for the cluster analysis procedures described in this section, a rigorous statistical treatment is lacking. More satisfactory methods for estimating missing values in cluster analysis are described by Dixon (1979).

Non-uniqueness and Inversions

Non-uniqueness and inversions are two problems that may occur in hierarchical cluster analysis. Although known for some time, they received relatively little attention until a recent paper by Morgan and Ray (1995). *Non-uniqueness* a problem with several hierarchical methods, is related to the occurrence of ties in the data. When several pairs of objects have the same level of similarity (to the level of accuracy used in the study), several sets of hierarchical cluster structures may be consistent with the data, depending on which rule is used to break the ties. With single linkage, non-uniqueness, is not a problem.

Inversion is the problem of two clusters being joined at a certain step of the algorithm at a level of similarity that is lower than the level of similarity at which clusters were joined at a preceding step. For example, in Figure 5.2, the clusters AE and HD are joined at a similarity level of 10.0. In the next step, KB is added with a level of similarity of 14. Had the similarity of KB with AEHD been 8.0 (while its similarity with both AE and HD was > 10.0), an inversion would have occurred, because KB would be joined to AEHD at a lower level of similarity (8.0) than the similarity of AE and HD (10.0). Such inversions may occur as the consequence of working with moving centroids, and consequently the problem arises only with the median and centroid linkage algorithms.

Morgan and Ray (1995) studied the occurrence of inversions and non-uniqueness for seven hierarchical methods on 25 datasets. It appeared that the non-uniqueness problems were worst for complete linkage and less severe for (weighted) average linkage algorithms. Non-uniqueness certainly did not appear to be a rare event, and the authors advise researchers to check for non-uniqueness when using standard hierarchical clustering methods. They suggest the use of clustering criteria to compare the adequacy of the resulting multiple solutions. Another approach is to construct a common, pruned tree from the multiple solutions (for this technique, refer to the Morgan and Ray article).

Inversions occurred frequently with both the median and centroid linkage methods. Because the resulting trees are useless and no remedies are known, Morgan and Ray strongly advise against the use of those clustering methods.

Cluster Validation

A final issue is cluster validation, or replication. Strategies for validation may be based on *external, internal* and *relative* criteria. External criteria measure performance by matching a cluster structure to exogenous information. Internal criteria assess the fit between the cluster structure and the data, using only the data themselves. Relative criteria involve the comparison of two cluster structures.

Beckenridge (1989) provides evidence that replication analysis may be useful in the validation of identified segments. Milligan (1994) summarizes the major steps: (1) the data are split into two random samples, (2) a cluster analysis is applied to the first sample, (3) the objects in the second sample are assigned to the clusters on the basis of distances to the cluster centroids, (4) the second sample is cluster analyzed, (5) the agreement between the two cluster solutions is computed. For that purpose, Hubert and Arabie's (1985) corrected Rand index is recommended. The level of agreement between the two classifications of the second (holdout) sample reflects the stability of the cluster solution across independent samples. Another important criterion is the interpretability of the cluster structure, that is, its face validity. To be managerially relevant, the number of clusters must be small enough to allow strategy development. At the same time, each segment must be large enough to warrant strategic attention (Arabie and Hubert 1994).

Cluster Analysis Under Various Sampling Strategies

In this section we address the issue of how to apply market segmentation methods to data that arise from various complex-sampling strategies. Complex sampling strategies include, for example, stratified, cluster and multistage sampling. We are concerned with statistical inferences about segments in the population identified on the basis of a sample obtained with a complex sampling strategy. Good surveys use the structure of the population and employ sampling designs that incorporate stratification and different levels of clustering of the observations to yield more precise estimates. Such sampling strategies are commonly used in marketing research. The emphasis in survey research has traditionally been on description, which still appears to be its dominant use in applied marketing research. The theory of probability-weighted estimation for descriptive purposes has been very well established; the book by Cochran (1957) remains a valuable source almost 20 years after its first publication. We refer to that book for details on sampling and estimation methods. However, surveys are being increasingly used for analytic purposes. In traditional inference for descriptive purposes the complexities in sample design are often intimately connected to the estimation procedures employed, but many statistical methods for data analysis do not take the complexity of the sampling strategy into account. Recently, that problem has been addressed by, for example, Skinner, Holt and Smith (1989).

In applications of clustering-type techniques to survey data, the sampling strategy used for data collection is mostly ignored. The clustering methodologies described heretofore all assume that the data come from a simple random sample of the population. If that is the case, the derived segment structure represents the segments present in the population. However, when more complex designs are used, that is no longer the case. The general solution we propose is to employ the segments obtained from the cluster analysis as a post-stratification factor in the sampling design. We describe it in detail below for several sampling designs, in particular stratified, cluster and two-stage samples.

Stratified samples

Stratified samples arise when the population can be divided into, say, G mutually exclusive strata (e.g., regions), from which separate random samples are drawn. Thus the sample of size N is stratified into G groups. Now a clustering technique is applied to the (K × 1) observation vectors $\mathbf{y}_n$ of the N subjects. We assume an appropriate clustering method (e.g., Ward's) is used to arrive at initial cluster centers, followed by a nonhierarchical procedure minimizing the det($\mathbf{W}$) criterion. Assume the Calinski and Habarasz criterion points to S segments. Under a stratified sampling strategy, the sample means of the $\mathbf{y}_n$ within each cluster are not unbiased estimates of the segment-level means in the population, and the segment sizes estimated from the sample do not conform to the sizes of the segments in the population. The reason is that the subjects composing the segments had unequal probabilities of being selected into the sample. The proper treatment of this problem is based on the fact that the segment structure of the sample presents a post-stratification on the basis of the unobserved grouping variable that identifies the segments. Hence, the appropriate estimate of the sample mean for each variable is a weighted estimate, where each subject is weighted by its selection probability. That selection probability equals $W_g = N_g^{(p)}/N^{(p)}$, where g is the stratum from which the respondent was drawn (we assume the stratum sizes in the population are known). Thus, the problem becomes one of estimating the domain mean for each of the S domains, given the stratification into G other strata. Cross-tabulating the S segments and the G strata yields estimates y_{sg}. Then the estimate of the population mean vector for segment s (or the vector of proportions if Y is discrete) is:

$$\overline{Y}_s = \sum_{g=1}^{G} W_g \, \overline{y}_{sg\bullet} \quad . \tag{5.1}$$

Unfortunately, the variance of an estimate of y_{ks} cannot be computed in the usual way. The reason is that a clustering procedure is used to minimize the within-segment variance, and the statistical properties of that variance are no longer known. Bootstrap procedures could be used to obtain correct estimates of the variance.

Given that we calculate the segment sizes within each of the G strata, $\pi_{s|g}$, the segment sizes are equal to:

$$\hat{\pi}_s = \sum_{g=1}^{G} W_g \, \hat{\pi}_{s|g} \quad . \tag{5.2}$$

By the standard formulas for stratified sampling, the variance of the estimate of π_s is:

$$Var(\hat{\pi}_s) = \sum_{g=1}^{G} W_g^2 (1 - f_g) \frac{\hat{\pi}_{s|g}(1 - \hat{\pi}_{s|g})}{N_g} \quad . \tag{5.3}$$

Using that formula, we can evaluate the effects of the sampling strategy on the precision of the estimates of the segment sizes. It is well known that in this case the

Part 2 Segmentation methodology

gain in precision from stratified random sampling over simple random sampling is small (< 30%) unless the π_{sg} vary greatly across the G strata and are outside the interval (0.2, 0.8). In effect what this tells us is that when we expect very small segments to be present, it would be efficient to employ a stratified sampling strategy, with stratification variables that are maximally correlated with the (yet unknown) underlying segments That may in fact be achieved by some of the hybrid segmentation procedures discussed in chapter 3. Thus, an additional advantage of such hybrid procedures is increased precision of the estimates. Optimal allocation provides little gain over proportional allocation. The presumption is that interest focuses on accurately identifying segment *sizes*. If we have prior estimates of the segment proportions, we can calculate the required size of the stratified sample. For a proportional allocation, an initial estimate (assuming the sampling fraction to be negligible) is provided by:

$$N_0 = \sum_{g=1}^{G} \frac{W_g \pi_g (1 - \pi_g)}{(d/t)^2},$$

where d is the desired error level and t is the value of the t-distribution corresponding to the desired significance level. The corrected value, taking account of the finite population correction, is obtained as above.

Cluster samples

For cluster samples, population estimates can be obtained if we again consider the segments arising from the application of the clustering algorithms to the sample as a post-stratification factor. A cluster sample is used when the elements in the population occur naturally in groups, such as households, neighborhoods or firms, where all members of the group are included in the sample. The purpose is to provide descriptive statistics for the population in each of the S clusters. A distinction should be made as to whether the segments need to be identified at the level of the primary units (i.e., sampling clusters) or at the level of the secondary units (i.e., consumer or firm). The latter situation is most common. If segments across the primary units are sought, the clustering method should be applied to measures at the primary-unit level, such as the means or proportions of the variables across the secondary units for each primary unit. Thus the data are first aggregated for each primary unit before the clustering procedures are applied. An example of this situation is a cluster sample of firms in which the board members of each firm are interviewed. If the purpose is to identify segments of firms, the data of the individual board members are aggregated before clustering. In this case, if a simple random sample is applied to sample the clusters of elements, the ordinary means calculated for each segment provide direct estimates of the population values. Unfortunately, variances of the estimates cannot be computed because the standard estimates do not take into account the fact that the data have been subjected to a clustering routine. Bootstrap procedures can be applied to provide the correct estimates for the variances of the population means.

If segments are assumed to be present at the level of the secondary units (i.e., the individual members of the board of each firm in the example, denoted by m = 1,...,M$_n$), a clustering routine is applied to the measurements at the secondary-unit level. To obtain population estimates, the derived segment structure of the sample is considered as a post-stratification factor, so the calculation of population estimates for most sampling procedures is straightforward. The estimates of the population mean of variable k within segment s, when for example the primary units are of unequal size and are selected with equal probabilities, are:

$$\overline{Y}_{ks} = \frac{\sum_{n=1}^{N}\sum_{m=1}^{M_n} y_{knms}}{\sum_{n=1}^{N} M_n}. \qquad (5.4)$$

The estimates of the proportions of the segments in the population can be calculated as follows. For example, suppose the primary units have unequal sizes M_n, and a sample of N primary units is drawn with probabilities proportional to size. Now for each primary unit, an estimate π_{ns} is available, by dividing the total number of secondary units in segment s by M_n. The estimator for the segment size in the population is:

$$\hat{\pi}_s = \frac{1}{N}\sum_{n=1}^{N}\hat{\pi}_{ns}. \qquad (5.5)$$

An estimate of the variance of the estimated segment proportion can be calculated as:

$$Var(\hat{\pi}_s) = \sum_{n=1}^{N}\frac{(\hat{\pi}_{ms} - \hat{\pi}_s)^2}{N(N-1)}. \qquad (5.6)$$

The variance estimate in equation 5.6 can be used, for example, in sample size calculations if initial estimates of the π_s and π_{ns} are available. An initial estimate of the sample size (ignoring the finite population correction) is then provided by

$$N_0 = (\frac{t}{d})^2 Var(\hat{\pi}_s),$$

with *d* and *t* defined as before.

Two-stage samples

The results for two-stage samples are rather straightforward extensions of the results for cluster samples, based on the standard results. Two-stage samples arise when the units in the population occur in groups as above, but the sample of secondary units (individuals) is drawn from each primary unit (households). Again, we use the segment structure derived by some clustering procedure as a post-stratification factor. If the clusters of elements (primary units) in the population have the same size $M^{(p)}$ and the N primary units and the M secondary units (population elements) in the sample are selected by simple random sampling, the population mean is estimated by the sample average. Note that here a constant fraction $M/M^{(p)}$ is sampled from each secondary unit and that the sample is therefore self-weighting.

If the secondary units and primary units are selected by simple random sampling, there are two types of estimators. In the situation where the sampling fraction within each secondary unit is constant $(M_n/M_n^{(p)} = f_2)$, the estimator is self-weighting:

$$\overline{Y}_{ks} = \frac{\sum_{n=1}^{N} \sum_{m=1}^{M_n} y_{knms}}{\sum_{n=1}^{N} M_n} . \tag{5.7}$$

That estimator is the standard estimator for this sampling strategy, but accounting for the post-stratification by segments. Variances of the estimator can be derived only by bootstrapping. Another simple situation occurs when the primary units are selected with probabilities proportional to size,

$$M_n^{(p)} / \sum_{n=1}^{N^{(p)}} M_n^{(p)} .$$

and the size of the sub-sample is constant, $M_n = M$. In this case, the sample is self-weighting. Cochran provides further extensions for selection without replacement, estimators for proportions and totals, and other situations.

The estimates of the segment proportions in the population for two-stage sampling can be calculated as in the case of cluster sampling. As an example, we assume the primary units have unequal sizes M_n, and a sample of N is drawn with probabilities proportional to size. The estimator for the population segment size is now simply the mean of the proportions of the segments across the primary units, as in equation 5.5, and its variance is provided by 5.6. Estimators for the population mean and variance of the mean under different sampling strategies follow directly from the results presented by Cochran (1957).

Overlapping and Fuzzy Methods

Overlapping Clustering

As indicated in part 1, the overlapping clustering model ADCLUS (Shepard and

Arabie 1979) and its generalizations INDCLUS (Carroll and Arabie 1983), GENNCL-US (DeSarbo 1982) and CONCLUS (DeSarbo and Mahajan 1984) provide useful approaches to overlapping clustering on the basis of similarities. Overlapping clusters arise in the models because objects can have zero-one membership in more than one cluster. To yield overlapping consumer segments, the methods are applied to similarities between subjects, where the similarities can be calculated according to the measures summarized in Table 5.1. The computational requirements for some of the iterative algorithms may be excessive for real-life problems with large sample sizes (Arabie et al. 1981), which limits the applicability of these methods to segmentation problems. Moreover, the zero-one measures of overlapping segment membership that these methods provide, are relatively coarse measures in comparison with the degrees of membership provided by the fuzzy methods, discussed below. Hruschka (1986) reported empirical evidence that fuzzy clustering procedures outperform overlapping procedures such as the ADCLUS algorithm.

Overlapping Clusterwise Regression

Whereas the ADCLUS and related algorithms are descriptive, DeSarbo, Oliver and Rangaswamy (1989) propose an overlapping clusterwise regression procedure that allows for the prediction of one or more dependent variables from independent variables in a regression context. The procedure is similar to that of Wedel and Kistemaker (1989) described previously in this chapter, but it provides overlapping clusters and uses a simulated annealing algorithm for optimization. The simulated annealing procedure was devised as a method to overcome problems of local optima. It starts from a random initial partition of the sample, and iteratively specifies steps in a random direction of the parameter space. The new solution thus obtained is accepted if it improves the criterion being optimized; if it does not, it is rejected with a probability proportional to the decrease in the criterion value. The computational effort is reported to be considerable, which in practical applications also limits the subspace of the parameters that can be searched. The method uses the more coarse zero - 1 measure of segment membership, which is shown by Wedel and Steenkamp (1989) to be outperformed by fuzzy clusterwise regression methods.

Fuzzy Clustering

The concepts of fuzzy sets and partial set membership were introduced by Zadeh (1965) as a way to handle imprecision in mathematical modeling. Dunn (1974) and Bezdek (1974) recognized the applicability of Zadeh's concepts to clustering problems, and developed the *fuzzy c-varieties* (FCV) family of clustering algorithms. In the FCV clustering methods based on fuzzy-set theory, the traditional assumption that every object is to be assigned to one and only one cluster is relaxed and replaced with two assumptions:(1) an object can belong to more than one cluster, where its degree of membership in a cluster is represented by a number in the interval [0,1], that is, $0 \leq u_{ns} \leq 1$, and (2) the total membership of an object in a cluster must sum to one

Part 2 Segmentation methodology

(i.e., $\Sigma_s\, u_{ns} = 1$) for all n. Four major methods of fuzzy clustering have been proposed (the computational details of the methods are provided in Table 5.4):

1. The fuzzy c-means cluster algorithm, FCM (Bezdek 1974; Dunn 1974), is a descriptive clustering algorithm, and derives clusters of a round shape.

2. The fuzzy c-lines algorithm, FCL (Bezdek et al. 1981a,b), is also a descriptive algorithm, and derives clusters of a linear or planar shape.

3. The fuzzy clusterwise regression algorithm, FCR (Wedel and Steenkamp 1989, 1991), is a predictive algorithm, and derives clusters that differ in their regression weights.

4. The fuzzy grade of membership model, GoM, for contingency tables and discrete data, proposed by Manton, Woodbury and Tolley (1994).

Fuzzy clustering procedures provide a flexible approach to clustering that allows for the identification of clusters of various shapes. Figure 5.4 provides an example.

The purpose of the FCM algorithms proposed by Dunn (1974) and Bezdek (1974) is to estimate the cluster centroids of a pre-specified number of clusters and the degree of membership of objects in those clusters. The algorithms arrive at a partitioning of the data by minimizing a sum of squared errors, analogous to the K-means algorithm or Ward's hierarchical method. That function is the same objective function minimized by the FCM, FCL and FCR algorithms. At the first step, the user provides a starting partition for the algorithm, which may be obtained by some other method (e.g., Ward's minimum variance). Next the algorithm iteratively arrives at a partitioning of the data by iterating between the two equations provided in Table 5.4 calculating respectively the memberships of subjects in the segments, p_{ns}, and the means of the variables characterizing each segment, y_{ks}. In other words, given initial memberships, the cluster means are computed. Those fuzzy means of all K variables for each of the classes are computed as simple weighted means, where the fuzzy memberships present the weights. The distances of subjects to these new cluster means are then calculated, as well as their degree of membership in the new clusters, which is based on their distance to the clusters. Some constant $r > 1$, must be pre-specified by the user, influencing the degree of separation of the clusters obtained. The iterations proceed until no improvement occurs (it can be shown that the iterations always converge).

The FCM algorithm can be seen as the fuzzy variant of the K-means nonhierarchical algorithm. Hruschka (1986) compared FCM empirically with a traditional nonoverlapping (k-means) and an overlapping (ADCLUS) procedure. The superiority of FCM was clearly demonstrated.

Figure 5.4: An Example of Cluster Structures Recovered with the FCV Family

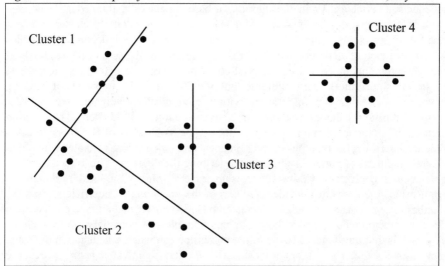

A generalization of the FCM algorithm developed by Bezdek et al. (1981a,b), called fuzzy c-lines (FCL), identifies clusters with a linear shape. The algorithm is very similar to the FCM algorithm, but it differs in how the cluster structure and the distances between clusters and objects are defined. Here, each cluster is characterized by a linear cluster structure described by a set of Q principal components (obtained by a PCA of the weighted within-cluster covariance matrix). A linear cluster structure is obtained if one principal component is used (Q = 1), a planar cluster structure if two principal components are used for that cluster (Q = 2), and so on. Thus, instead of calculating the cluster means on the raw data as in the FCM algorithm, the analyst first performs a principal components analysis within each cluster. The distance of an object to a cluster is now defined as the distance of the object to the principal components of that cluster (Table 5.4). Again, the clustering of the data is obtained from an iteration of the two equations for calculating the cluster principal components and the subject memberships (provided in the Table 5.5), starting from some initial partition of the sample. Convergence of the iterations is ensured. However, the maximum obtained may be a local maximum, and the algorithms may need to be run from different starts to investigate the seriousness of the local optima problem. The FCL algorithm was demonstrated to perform well in the recovery of linear cluster structures. Disadvantages of the algorithm are that it requires several choices to be made by the researcher: first, S, the number of segments, must be decided, second, r, the fuzzy weight parameter, must to be chosen, and third the dimensionality of the linear cluster structure, Q, must be determined. For the third problem, a solution has been provided by Gunderson (1982), which involves an adaptive estimation algorithm that may yield clusters of different (round and linear) shapes.

A third member of the FCV family is the fuzzy clusterwise regression (FCR) algorithm. Contrary to the FCM and FCL methods, this algorithm is an interdependence method that seeks to explain, within each cluster, a dependent variable from a set of independent variables. Again, the algorithm is based on the estimation equation derived from minimizing a sum of squared residuals. In the FCR algorithm the clusters are defined from regressions of the dependent variables **y** on a set of P independent variables **X**. The distance of an object to a cluster is here defined as the sum of the residuals of the measurements of the object relative to their predicted values for that cluster. The clustering algorithm is again similar to that of FCM and FCL: it consists of iterating between calculating the regressions within each cluster (weighted with cluster membership) and calculating the memberships of subjects in clusters. The solution consists of a characterization of each of the segments by a set of P regression coefficients describing the relations of the dependent and independent variables within that cluster, and a set of membership values describing the degrees of membership of the objects in the clusters. Convergence is guaranteed only to a local optimum. In applications the user must decide on values of the fuzzy weight coefficient r and the number of segments S. Wedel and Steenkamp (1989) provide a heuristic procedure to determine the optimal values of those numbers from plots. A potential problem with the approach is that the standard errors of the regression estimates, calculated in the usual way, are not valid. The authors propose a Monté Carlo significance testing procedure for that purpose. Here, the null hypothesis is to be rejected if the test-criterion from the observed data (e.g., the t-value of the regression coefficient) is greater (or smaller) than at least $M - M(\alpha/2)$ values of the same test criterion calculated from $M - 1$ random samples (α is the level of significance and M is an integer, usually $M = 20$). The random samples are obtained by randomly permuting the dependent variable across n and k. Empirical comparisons with overlapping clusterwise regression procedures and Monté Carlo studies support the performance of the FCR algorithm. Wedel and Steenkamp (1991) extend the FCR procedure to a two-mode clustering context; the discussion of that procedure is beyond the purpose of this chapter.

A final method for fuzzy set clustering, the grade of membership mode, GoM, was proposed by Manton, Woodbury and Tolley (1994). Their model is tailored to the analysis of discrete data, where y_{nkl} is zero or one (here, $l = 1,...,L_k$ denotes the categories of the outcome of variable k). The probability that $y_{nkl} = 1$ (denoted by λ_{nkl}, where $\Sigma_l \lambda_{nkl} = 1$ for all n and k) is expressed as a bilinear form, as the product of a segment-specific probability (λ_{skl}) and the grade of membership of subject n in segment s (p_{ns}). The model is estimated from the data by using the method of maximum likelihood. The purpose of likelihood estimation is to find parameters such that the observations **y** are more likely to have come from the multinomial distribution describing the data with that set of parameters than from a multinomial distribution with any other set of parameters. It is accomplished by maximizing the likelihood function, which measures the relative likelihood that different sets of parameters have given rise to the observed vector **y**. For the GoM model, the likelihood, L, is specified on the basis of the multinomial distribution for the data. The method estimates the grade of membership of each observation vector for subject n in segment s, and the segment-specific response probabilities for the categories of the K variables within each segment. Maximizing the likelihood amounts to alternating between two estimation equations: one for the segment-specific probabilities of a category l of the discrete variable k, and the other for the grade of membership of subject n in each class (the likelihood expression for the mode and the two estimation equations are provided in Table 5.4).[1]

A complete family of models has spawned from the basic GoM model, including models for longitudinal discrete data and empirical Bayes approaches. Discussion of those approaches is beyond the purpose of this chapter. Manton, Woodbury and Tolley (1994) show that some latent class methods, which arise as special cases of mixture models for categorical data, are in fact special cases of the GoM model. They demonstrate theoretically and empirically that the GoM model represents individual differences better because of the grade of membership representation. Mixture and latent class models are described in more detail in the next chapters.

Market Segmentation Applications of Clustering

The number of applications of clustering methods in marketing is enormous, making it virtually impossible to review all applications to date. In Table 5.5 we summarize several illustrative applications. The relatively older references, drawn from the review by Punj and Stewart (1983), are supplemented with some more recent references. This incomplete listing shows both the wide array of problems addressed and the large variety of clustering methods used. In many segmentation studies, however, the authors are not specific about the type of clustering method applied to their segmentation problem.

---

[1] The authors note that p_{ns} cannot be estimated consistently. Rather, they show that their procedure provides consistent estimates of the moments of the distribution of the p_{ns} to the order K (consistency for each p_{ns} is obtained asymptotically if K increases much faster than S)

Part 2 Segmentation methodology

Table 5.4: Fuzzy Clustering Algorithm Estimating Equations

	Criterion	Memberships
FCM	$J_{FCV} = \sum_{s=1}^{S} \sum_{n=1}^{N} p_{ns}^{r} D_{ns}$	$p_{ns} = \dfrac{1}{\sum_{t=1}^{S} (D_{ns} / D_{nt})^{2/(r-1)}}$
FCL	$J_{FCV} = \sum_{s=1}^{S} \sum_{n=1}^{N} p_{ns}^{r} D_{ns}$	$p_{ns} = \dfrac{1}{\sum_{t=1}^{S} (D_{ns} / D_{nt})^{2/(r-1)}}$
FCR	$J_{FCV} = \sum_{s=1}^{S} \sum_{n=1}^{N} p_{ns}^{r} D_{ns}$	$p_{ns} = \dfrac{1}{\sum_{t=1}^{S} (D_{ns} / D_{nt})^{2/(r-1)}}$
GoM	$L = \prod_{n=1}^{N} \prod_{k=1}^{K} \prod_{l=1}^{L_k} (\sum_{s=1}^{S} p_{ns} \lambda_{skl})^{Y_{nkl}}$	$p_{ns} = \dfrac{1}{y_{n++}} \sum_{n=1}^{N} \sum_{l=1}^{L_k} y_{nkl} \dfrac{p_{ns}^{*} \lambda_{skl}^{*}}{\sum_{s=1}^{S} p_{ns}^{*} \lambda_{skl}^{*}}$

Table 5.4: Fuzzy Clustering Algorithm Estimating Equations - continued

	Distance	Segment-structure
FCM	$D_{ns} = \sum_{k=1}^{K} (y_{nk} - \bar{y}_{ks})^2$	$\bar{y}_{ks} = \dfrac{\sum_{s=1}^{S} p_{ns}^r y_{nk}}{\sum_{s=1}^{S} p_{ns}^r}$
FCL	$D_{ns} = \sum_{q=1}^{Q}\sum_{k=1}^{K} (y_{nk} - z_{kqs})^2$	$S_s = \sum_{n=1}^{N} p_{ns}^m (y_n - \bar{y}_s)(y_n - \bar{y}_s)' = Z_s \Lambda_s Z_s$
FCR	$D_{ns} = (y_n - X_n\beta_s)'(y_n - X_n\beta_s)$	$\beta_s = (X' P_s X)^{-1} X' P_s y,$
GoM	-	$\lambda_{skl} = \dfrac{\sum_{n=1}^{N} y_{nkl} \dfrac{p^*_{ns}\lambda^*_{skl}}{\sum_{s=1}^{S} p^*_{ns}\lambda^*_{skl}}}{\sum_{n=1}^{N} y_{nk} + \sum_{l=1}^{L_k} \dfrac{p^*_{ns}\lambda^*_{skl}}{\sum_{s=1}^{S} p^*_{ns}\lambda^*_{skl}}}$

Table 5.5: Cluster Analysis Applications to Segmentation

Authors	Year	Purpose of research	Clustering method
Anderson, Cox and Fulcher	1976	Segmenting commercial bank customers on the basis of bank selection attribute scores	Iterative partitioning
Bass, Pessemier and Tigert	1969	Segmenting consumers on the basis of media exposure variables	Average linkage
Calantone and Sawyer	1978	Identification of segments in the retail banking market on the basis of bank selection variables	K-means
Claxton, Fry and Portis	1974	Segmentation of furniture and appliance buyers on the basis of information search behavior	Complete linkage
Greeno, Sommers and Kernan	1973	Identification of consumer market segments on the basis of personality and implicit behavior	Ward's method
Kernan	1986	Identification of segments on the basis of personality and decision behavior	Ward's method
Kiel and Layton	1981	Segmentation of car buyers on the basis of search behavior	Average linkage
Landon	1974	Consumer segmentation on the basis of buying behavior	Ward's method
Montgomery and Silk	1972	Identify opinion leadership and consumer interest segments	Complete linkage
Moriarty and Venkatesan	1978	Segmentation of educational institutions on importance of financial aid services	K-means
Myers and Nicosia	1968	Consumer segmentation on the basis of supermarket image	Iterative partitioning
Sethi	1971	Country segmentation on the basis of macro-level data	Iterative partitioning
Schaninger, Lessig and Panton	1980	Segmenting consumers on the basis of product usage	K-means
Assael & Roscoe	1976	Socioeconomic and demographic segmentation	AID

Table 5.5 – continued

Authors	Year	Purpose of research	Clustering method
Currim	1981	Benefit segmentation using conjoint	Ward's method
Singh	1990	Segmenting consumers on the basis of complaint behavior for services	Ward's, K-means
Maier and Saunders	1990	Segmentation of doctors on the basis of drug prescription behavior	A-priori and Ward's method
Hruschka	1986	Usage situation segmentation	FCM
Steenkamp and Wedel	1993	Segmentation of consumers on the basis of benefits	FCR
Steenkamp and Wedel	1991	Identify consumer segments on the basis of retail image	FCR
Magdison	1994	Identification of consumer segments in direct mail on the basis of demographics	CHAID

6
MIXTURE MODELS

We now discuss the main statistical approach to clustering and segmentation: mixture models. We examine a general form of such mixtures and describe the EM estimation algorithm. We consider problems of the model and its estimation that are related to identification, local optima, standard errors and the number of segments. Those issues provide the basic foundation for the chapters that follow.

The development of models for mixtures of distributions dates back to the work of Newcomb (1886) and Pearson (1894). In finite mixture models, the observations in a sample are assumed to arise from two or more groups that are mixed in unknown proportions. The purpose is to "unmix" the sample, that is, to identify the groups or segments, and to estimate the parameters of the density function underlying the observed data within each group. Mixture models assume a specific density function for the observations within each of the underlying populations. The density function is used to describe the probabilities of occurrence of the observed values of the variable in question, conditional on knowing the group from which those values were drawn. The normal distribution is probably the most frequently used distribution for continuous variables, whereas the multinomial distribution is commonly used for multichotomous variables. The unconditional likelihood of the observed data is then defined as a mixture of the group-level densities, with the mixing weights defining the relative size of each group. To illustrate the mixture approach, we provide two examples taken from the literature.

Mixture Model Examples

Example 1: Purchase Frequency of Candy

Dillon and Kumar (1994) provide an illustrative application of mixture models in marketing. They describe data on the number of packs of a new hard-candy product purchased by a sample of 456 consumers. Figure 6.1 shows part of the frequency distribution of the number of packs purchased. The frequency distribution can be modeled by a Poisson distribution. The Poisson distribution is characterized by a single parameter: the mean, λ, which represents the mean purchase rate for the category. The (maximum likelihood) estimate of that parameter is simply the average of the observations, which is 3.99 purchases per week in the example. The figure also shows the frequencies of purchases predicted by the Poisson distribution with that mean. It is obvious that the Poisson distribution does not fit the data very well. Dillon and Kumar attribute the lack of fit to the fact that the data are too heterogeneous, and the Poisson

Part 2 Segmentation bases

distribution cannot account for the heterogeneity. They subsequently assume that the heterogeneity arises from the presence of a number of segments of consumers in the sample that purchase the product at different rates. Those segments are unobserved, but can be recovered from the data by using a mixture Poisson model. Dillon and Kumar estimate such a model, which assumes that within each unobserved segment, consumers purchase the product according to a Poisson process with a segment-specific rate. They estimate the mixture Poisson model on the data and identify six underlying segments.

Two segments consist of relatively light users: segment 1 (15.6%), non-buyers (average purchase rate of 0.00), and segment 2 (13.4%), consumers purchasing 1.08 packs per week on average. The largest segment, segment 3 (47.4%), purchased 3.21 packs on average, and segment 4 (12.8%), consisting of heavy users, purchased 7.53 packs on average. The remaining two segments, segment 5 (3.0%) and segment 6 (7.9%), are small and are made up of extremely heavy users, purchasing 11.1 and 13.6 packs per week, respectively. Dillon and Kumar showed that the six-segment solution provided a much better fit to the data (Figure 6.1). The application provides an example of usage frequency segmentation, enabling managers to identify and target segments that are light and heavy users to varying degrees.

Figure 6.1: Empirical and Fitted Distributions of Candy Purchases (adapted from Dillon and Kumar 1994)

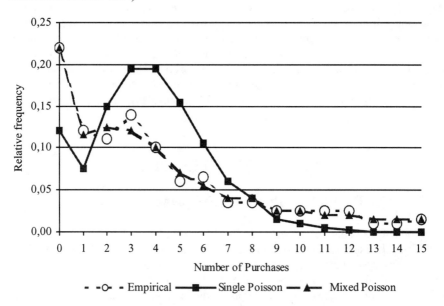

Example 2: Adoption of Innovation

A second example of an application of mixture models to market segmentation has been provided by Green, Carmone and Wachpress (1976). It actually represents the first application of mixture models in marketing. The authors consider a contingency table, where 10,524 consumers are classified according to five variables collected in a mail survey executed for AT&T after the introduction of a new telecommunications service.

- education: low: high-school graduate or less; high: at least some college
- mobility: non-motile: no change in residence in five years; mobile: one or more moves
- age: young: < 35, middle: 35-54, older: > 54
- income: low: < $12,500; middle: 3 = $12,500-$20,000; high > $20,000
- reaction to Innovation: adopters or non-adopters of a new telecommunications service

The five-way contingency table consists of 72 cells showing various statistically significant dependencies among the variables. Green and his co-authors assume that the associations found among the variables result from an unobserved discrete variable, which is associated with those observed in the contingency table. The unobserved variable represents the segmentation variable, classifying the respondents into unobserved segments. Had the segmentation variable been observed, computing the five-way cross-classification table within each of the segments would have revealed no association among the five measured variables for each segment. The observed associations among the five classification variables are assumed to be caused by the differences across the various levels of the unobserved discrete variable (i.e., segments). However, because the discrete variable is unobserved, must to be estimated.
That was accomplished by using latent class analysis, a special type of mixture model in which the observed variables have binomial and/or multinomial distributions within each segment (i.e., a mixture of multinomials). The results indicate that a three-segment solution provides a good fit to the data. The profile of probabilities for the categories within each variable for each segment are used to aid in interpretation of the segments. The results are reproduced in Table 6.1. Segment 1 (37%) has a high probability of adopting the new service, 0.25. It is characterized as young, highly educated, mobile and of relatively low income. Probability of adopting is 0.16 for the sample as a whole. Segment 2 (34%) also contains a higher than average proportion of adopters. However, it differs from segment 1 in being less mobile, older and higher in income. Segment 3 (29%) tends to consist of non-adopters. They have lower education and lower mobility, are older and have lower income than the other two segments. The segmentation scheme supports differential promotion targeted, for example, at segments 1 and 2.

Mixture Distributions (MIX)

Mixture distributions have been widely used in the field of cluster analysis. Here, such mixtures are used as a clustering or segmentation technique. Unlike most of the procedures described in the preceding chapter, which merely present convenient heuristics for deriving groups or segments in a sample, mixture distributions are a

Part 2 Segmentation bases

statistical-model-based approach to clustering. They allow for estimation within the framework of standard statistical theory. Moreover, the mixture approach to clustering provides a very flexible class of clustering algorithms that can be tailored to a very wide range of substantive problems, as demonstrated by the two examples above. Thus, mixture models enable marketing researchers to cope with heterogeneity in their samples and to identify segments by using a model-based approach that connects classical clustering to conventional statistical estimation methods. We demonstrate those capabilities in this and the subsequent chapters. Here we predominantly use the books by Everitt and Hand (1981), Titterington, Smith and Makov (1985), McLachlan and Basford (1988) and Langeheine and Rost (1988) as well as the reviews papers by Dillon and Kumar (1994) and Wedel and DeSarbo (1994) as the basic sources.

Table 6.1: Results of the Green et al. Three-Segment Model (adapted from Green et al. 1976)

Variable	Segment 1 (%)	Segment 2 (%)	Segment 3 (%)
1. Low education	0.36	0.39	0.76
High education	0.64	0.61	0.24
2. Non-motile	0.13	0.78	0.81
Mobile	0.87	0.22	0.19
3 Young age	0.83	0.07	0.11
Middle age	0.16	0.59	0.29
Older age	0.01	0.34	0.60
3. Low income	0.57	0.05	0.85
Middle income	0.31	0.44	0.14
High income	0.12	0.51	0.01
4. Adopter	0.25	0.18	0.06
Non-adopter	0.75	0.82	0.94

To formulate the finite mixture model, which we call MIX, assume the objects (consumers or firms) on which the variables $\mathbf{y_n} = (y_{nk})$ are measured arise from a population that is a mixture of S segments, in proportions $\pi_1,....,\pi_S$. We do not know in advance from which segment a particular object arises. The probabilities π_s are subject to the following constraints.

$$\sum_{s=1}^{S} \pi_s = 1, \quad \pi_s \quad \pi_s \geq 0, \quad s = 1,\ldots, S. \tag{6.1}$$

Given that y_{nk} comes from segment s, the conditional distribution function of the vector $\mathbf{y}_n$ is represented by the general form $f_s(\mathbf{y}_n|\boldsymbol{\theta}_s)$, where $\boldsymbol{\theta}_s$ denotes the vector of all unknown parameters associated with the specific form of the density chosen. For example, in the case that the y_{nk} within each segment are independent normally distributed, $\boldsymbol{\theta}_s$ contains the means, μ_{ks}, and variances, σ_s^2, of the normal distribution within each of the S segments. In the two examples above, $f_s(\mathbf{y}_n|\boldsymbol{\theta}_s)$ represents respectively the Poisson and the binomial (multinomial) distributions, and $\boldsymbol{\theta}_s$ the segment-specific purchase rates or multinomial probabilities. The simple idea behind the mixture distributions is that if the distributions conditional upon knowing the segments have been formulated, the unconditional distribution of $\mathbf{y}_n$ is obtained as:

$$f(\mathbf{y}_n | \phi) = \sum_{s=1}^{S} \pi_s f_s(\mathbf{y}_n | \theta_s), \tag{6.2}$$

where $\phi = (\boldsymbol{\pi}, \boldsymbol{\theta})$. That can be seen quite easily from the basic principles of probability theory: the unconditional probability is equal to the product of the conditional probability given s times the probability of s and that expression summed over all values of s.

The conditional density function, $f_s(\mathbf{y}_{nk}|\boldsymbol{\theta}_s)$, can take many forms, including the normal, Poisson and binomial used above, and other well-known functions such as the negative binomial, Dirichlet, exponential gamma, and inverse Gaussian distributions. All of those more commonly used distributions are members of the exponential family of distributions, a general family that encompasses both discrete and continuous distributions. The exponential family is a very useful class of distributions, and the common properties of the distributions in the class allows them to be studied simultaneously, rather than as a collection of unrelated cases. The distributions are characterized by their means, μ_{ks}, and possibly a dispersion parameter, λ_s. Those parameters are assumed to be common to all observations in segment s.

Another useful property of the exponential family is that a conjugate distribution always exists within the family, so that distributions resulting from compounding are also members of the family. Examples of such compounding distributions are the beta-binomial arising from compounding the beta and binomial distributions, the negative binomial arising from compounding the gamma and Poisson, and the Dirichlet multinomial arising from compounding the Dirichlet and multinomial distributions. Such compound distributions will prove to be useful later to account for within-segment heterogeneity.

Table 6.2 presents several characteristics of some of the most well known univariate distributions in the exponential family. The univariate distributions are listed in Table 6.2. Other distributions (such as the Bernoulli or Geometric) are special cases of distributions listed (binomial and negative binomial, respectively). Table 6.2 is not exhaustive; for example the beta (a special case of the Dirichlet) and inverse Gaussian distributions are also members of the exponential family. The table gives a short notation for each distribution, its distribution function as a function of the parameters, the mean or expectation of the distribution, its variance (or variance-covariance matrix) and the canonical link function, which are discussed more extensively in the next section. McCullagh and Nelder (1989) provide more detailed discussions.

If the observations cannot be assumed to be conditionally independent given knowledge of the segments or mixture components, one of the members of the multivariate exponential family may be appropriate. The two most important and most frequently used distributions in the family are the multinomial and the multivariate normal (shown in Table 6.2). In the latter case, the distribution of y_n takes the well-known form, where μ_s is the (K x 1) vector of expectations and Σ_s is the (K x K) covariance matrix of the vector y_n, for a given mixture component s.

Maximum Likelihood Estimation

The purpose of the analysis of mixture models is to estimate the parameter vector $\phi = (\pi, \theta)$. That is accomplished by using the method of maximum likelihood, which provides a statistical framework for assessing the information available in the data about the parameters in the model. We formulate the likelihood for ϕ as:

$$L(\phi; y) = \prod_{n=1}^{N} f(y_n | \phi). \qquad (6.3)$$

An estimate of ϕ can be obtained by maximizing the likelihood equation (6.3) with respect to ϕ subject to the restrictions in equation 6.1. The purpose of likelihood estimation is to find a parameter vector ϕ_0 such that the observations y are more likely to have come from $f(y|\phi_0)$ than from $f(y|\phi)$ for any other value of ϕ. The estimate ϕ_0 thus maximizes the likelihood of the observed data, given the model. That is accomplished by maximizing the likelihood function L(.) above, which measures the likelihood that the parameter vector ϕ could have produced the observed vector y. The likelihood is simply the product over the densities of the N individual observations (or observation vectors), as those are assumed to be independent. Hasselblad (1966, 1969) was among the first to use maximum likelihood estimation for mixtures of two or more distributions from the exponential family. The likelihood approach for finite normal mixtures has become increasingly popular (e.g., Day 1969; Wolfe 1970; Fowlkes 1979; Symons 1981; McLachlan 1982; Basford and McLachlan 1985; Wedel and DeSarbo 1995). A review of maximum likelihood algorithms for mixture models is provided by Böhning (1995).

The likelihood of finite mixtures can be maximized in basically two ways: by using standard optimization routines such as the Newton-Raphson method (McHugh 1956, 1958) or by using the expectation-maximization (EM) algorithm (Dempster, Laird and Lubin 1977). Both methods find an approximate solution of the likelihood equations in an iterative fashion. The Newton-Raphson method requires relatively few iterations to converge, and provides the asymptotic variances of the parameter estimates as a byproduct, but convergence is not ensured (McLachlan and Basford 1988). In the EM algorithm, iterations are computationally attractive, the algorithm can usually be programmed easily, convergence to a local optimum is ensured, but the algorithm requires many iterations (Titterington, Smith and Makov 1985; McLachlan and Basford 1988). Both types of procedures may converge to local optima. As stated in the preceding chapter, this problem is related to the fact that the likelihood may have several local optima, and the algorithm may end up in one of those optima depending on the values of the parameters used to start the iterative search process (note that a similar problem occurs in traditional cluster analysis). The optimum that is reached from a particular start may not be the global (i.e., the "absolute") optimum.

It is not yet clear which of the two methods, the EM algorithm or numerical optimization, is to be preferred in general (cf. Everitt 1984; McLachlan and Basford 1988; Mooijaart and Van der Heijden 1992), but the EM algorithm has apparently been the most popular (Titterington 1990) because of its computational simplicity. Throughout this book, we use the EM framework for estimating a variety of mixture models. Although many of the mixture likelihood approaches (beginning with Newcomb 1886, and later Hasselblad 1966, 1969, and Wolfe 1970) have used iterative schemes corresponding to particular instances of the EM algorithm, the formal applicability of the EM algorithm to finite mixture problems was recognized only after the developments of Dempster, Laird and Rubin (1977), which were later supplemented by Boyles (1983) and Wu (1983). (See also Redner and Walker 1984; Titterington 1990.) We show below that the mixture likelihood can be maximized by using an EM algorithm (Dempster, Laird and Rubin 1977).

Once an estimate of ϕ has been obtained, estimates of the posterior probability p_{ns} that observation n comes from mixture component s can be calculated for each observation vector y_n by means of Bayes' theorem. The posterior probability is given by:

$$p_{ns} = \frac{\pi_s f_s(y_n|\theta_s)}{\sum_{s=1}^{S} \pi_s f_s(y_n|\theta_s)}. \tag{6.4}$$

The p_{ns} estimates provide a probabilistic allocation of the objects to the mixture components (see chapter 5), and can be used to classify a sample into segments. The Bayesian updating in equation 6.4 will lead to a more precise allocation of consumers/firms into the segments when more information is available from each consumer/firm n. As shown below, the posterior probabilities p_{ns} play a crucial role in the estimation of the parameters for each segment, so that the more information is available from each consumer/firm n, the more accurate the segment-level estimates will be.

Part 2 Segmentation bases

Table 6.2: Some Distributions from the Univariate Exponential Family

Distribution	Notation	Distribution. Function
Discrete		
Binomial	$B(K,\mu)$	$\binom{K}{y} (\frac{\mu}{K})^y (1-\frac{\mu}{K})^{(K-y)}$
Poisson	$P(\mu)$	$\dfrac{e^{-\mu} \mu^y}{y!}$
Negative Binomial	$NB(\mu,v)$	$\left(\dfrac{v}{v+\mu}\right)^v \dfrac{\Gamma(v+y)}{y!\,\Gamma(v)} \left(\dfrac{\mu}{v+\mu}\right)^y$
Continuous		
Normal	$N(\mu,\sigma)$	$\dfrac{1}{\sqrt{2\pi}\sigma} \exp\left[\dfrac{-(y-\mu)^2}{2\sigma^2}\right]$
Exponential	$E(\mu)$	$\dfrac{1}{\mu} \exp\left[-\dfrac{y}{\mu}\right]$
Gamma	$G(\mu,v)$	$\dfrac{1}{y\Gamma(v)} \left(\dfrac{yv}{\mu}\right)^{v-1} \exp\left[-\dfrac{vy}{\mu}\right]$
Discrete		
Multinomial	$M(\boldsymbol{\mu})$	$\prod_{k=1}^{K} \mu_k^{y_k}$
Continuous		
Multivariate Normal.	$MVN(\boldsymbol{\mu},\Sigma)$	$\dfrac{1}{(2\pi)^{K/2} |\Sigma|^{1/2}} \exp[-1/2(y_n-\mu)'\Sigma^{-1}(y_n-\mu)]$
Dirichlet	$D(\boldsymbol{\mu})$	$\dfrac{\Gamma(\sum_{k=1}^{K}\mu_k) \prod_{k=1}^{K} y_k^{\mu_k-1}}{\prod_{k=1}^{K}\Gamma(\mu_k)}$

Table 6.2 – continued

Distribution	Domain	Mean	Variance	Link
Discrete				
Binomial	(0,1)	μ	$\mu(1-\mu/K)$	$\ln(\mu/(K-\mu))$
Poisson	(0,∞)	μ	μ	$\ln(\mu)$
Negative Binomial	(0,∞)	μ	$\mu+\mu^2/v$	$\ln(\mu/(v+\mu))$
Continuous				
Normal	(-∞,∞)	μ	σ^2	μ
Exponential	(0,∞)	μ	μ^2	$1/\mu$
Gamma	(0,∞)	μ	μ^2	$1/\mu$
Discrete				
Multinomial	(0,1)	$\boldsymbol{\mu}$	$\mu_k(1-\mu_k)$, $-\mu_k\mu_l$	$\ln(\mu_k/\mu_K)$
Continuous				
Multivariate Normal.	(-∞,∞)	$\boldsymbol{\mu}$	Σ	$\boldsymbol{\mu}$
Dirichlet	(0,1)	$\mu_k/\Sigma\mu_k$	$\dfrac{\mu_k(\mu-\mu_k)}{\mu^2(\mu_k+1)}$	$\ln(\mu_k/\mu_K)$

The EM Algorithm

The basic idea behind the EM algorithm is the following. Rather than applying a complicated maximization routine to maximize the likelihood across the entire parameter space, one augments the observed data (y) with additional information that replaces unobserved data (z), which greatly simplifies the maximization of the likelihood. In the context of mixture models, this unobserved data constitute the membership of subjects in segments. That membership is unobserved, but were it observed, it would facilitate the computation of the segment-level estimates. The information that is substituted for the missing data is the expected membership of subjects, given a set of preliminary estimates of the parameters of the model. A detailed description of the EM algorithm in general is provided, in for example, Tanner (1993).

To derive the EM algorithm for mixture models, we introduce unobserved data, z_{ns}, indicating whether observation n belongs to mixture component s: $z_{ns} = 1$ if n comes from segment s, and $z_{ns} = 0$ otherwise. The z_{ns} are assumed to be independently multinomially distributed, arising from one draw of the S segments with probabilities π_s. Further, the observed data y_n, given z_n, are assumed to be independent. With z_{ns} considered as missing data, the log-likelihood function for the complete data X and Z = (z_{ns}) can be formed:

$$\ln L_c(\phi) = \sum_{n=1}^{N} \sum_{s=1}^{S} (z_{ns} \ln f_s(y_n | \theta_s) + z_{ns} \ln \pi_s). \tag{6.5}$$

That complete log-likelihood is maximized by using an iterative EM algorithm. In the E-step, the expectation of ln L_c is calculated with respect to the unobserved data Z, given the observed data y and provisional estimates of ϕ. It can be easily seen that this expectation $E(\ln L_c(\phi;y,Z))$ is obtained by replacing z_{ns} in the likelihood by their current expected values, $E(z_{ns}|y, \phi)$, which can be shown to be identical to the posterior probability that object n belongs to segment s: p_{ns}.

In the M-step, the expectation $E(\ln L_c(\phi;y,Z))$ is maximized with respect to π_s under the constraints on those parameters, producing:

$$\hat{\pi}_s = \sum_{n=1}^{N} \hat{p}_{ns} / N. \tag{6.6}$$

Thus, the estimates of the prior probabilities at each step of the algorithm are simply the averages of the posterior probabilities in each segment. Maximizing $E(\ln L_c(\phi;y,Z))$ with respect to θ leads to independently solving each of the S expressions:

$$\sum_{n=1}^{N} \hat{p}_{ns} \frac{\partial \ln f_s(y_{nk} | \theta_s)}{\partial \theta_s} = 0, \tag{6.7}$$

where the p_{ns} terms are treated as known constants.

The maximum likelihood equations of θ_s are obtained by weighting the contribution from each sample unit (i.e., consumer/firm) with the posterior probabilities of segment membership. Titterington, Smith and Makov (1985) discuss the general form of the equations (6.7) for mixtures of distributions within the exponential family. An attractive feature of the EM algorithm is that the above equations have a closed form for some models. For example, if $f_s(y_{nk}|\theta_s)$ is the density of the normal or Poisson distributions, the mixture component means are estimated by $\mu_{ks} = \Sigma_n p_{ns} y_{nk}/\Sigma_n p_{ns}$, which is the weighted mean of the observations. Similarly, the component variances for the normal distribution are calculated by using the posterior membership probabilities as weights. For binomial mixtures a closed-form expression for its parameters also exists in the M-step. In other cases, numerical optimization methods may have to be used in the M-step to find the solution of equation 6.7. Such numerical methods involve the use of (conjugate) gradients, Newton-Raphson, quasi-Newton methods and Fisher scoring. A large literature is available on those techniques for numerical optimization (e.g., Dennis and Schnabel 1983; Scales 1985).

The E- and M-steps are alternated until the improvement in the likelihood function is possible. Schematically, the EM algorithm consists of the following steps:

1. At the first step of the iteration, $h = 1$, initialize the procedure by fixing the number of segments, S, and generating a starting partition $p_{ns}^{(1)}$. A random starting partition can be obtained or a rational start can be used (e.g., using K-means cluster analysis).
2. Given $p_{ns}^{(h)}$, obtain estimates of π_s by using equation 6.6 and of θ_s by using either closed-form expressions or a numerical optimization procedure.
3. Convergence test: stop if the change in the log-likelihood from iteration (h - 1) to iteration (h) is sufficiently small.
4. Increment the iteration index, $h \leftarrow h + 1$, and calculate new estimates of the posterior membership, $p_{ns}^{(h+1)}$ by using equation 6.4.
5. Repeat steps 2 through 4.

An attractive feature of the EM algorithm is that it provides monotonically increasing values of the log-likelihood. Under moderate conditions, the log-likelihood is bounded from above, thus assuring that the algorithm converges to at least a local optimum. The EM algorithm provides a general framework for estimating a large variety of mixture models, including those in the next two chapters. As can be seen in equation 6.7, the M-step amounts to the maximum-likelihood estimation of the segment-level model for each segment s, in which the contribution by each consumer/firm n to the likelihood is weighted by the posterior probabilities p_{ns} obtained in the E-step. The EM algorithm makes the extension of aggregate models to the segment level quite straightforward, as long as maximum-likelihood estimates can be obtained for the aggregate model either analytically or through numerical optimization. The estimation for those models involves only modifications of the likelihood equations in the M-step, weighting each consumer/firm by the posterior probabilities computed in the E-step. The EM algorithm for mixture models is a specific instance of the EM algorithm for mixture regression models (where only constants are included in the regression equation). The latter algorithm is provided in detail in Appendix A1 to the next chapter.

EM Example

We provide a simple example of the EM algorithm in the case of a two-segment Poisson model. We generated data from a mixture of two Poissons ($S = 2$), representing the number of purchases in a product category during a fixed time interval, for 20 hypothetical subjects. We assume one observation per subject, (i.e., $K = 1$). Subjects 1 through 10 were in segment 1 with a purchase rate of $\mu_1 = 1.0$ and subjects 11 through 20 were in segment 2 with a purchase rate of $\mu_2 = 10.0$. The data generated are displayed in the second column of Table 6.3. The EM-algorithm was applied to this hypothetical small data set to illustrate the procedure. In the estimation we used $S = 2$, according to the true number of segments. The table displays steps 1 to 4 of the algorithm, as well as the results of the final step, step 11.

In step 1 random probabilities were generated representing each subject's membership in the two segments as starting values for the EM algorithm. They are displayed in the third column of Table 6.3. Given those initial posterior values, we computed the segment means in the M-step. The M-step for this model entails only closed-form expressions; the estimates of the prior probabilities π are equal to the mean of the posteriors for each segment, according to equation (6.6), and the segment means are simple weighted means of the data, where the weights constitute the posterior probabilities at step 1: $\mu_s = \Sigma_n\, p_{ns}\, y_n / \Sigma_n\, p_{ns}$, (note that the canonical parameter $\theta_s = \log(\mu_s)$). In step 2, new posteriors are calculated from the current estimates of the priors and the segment means, according to equation 6.4. In the case of the Poisson distribution, the kernel of log-likelihood takes a simple form and equals $y_n \ln(\mu_s) - \mu_s$. Thus, the posteriors in step 2 are calculated on the basis of the segment means and the posteriors in column 2. New segment means are estimated as a weighted average on the basis of the new priors in step 2, and so on.

Note that in the example the structure of the data is already well recovered after a few iterations because the segments are very well separated. The bottom row of the table provides the value of the log-likelihood during the iterations. Observe that it decreases considerably during the first two steps, but the subsequent decrease is much smaller. The EM algorithm converged after 11 iterations; the change in the likelihood from step 10 to step 11 was less than 0.0001%. The value of the log-likelihood was -21.966 at convergence. At the final step of the iterations, the value of $\lambda_1 = \ln(\mu_1) = 0.335$ (se = 0.269), and $\lambda_2 = 2.282$ (se = 0.100). The true means appear to be quite well recovered despite the small number of observations.

Chapter 6 Mixture models

Table 6.3: Application of the EM Algorithm to Synthetic S = 2 Mixture Poisson Data

n	y_n	p_{1n}^1	p_{2n}^1	p_{1n}^2	p_{2n}^2	p_{1n}^3	p_{2n}^3	p_{1n}^4	p_{2n}^4	p_{1n}^{11}	p_{2n}^{11}
1	1	0.835	0.165	0.755	0.245	0.976	0.024	0.998	0.002	0.998	0.002
2	1	0.520	0.480	0.755	0.245	0.976	0.024	0.998	0.002	0.998	0.002
3	1	0.221	0.779	0.755	0.245	0.976	0.024	0.998	0.002	0.998	0.002
4	3	0.486	0.514	0.661	0.339	0.887	0.113	0.959	0.041	0.926	0.074
5	0	0.464	0.536	0.795	0.205	0.989	0.011	1.000	0.000	1.000	0.000
6	2	0.623	0.377	0.710	0.290	0.947	0.053	0.992	0.008	0.989	0.011
7	2	0.198	0.802	0.710	0.290	0.947	0.053	0.992	0.008	0.989	0.011
8	0	0.766	0.234	0.795	0.205	0.989	0.011	1.000	0.000	1.000	0.000
9	3	0.668	0.332	0.661	0.339	0.887	0.113	0.959	0.041	0.926	0.074
10	1	0.840	0.160	0.755	0.245	0.976	0.024	0.998	0.002	0.998	0.002
11	13	0.203	0.797	0.167	0.833	0.002	0.998	0.000	1.000	0.000	1.000
12	13	0.062	0.938	0.167	0.833	0.002	0.998	0.000	1.000	0.000	1.000
13	7	0.226	0.774	0.440	0.560	0.233	0.767	0.033	0.967	0.005	0.995
14	13	0.640	0.360	0.167	0.833	0.002	0.998	0.000	1.000	0.000	1.000
15	11	0.374	0.626	0.240	0.760	0.012	0.988	0.000	1.000	0.000	1.000
16	6	0.495	0.505	0.496	0.504	0.406	0.594	0.149	0.851	0.035	0.965
17	7	0.511	0.489	0.440	0.560	0.233	0.767	0.033	0.967	0.005	0.995
18	12	0.504	0.496	0.201	0.799	0.005	0.995	0.000	1.000	0.000	1.000
19	9	0.814	0.186	0.332	0.668	0.056	0.944	0.001	0.999	0.000	1.000
20	8	0.882	0.118	0.385	0.615	0.118	0.882	0.007	0.993	0.001	0.999
π_s		0.517	0.483	0.519	0.481	0.531	0.469	0.506	0.494	0.493	0.507
μ_s		5.028	6.315	3.522	7.949	1.925	9.865	1.496	9.905	1.397	9.796
LL		-42.3681		-31.0035		-22.6753		-22.0026		-21.9657	

87

Limitations of the EM Algorithm

In comparison with direct numerical optimization of the likelihood, the convergence of the EM algorithm may be slow. Sometimes more than 100 iterations may be necessary because convergence is quadratic in the number of parameters. Several procedures have been proposed to improve the rate of convergence (e.g., Peters and Walker 1978; Louis 1982).

Local maxima

Another potential problem associated with the application of the EM algorithm (as well as direct optimization) to mixture problems is its convergence to local maxima. The problem is well documented (McLachlan and Basford 1988, p. 16; Titterington, Smith and Makov 1985, p. 84). It is caused by the likelihood being multimodal, so that the algorithm becomes sensitive to the starting values used. The convergence to local optima seems to be exacerbated when the component densities are not well separated, when the number of parameters estimated is large, and when the information embedded in each observation is limited, leading to a relatively weak posterior update of the membership probabilities (p_{ns}) in the E-step.

Among the solutions to reduce the occurrence of local optima are to (1) start the EM algorithm from a wide range of (random) starting values, (2) start the algorithm from a larger number of classes and work down to a smaller number, or (3) use some clustering procedure such as K-means applied to the dependent variable to obtain an initial partition of the data (e.g., Banfield and Bassil 1977; McLachlan and Basford 1988). If different starting values yield different optima, the solution with the maximum value of the likelihood is recommended as the best solution. Figure 6.2. is a hypothetical illustration of the situation of local maxima of the likelihood.

The choice of a good starting partition is also related to the problem of when to stop the algorithm. The EM algorithm is stopped when the likelihood hardly changes from one iteration to the next. However, some researchers have argued that this is a measure of lack of progress rather than a measure of convergence, and there is evidence that the algorithm is often stopped too early. The choice of a good starting partition can positively affect algorithm performance and model identification. Hamilton (1991) proposed a quasi-Bayesian EM estimation approach for mixtures of normal distributions, which consistently improves on the maximum likelihood approach and is less prone to local optima. Another approach, nonparametric maximum likelihood estimation (NPMLE), treats the number of segments S as a random variable rather than a fixed number (cf. Dillon and Kumar 1994).

Standard errors

A third problem is that the EM algorithm does not require an information matrix in the search for the maximum likelihood estimates, and consequently standard errors

Figure 6.2: Local and Global Maxima

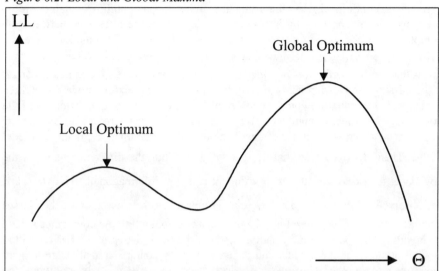

of the estimates are not known at convergence. Several procedures for computing standard errors have been proposed, however. Under certain regularity conditions, the estimators of θ_s, being maximum likelihood estimators, are asymptotically (i.e., for large N) normal. The asymptotic covariance matrix of the estimates of θ_s can be calculated from the inverse of the observed Fisher information matrix or from the expected information matrix (Louis 1982; McLachlan and Basford 1988). From a frequentist point of view, the observed information matrix, $I(\phi)$, is a preferred method for calculating the standard errors, and it is also easier to obtain as a function of the gradient (the vector of first derivatives) of the complete log-likelihood. Alternatively, the covariance matrix of the estimates, can be calculated by using the mispecification-consistent method proposed by White (1982). The mispecification-consistent method yields the correct variances and covariances if the model, and in particular the distribution function, are mispecified. If the model is correctly specified, the estimates of the asymptotic (co)variances from this method and those based on the information matrix are the same. The mispecification-consistent covariance matrix is calculated from the information matrix (I) and the cross-product matrix (H) of the first derivatives:

$$\Sigma(\phi) = NK\, I(\phi)^{-1} H\, I(\phi)^{-1}. \tag{6.8}$$

The information matrix $I(\phi)$ referred to in equation (6.8) is the observed information matrix, also denoted as: $I_O(\phi)$. The complete information matrix is the information matrix that would be obtained if the segment-memberships of individuals would be known: $I_C(\phi)$. It is also the information matrix that is produced in the M-step of the EM algorithm, if a numerical search procedure is used to obtain estimates of the parameters, conditional upon the segment membership probabilities obtained in the previous E-step. The missing information is defined as the difference between the complete information matrix and the observed information matrix: $I_M(\phi) = I_C(\phi) - I_O(\phi)$, (cf. McLachlan and Krishnan 1997, p.101). A consequence of this is that the standard errors provided by the EM algorithm are approximately correct if the missing information tends to zero. In applications of mixture models, one often observes that the mixture components are well separated, so that all estimates of the posterior probabilities (equation (6.4)) are very close to either zero or one. This occurs in particular when the number of observations on an individual, K, is large (Wedel 1999). Then, $I_M(\hat{\phi}) \to 0$ so that $I_O(\hat{\phi}) \to I_C(\hat{\phi})$. Thus the difference between the covariance matrices estimated by inverting the complete information matrix $I_C(\hat{\phi})$, and the observed information matrix $I_O(\hat{\phi})$, disappears as the separation of the classes becomes more perfect. Therefore, the procedure of computing the covariance matrix by inverting the complete information matrix in the final M-step of the EM algorithm is appropriate when the mixture components are well separated and all posteriors are close to zero or one. The question arises, however, under which conditions the complete data information matrix obtained in the M-step of the EM algorithm is an accurate approximation to the observed data information matrix. Wedel (1999) provides theoretical and empirical support for the fact that the approximation is reasonably accurate if the entropy of the segment memberships (cf. equation 6.11) is higher than 0.95.

Identification

Identification of a model refers to the situation in which only one set of parameter values uniquely maximizes the likelihood. The model is said to be unidentified when more than one set of values provide a maximum. The interpretation of parameters from unidentified models is useless because an infinite set of parameters yields the same solution. Dillon and Kumar (1994) recommend that more attention be paid to identification issues in the estimation of mixture models in marketing. Throughout the exposition of the mixture models above, ϕ was assumed to be identifiable. Titterington, Smith and Makov (1985) provide an extensive overview of the identifiability of mixtures, including a survey of the literature. Many mixtures involving members of the exponential family, including the univariate normal, Poisson, exponential and gamma distributions, are identifiable (see Titterington, Smith and Makov 1985). The lack of identifiability due to invariance of the likelihood under interchanging of the labels of the mixture components (Aitkin and Rubin 1985) is of no major concern. An example of that problem is to the situation in which, a three-segment solution with segments 1, 2 and 3 is exactly the same as a solution that recovers the same segments but in the order 2, 1, 3. We follow the solution of McLachlan and Basford (1988) in reporting results for only one of the possible arrangements of the segments.

For some more complex models, it is difficult to demonstrate that the parameters are globally identified (Dillon and Kumar 1994). In such cases, an investigation of the Hessian matrix of second-order partial derivatives can be performed. If the matrix is positive definite (e.g., all its eigenvalues are positive), the model is said to be locally identified. Local identification refers to the situation in which a model is identified for parameter values near the estimated values, but not necessarily for all possible parameter values.

Determining the Number of Segments

When applying the above models to data, the actual number of segments S is unknown and must be inferred from the data. The problem of identifying the number of segments is still without a satisfactory statistical solution. Suppose we want to test the null hypothesis (H_0) of S segments against the alternative hypothesis (H_1) of S + 1 segments. Unfortunately, the standard likelihood ratio statistic for this test does not apply, because it is not asymptotically distributed as chi-square. The likelihood ratio test statistic is simply the difference of the maximized likelihood of the two models. The statistic follows a chi-square distribution for nested models under the null hypothesis, given that certain regularity conditions are satisfied. In testing for the number of components in a mixture model, this asymptotic chi-square distribution is not valid, because H_0 corresponds to a boundary of the parameter space for H_1, a situation that violates one of the regularity conditions. A study into the distribution of the likelihood ratio test for the S = 1 versus the S = 2 model for mixtures of members of the exponential family has been reported by Böhning, et al. (1994). The main result of their study, obtained through Monté Carlo simulation, is that the asymptotic distribution of that LR test appears to be not very well approximated by the conventional chi-square distribution, and deviates from it in a way that depends on the actual distribution (e.g. normal, Poisson, binomial).

An alternative procedure for determining the number of segments is the Monte Carlo test procedure (Hope 1968) applied to mixture problems by Aitkin, Anderson and Hinde (1981), McLachlan (1987) and De Soete and DeSarbo (1991). It involves comparing the likelihood ratio statistic from the real data with a distribution of that statistic obtained from a number (usually 20) of synthetic datasets generated under the null hypothesis H_0. The procedure is computationally demanding. For example, in testing for a four-segment versus a three-segment solution, 20 datasets are generated according to a three-segment solution (using the parameters of the three-segment solution obtained from the real data). Those datasets are analyzed with the S = 3 and S = 4 models, and the likelihood-ratio test statistics are computed. The LR test statistic from the real data is compared with the 20 statistics from the simulated data, and if it exceeds 19 of the simulated statistics, the 3-segment solution is rejected.

Because of the high computational burden of that procedure, another class of criteria for investigating the number of segments is frequently used: information criteria. Those measures are justified by their attempt to balance the increase in fit obtained against the larger number of parameters estimated for models with more segments. Basically, such criteria impose a penalty on the likelihood that is related to the number of parameters estimated:

$$C = -2 \ln L + Pd. \tag{6.9}$$

Here, P is the number of parameters estimated and d is some constant. The constant imposes a penalty on the likelihood, which weighs the increase in fit (more parameters yield a higher likelihood) against the additional number of parameters estimated. The classical Akaike (1974) information criterion, AIC, arises when d = 2. Two criteria that penalize the likelihood more heavily are the Bayesian information criterion, BIC, and the consistent Akaike information criterion, CAIC. For those criteria, d = ln(N) and d = ln(N + 1), respectively. Note that the CAIC penalizes the likelihood even more than BIC, although the two criteria are quite close. Both statistics impose an additional sample size penalty on the likelihood, and are more conservative than the AIC statistic in that they tend to favor more parsimonious models (i.e., models with fewer segments). Studies by Bozdogan (1987) indicate that CAIC is preferable in general for mixture models. Bozdogan also proposed the modified AIC (MAIC) criterion, for which d = 3.

The major problem with those criteria is that they rely on the same properties as the likelihood ratio test, and can therefore be used only as indicative for the number of segments; the number of segments for which the statistics reach a minimum value is chosen. Bozdogan (1994) used the information theoretic measure ICOMP, which is based on the properties of the estimated information matrix. Here d equals:

$$d = -tr[\hat{I}(\phi)^{-1}] - \frac{1}{P} \det[\hat{I}(\phi)^{-1}], \tag{6.10}$$

where P is the effective number of parameters estimated. That measure tends to penalize the likelihood more when more parameters are estimated, but also when the Hessian becomes near-singular because of an increasing number of parameters, in which case the term involving the determinant of the information matrix increases. The ICOMP criterion also penalizes models that produce high variances in the parameter estimates.

The preceding heuristics account for over-parameterization as large numbers of segments are derived, but one must also ensure that the segments are sufficiently separated for the solution that is selected. To assess the separation of the segments, an entropy statistic can be used to investigate the degree of separation in the estimated posterior probabilities:

$$E_S = 1 - \frac{\sum_{n=1}^{N} \sum_{s=1}^{S} p_{ns} \ln p_{ns}}{N \ln S}. \tag{6.11}$$

E_S is a relative measure and is bounded between zero and one. Values close to one indicate that the derived segments are well separated. A value close to zero, indicating that all the posteriors are equal for each observation, is of concern as it implies that the centroids of the segments are not sufficiently well separated.

Recently, Celeux and Soromenho (1996) proposed a normed entropy criterion, NEC, for the selection of the number of segments in mixture models. It is defined as:

$$NEC(S) = \frac{Es}{lnL(S) - lnL(1)}.$$ (6.12)

Here, lnL(S) is the likelihood for S segments. A Monté Carlo comparison of this criterion shows favorable results relative to AIC and CAIC, and performance comparable to that of ICOMP. However, it is computationally more attractive than the latter A potential problem with the measure is that it is not defined for $S = 1$. A solution to that problem has been proposed by the authors. However, it applies only to special cases within the class of mixtures of normal distributions, and therefore cannot be considered as general. Hence, NEC should preferably be applied in conjunction with one of the other informational criteria in determining the number of segments for S>2.

Another set of diagnostic tools that are useful for mixture models and the selection of the number of segments has been developed by Lindsay and Roeder (1992), and applied by Deb and Trivedi (1997) in the context of mixture regression models (see chapter 7). Lindsay and Roeber establish the properties of the tools for mixtures of distributions from the exponential family. They propose two diagnostic tools, a residual function and a gradient function, which are closely related. The residual function appears to be rather uninformative in practical applications, as the gradient function is smoother (Deb and Trivedi 1997). We therefore limit our discussion to the gradient function, defined as:

$$d(y, S^*) = \left(\frac{f(y|\hat{\theta}, S = 1)}{f(y|\hat{\theta}, S = S^*)} - 1 \right),$$ (6.13)

where, for example, for a Poisson or negative binomial model, $f(y|\hat{\theta}, S = 1)$ is the fitted probability at support point y for $S = 1$. The graph of 6.13 is used to investigate the presence of a mixture: the convexity of the graph is interpreted as evidence in favor of a mixture. However, a convex graph may be observed for more than one value of S^*, so AIC or other information criteria still need to be used. An additional statistic suggested by Lindsay and Roeder (1992) is a weighted sum of equation 6.13:

$$D(S^*) = \sum_{i=1}^{n} \sum_{j=1}^{J} d(y, S^*) f(y),$$ (6.14)

with f(y) the sample frequency of y. The statistic should rise above zero if additional segments are required to fit the data. However, Deb and Trivedi (1997) found this statistic to be of relatively little use in comparison with the gradient function presented in equation 6.13.

Some Consequences of Complex Sampling Strategies for the Mixture Approach

Mixture models are estimated by maximum likelihood. However, a sampling strategy other than simple random sampling may require the modification of the likelihood equations provided in the previous chapter. Here, we explain a possible way to modify the likelihood equations to accommodate the sampling strategy. The approach to take with complex designs is to introduce for each subject a weight into the likelihood function that is proportional to the inverse of that subject's selection probability under the sampling design. This is called an approximate, or pseudo-maximum likelihood (PML) estimation approach, and it requires the knowledge of the inclusion probabilities of each of the units selected in the sample. The discussion of the PML approach is based on Skinner (1989), its application to the estimation of mixture models was explored by Wedel, ter Hofstede and Steenkamp (1998). Assume a general complex sampling that may involve combinations of sampling schemes, for example, stratified and two-stage sampling. Suppose that under the sampling design a subject n, has an inclusion probability of P_n. Complex samples result in different inclusion probabilities for different n. Mixture models are usually estimated by maximizing the log likelihood, provided by (6.2) and (6.3). The ML estimator solves the equations:

$$\sum_{n=1}^{N} J_n(\phi) = \sum_{n=1}^{N} \frac{\partial \ln f(y_n | \phi)}{\partial \phi} = 0. \qquad (6.15)$$

This standard formulation of the log likelihood applies under simple random sampling, in which each sampling unit n receives the same weight. The PML approach involves a modification of the Equations (6.15) to include consistent estimates of the selection probabilities, P_n, yielding the sample estimating equations:

$$\sum_{n=1}^{N} \omega_n J_n(\phi) = 0, \qquad (6.16)$$

in which the weights ω_n are inversely proportional to the P_n: $\omega_n = N / (P_n \sum 1/P_n)$, so that they sum to N across the sample. Solving equation (6.16) yields an approximate or pseudo- maximum likelihood estimator (PML) for ϕ, which is consistent. PML estimates can be obtained by solving Equation (6.16) using either numerical optimization or the EM algorithm. The PML estimator, however, is not efficient. The reason is that introducing as weights the selection probabilities decreases the efficiency of the estimator. (This points to the advantages of using self-weighting samples.) A consistent and robust estimator of the asymptotic variance can be obtained, based on the method proposed by White (1982), see also equation (6.8) and Wedel, ter Hofstede and Steenkamp (1998). For model selection, Wedel, ter Hofstede and

Steenkamp (1998) suggest the use of information statistics based on the log pseudo likelihood, instead of the likelihood, which involves a modification of equation (6.9).

An advantage of the PML approach is that it provides a unifying framework for dealing with a large variety of complex sample designs, thus alleviating the problem of building mixture models that accommodate the specific sample design and the structure of the observations on a case by case basis. Complex sampling designs for which all selection probabilities are equal are called self- weighting. For such sampling designs, the ML and PML estimators coincide. Neglecting the sampling design for samples that are not self-weighting leads to inconsistent estimators. We provide the weights for a few common complex sampling designs.

In stratified sampling it is assumed that the population is grouped into $g = 1,...,G$ strata. Within stratum g, N_g subjects are sampled. Let $N_g^{(p)}$ indicate the corresponding number of units in stratum g in the population. The sample size is $N = \sum N_g$ and the population size is $N^{(p)} = \sum N_g^{(p)}$. A mixture model is to be applied to the N (K×1) observation vectors y_n. The appropriate PML estimates of the parameters are weighted estimates obtained from Equation (6.16), in which case the selection probabilities equal: $P_n = N_{g(n)} / N_{g(n)}^{(p)}$, where g(n) is the stratum from which subject n comes. If the sampled fraction is constant across strata $P_n = P$, the sample is self-weighting and the ML and PML estimators coincide.

In cluster sampling a sample of primary units m=1,...,M (often referred to as clusters) is drawn from a population of $M^{(p)}$ units. Within each primary unit observations on all secondary units $n = 1,...,N_m$ are obtained, where the sample size is $N = \sum N_m$ and the size of the population $N^{(p)} = \sum N_m^{(p)}$. Since cluster samples contain all secondary units in the selected primary units, the inclusion probabilities of the secondary units equal those of the primary units ($P_{nm} = P_m$). For example, when a simple random sample is drawn from primary units, then $P_{nm} = M/M^{(P)}$. The PML and ML estimators coincide in this case. If the selection probabilities of the primary units differ, the sampling design is not self-weighting and should be accommodated in estimation.

Like in cluster sampling, in two-stage sampling a sample of size M is drawn from the primary units in the population. However, in two-stage sampling, a *sample* of secondary units of size N_m is drawn from each primary unit that has been drawn in the first stage. If, for example, (a) the primary units in the population have the same size, and (b) the primary and the secondary units are selected by simple random sampling in a constant fraction, the sample is self-weighting. If the primary and secondary units are selected by simple random sampling and the sizes of the primary units differ, the selection probabilities are: $P_{nm} = MN_m / M^{(p)} N_m^{(p)}$. If the sampling fraction within each primary unit is constant, the sampling strategy is again self-weighting.

In an empirical application to values segmentation, Wedel, ter Hofstede and Steenkamp (1998) showed that not accounting for the sample design may yield incorrect conclusions on the appropriate numbers of segments, as well as biased estimates of parameters within segments. Their data were collected within an international segmentation study. The sample was stratified by country. From each

Part 2 Segmentation bases

country a sample of approximately the same size was drawn, although the population sizes in the countries differ substantially. Stratified samples are the rule rather than the exception in international marketing research. The PML approach was also applied to international market segmentation using mixture models by ter Hofstede, Steenkamp and Wedel (1999).

Marketing Applications of Mixtures

Here, we briefly review applications of mixture models in marketing. Such applications have not been very numerous but applications of mixture regression models and mixture unfolding models, described in the following chapters, have been abundant. The applications are classified in Table 6.4 according to the distribution (from the exponential family) assumed for the data.

The first application, by Green, Carmone and Wachspress (1976), was extensively described previously in this chapter. It was a relatively standard application of existing latent class algorithms to a marketing problem of describing segments of adopters of a telecommunication service. Poulsen (1993, cf. Poulsen 1990) was been the first to recognize the applicability of mixture models to choice modeling. He identified segments of consumers on the basis of stationary and first-order Markov choice behavior, using binomial and multinomial models of purchase. Grover and Srinivasan (1987) used a similar but simpler approach, where segments of coffee brand switchers were identified through latent-class decomposition of a two-way contingency table. The approach was based on a decomposition of the contingency table of brand switching on two consecutive purchase occasions. In contrast to Poulsen's, the model was based on the traditional assumption of zero-order switching behavior within each latent class. Dillon and Kumar (1994) pointed to identification problems with those applications of latent-class models. Later, Grover and Srinivasan (1989) extended their approach to track choice behavior of segments over time by modeling repeated cross-classification tables of brand switching. In a second step of the analysis, they related the choice probabilities over time to marketing mix instruments.

Another early application of a mixture model was that of Lehman, Moore and Elrod (1982). Those authors identified segments on the basis of the information acquisition process prior to purchase, and identified segments of consumers with limited problem-solving and routinized response behavior. The Poisson mixture application provided by Dillon and Kumar (1994) is described in detail above. Dillon and Kumar also described a mixture model application to the analysis of paired-comparison data, according to the Bradley-Terry-Luce model. Using a binomial distribution for paired-comparison choices of distinct types of deodorants, they identified four segments with different profiles of preference scale values for those product forms.

Table 6.4: Mixture Model Applications in Marketing

Authors	Year	Application	D[a]
Green, Carmone and Wachspress	1976	Identification of segments of adopters	M
Lehman, Moore and Elrod	1982	Identification of segments of limited and routinized problem solvers	N
Poulsen	1983	Identification of segments on the basis of brand purchase probabilities	M
Grover and Srinivasan	1987	Identification of segments of brand loyals and switchers	M
Grover and Srinivasan	1989	Identification of segments of brand loyals and switchers over time	M
Kamakura and Mazzon	1991	Identification of value segments on the basis of LOV	M
Dillon and Kumar	1994	Identification of light, medium, and heavy buyers segments on the basis of purchase frequencies	P
Dillon and Kumar	1994	Identification of segments on the basis of BTL model for paired comparisons of brand forms	B
Kamakura and Novak	1992	Identification and validation of LOV segments	M
Böckenholt	1993	Purchase frequency and brand choice segmentation	DNB, DP, MNB, MP

[a] Distribution: B = binomial, M = multinomial, P = poisson, N = normal, DNB = Dirichlet-negative binomial, DP = Dirichlet-poisson, MNB = multinomial-negative binomial, MP = multinomial-poisson

Part 2 Segmentation bases

Kamakura and Mazzon (1991) developed a multinomial mixture model for the analysis of rank-ordered data to identify groups of consumers sharing the same value system based on the observed ranking of Rokeach's (1973) terminal values. The same model was applied by Kamakura and Novak (1990) to partial rankings (with ties) of Kahle's (1983) list of values; they also obtain a graphical representation of the value systems using a singular-value decomposition of the parameter estimates in the model. A very general class of mixture models with compound distributions that characterize within-segment heterogeneity has been provided by Böckenholt (1993). He developed a mixture model framework for purchase frequency data. Those data are contained in an ($N \times K$) subjects by brands purchase-frequency data matrix. Böckenholt assumes that the overall purchase rate follows a Poisson distribution (conditional on the segments). Further, he assumes that the parameter of the Poisson distribution (the expected purchase rate) follows a gamma distribution across respondents within each segment. As a result of compounding those two distributions, the overall purchase incidence follows a negative binomial distribution. Conditional on the overall purchase incidence, the brand choice frequencies are assumed to follow a multinomial distribution. Again, within-segment heterogeneity is accounted for by assuming that the multinomial purchase probabilities of subjects to follow a Dirichlet distribution. By compounding the Dirichlet and the multinomial, a Dirichlet-multinomial distribution of brand purchases is obtained. That model is a very general one accounting for within-segment heterogeneity in both overall purchase incidence and brand choice. Several models arise as special cases, specifically by assuming within-segment homogeneity in the brand selection probabilities and/or in the purchase incidence, or by assuming the absence of segments ($S = 1$ solution). Table 6.5 summarizes the special cases of the model. Note that the entire class of models presented by Böckenholt is subsumed under the exponential family of mixtures described in this chapter.

Table 6.5: Special Cases of the Böckenholt (1993) Mixture Model Family

	Within segment heterogeneity in:	
Model	Purchase incidence rate	Brand selection probabilities
Dirichlet gamma	Gamma	Dirichlet
Dirichlet Poisson	None	Dirichlet
multinomial gamma	Gamma	None
multinomial Poisson	None	None

Analysis of a synthetic data set supported the model. The results suggested that neglecting to account for within-segment heterogeneity, as for example in the multinomial-Poisson model, leads to identification of a much larger number of segments than are identified when within-segment heterogeneity is accounted for, as in the multinomial-gamma or the Dirichlet-Poisson models. There was substantial agreement of the segments identified (in terms of posterior probabilities) with the different models. The model was applied to a data set on choice frequencies of three tuna brands. The S = 2 Dirichlet-Poisson and the S = 3 multinomial-Poisson both fit the data well and give insights to the variations in consumers' purchase rates among the three brands.

Conclusion

We have provided a general framework for market segmentation based on mixture models. The statistical approach to segmentation affords several advantages over the heuristic clustering-based approach, although several issues related to local optima and testing for the number of segments remain to be resolved satisfactorily. By compounding distributions within the exponential family, within-segment heterogeneity can be accounted for.

One of the main limitations of mixture models, when applied to the analysis of consumer shopping behavior, is that they ignore the strong influences of price and sales promotions, which may lead to biased estimates of brand preferences and purchase rates within each segment. That important limitation is overcome by the mixture regression models discussed in the next chapter.

7
MIXTURE REGRESSION MODELS

We review mixture models that relate a dependent variable to a set of exogenous or explanatory variables. Also, we describe a generalized linear regression mixture model that encompasses previously developed models as special cases. The model allows for a probabilistic classification of observations into segments and simultaneous estimation of a generalized linear regression model within each segment. Previous applications of the approach to market segmentation are extensively reviewed.

The preceding chapter described "unconditional" approaches to mixtures. "Unconditional" refers to the situation in which there are no exogenous variables explaining the means and variances of each component in the finite-mixture distribution. The unconditional mixture models correspond to the descriptive clustering approaches reviewed in chapter 5, including the K-means and the FCM fuzzy procedure. In the unconditional mixture of normal distributions, for example, the mean and variance of each underlying segment (or component of the mixture) are estimated directly. In contrast, "conditional" mixture models allow for the simultaneous probabilistic classification of observations into underlying segments, and estimation of regression models explaining the means and variances of the dependent variable within each of those segments. The regression models relate a dependent variable, such as purchase frequencies or brand preferences, to explanatory variables, such as marketing mix variables or product attributes, within latent classes. Hence, the conditional regression mixtures represent the mixture analog to the clusterwise regression methods described in the preceding chapter.

In marketing, the fact that it is not very difficult to find reasons for the potential presence of heterogeneous groups or segments has led to the wide use of clustering and, more recently, of unconditional mixture procedures for market segmentation. In those applications, the main objective is mostly *descriptive*, that is, to form homogeneous groups of consumers on the basis of several *observed* characteristics. More recently, attention has shifted toward forming segments that are homogeneous in terms of their responsiveness to price, sales promotions, product features, etc. Such segments are formed on the basis of the *inferred* relationship between a response variable measuring behavior (purchase incidence and timing, brand choice, purchase volume, etc.), intentions, or stated preferences and a set of causal variables (product features, price, sales promotions, etc.) within each homogeneous group.

The identification of segments and simultaneous estimation of the response functions within each segment has been accomplished by a variety of mixture regression models, including mixtures of linear regressions (DeSarbo and Cron 1988), multinomial logits (Kamakura and Russell 1989), rank logits (Kamakura, Wedel and Agrawal 1994), Poisson regressions (Wedel et al. 1995; Wedel and DeSarbo 1995), nested logits (Kamakura, Kim and Lee 1996), and so on. Most importantly, these mixture regression models directly identify segments that are homogeneous in how they respond to the marketing mix, thus leading to actionable *normative* segmentation. We start with two examples of this valuable segmentation method.

Examples of the Mixture Regression Approach

Example 1: Trade Show Performance

DeSarbo and Cron (1988) propose a mixture regression model that makes possible the estimation of separate linear regression functions (and corresponding object memberships) of several of segments. The model specifies a finite mixture of normal regression models where the dependent variable is specified as linear functions of a set of explanatory variables. Thus their model generalizes to more than two classes the stochastic switching regression models developed previously in econometrics (Quandt 1972; Hosmer 1974; Quandt and Ramsey 1972, 1978).

DeSarbo and Cron used their model to analyze the factors that influence perceptions of trade show performance. A sample of 129 marketing executives were asked to rate their firm's trade show performance on the following eight performance factors, as well as on overall trade show performance.

1. Identifying new prospects
2. Servicing current customers
3. Introducing new products
4. Selling at the trade show
5. Enhancing corporate image
6. Testing of new products
7. Enhancing corporate moral
8. Gathering competitive information

An aggregate-level regression analysis of overall performance on the eight performance factors revealed that "identifying new prospects" and "new product testing" were related significantly to trade show performance. The results (reported in Table 7.1) were derived by a standard regression of overall performance ratings on the ratings of the eight factors. The mixture regression model revealed two segments (on the basis of AIC), composed of 59 and 70 marketing executives, respectively. The effects of the performance factors in the two segments are markedly different from those at the aggregate level, as shown in Table 7.1. Managers in segment A primarily evaluated trade shows in terms of non-selling factors, including servicing current customers and enhancing corporate image and moral. Managers in segment B evaluated trade shows primarily on selling factors, including identifying new prospects, introducing new products, selling at the shows and new product testing. Neither of the two segments considered gathering competitive information important.

Whereas the variance explained by the aggregate regression was only 37%, the variance in overall trade show performance explained in segments A and B was 73% and 76%, respectively, thus demonstrating the value of the mixture regression model. The models provide some clear indications for differentially targeting each segment of marketing executives.

Table 7.1: Aggregate and Segment-Level Results of the Trade Show Performance Study (adapted from DeSarbo and Cron 1988)

	Aggregate	Segment A	Segment B
Intercept	3.03*	4.093*	2.218*
1. New prospects	0.15*	0.126	0.242*
2. New customers	-0.02	0.287*	-0.164*
3. Product introduction	0.09	-0.157*	0.204*
4. Selling	-0.04	-0.133*	0.074*
5. Enhancing image	0.09	0.128*	0.072
6. New product testing	0.18*	0.107	0.282*
7. Enhancing morale	0.07	0.155*	-0.026
8. Competitive information	0.04	-0.124	0.023
Size (%)	1.00	0.489	0.511

*$p < 0.05$

Example 2: Nested Logit Analysis of Scanner Data

Many choice models assume that each consumer follows a choice process in which the choice probability for a given brand is *independent from irrelevant alternatives (IIA)*. That property implies a simple pattern of substitution among all available alternatives for a given consumer or a homogeneous segment. In mixture models based on such choice models (e.g., Kamakura and Russell 1989), attention is focused on consumer heterogeneity in preferences, so that segments are defined on the basis of their similarity in brand preferences. Kamakura, Kim and Lee (1996) argue that consumers may make choices according to more complex hierarchical processes, and might in fact differ in the nature of those hierarchical processes. They propose a model that accounts for both *preference* and *structural* heterogeneity in brand choice. They develop a finite mixture of nested logits that identifies consumer segments that may differ in their preferences for the available alternatives, in their responses to the marketing mix, and perhaps also in the type of choice structure. Some consumers may be modeled by a simple multinomial logit model with *independence from irrelevant alternatives*. Others, in the *brand-type* segments, may follow a hierarchical choice process, choosing the brand first and then making choices of product forms within the chosen brand. Still others, of the *form type*, may also follow a hierarchical choice process, but first choose the product form and then select the brand among the ones offering that product form. An important implication of the model is that it allows for a more flexible structure of brand competition *within* each segment, leading to cross-elasticity structures with nonproportional draw even when computed within a homogeneous consumer segment. In other words, the mixture of nested logits allows for a structural pattern of price/promotion competition within a market segment so that a particular stock keeping unit (SKU) would have more impact on SKUs that are in the same "branch" of the hierarchical choice structure of that segment. This integrated model gives managers a more realistic portrait of brand preferences, choice process and price/promotion response within each consumer segment.

Most importantly, the model *allows* for non-IIA structures, rather than *requiring* the elasticities to have any structure defined *a priori*. That feature is illustrated in Table 7.2, which shows the cross-elasticity matrices (i.e., effect of a price change by the column brand on the market share of the row brand) for three segments of brand-switching consumers for the leading brands of peanut butter in one metropolitan area in the United States. The cross-elasticities in *form-type* segment 1 are larger among different brands of Peanut Butter in the same product form (crunchy or creamy) than across product forms. For example, a price discount by creamy Peter Pan would have more impact on creamy Jif, Skippy and store brand than on their respective crunchy versions. Therefore, a promotion by any particular SKU (e.g., creamy Peter Pan) targeted to this segment is likely to draw shares from competitors in the same product form (e.g., creamy Jif, Skippy or store brand). The second part of Table 7.2 shows the typical cross-elasticity pattern in a *brand-type* segment (segment 2). In that type of segment, cross-elasticities are larger among the various forms of the same brand, therefore indicating that promotions targeted to such a segment are more likely to cannibalize sales of "sister" brands than to take sales from competitors. For example, a 1% discount on creamy Peter Pan is expected to draw 1.58% of the original shares of crunchy Peter Pan in segment 2, and only 0.81% of the original share of any other SKU. The cross-elasticities for the IIA-type segment 3 (bottom part of Table 7.2) show the proportional draw one would expect from a multinomial logit model. In this case, a 1% discount on creamy Peter Pan would produce approximately the same percentage loss in share for all SKUs. These results suggest that the benefits from a price promotion would vary across the consumer segments. In the product-form segments, a promotion is likely to draw shares from competitors, whereas in brand-type segments it is likely to cannibalize shares from other forms of the same brand.

Chapter 7 Mixture regression models

Table 7.2: *Average Price Elasticities within Three Segments (adapted from Kamakura et al. 1996)*

	Peter PanCreamy	Peter Pan Crunchy	Jif Creamy	Jif Crunchy	Skippy Creamy	Skippy Crunchy	Store Creamy	Store Crunchy
[1] Form-Type (Segment 1)								
PPR	-3.0	0.3	1.4	0.0	0.9	0.0	0.4	0.0
PPCrun	5.0	-7.6	1.0	0.2	0.6	0.0	0.3	0.2
Jcrea	6.7	0.3	-9.1	0.0	0.9	0.0	0.4	0.0
Jcrun	5.0	2.3	1.0	-10.2	0.6	0.0	0.3	0.2
SKCrea	6.7	0.3	1.4	0.0	-9.6	0.0	0.4	0.0
SKCrun	5.0	2.3	1.0	0.2	0.6	-10.4	0.3	0.2
STCrea	6.7	0.3	1.4	0.0	0.9	0.0	-6.4	0.0
STCrun	5.0	2.3	1.0	0.2	0.6	0.0	0.3	-6.6
[2] Brand-Type (Segment 2)								
PPR	-13.0	7.4	0.4	3.7	0.1	1.3	0.1	0.3
PPCrun	1.6	-7.4	0.4	3.7	0.1	1.3	0.1	0.3
Jcrea	0.8	3.9	-14.8	7.9	0.1	1.3	0.1	0.3
Jcrun	0.8	3.9	0.9	-7.8	0.1	1.3	0.1	0.3
SKCrea	0.8	3.9	0.4	3.7	-15.1	5.5	0.1	0.3
SKCrun	0.8	3.9	0.4	3.7	0.6	-10.2	0.1	0.3
STCrea	0.8	3.9	0.4	3.7	0.1	1.3	-9.3	2.5
STCrun	0.8	3.9	0.4	3.7	0.1	1.3	0.9	-7.8
[3] IIA-Type (Segment 3)								
PPR	-5.2	4.8	0.1	0.1	0.0	0.1	0.1	0.0
PPCrun	3.2	-3.8	0.1	0.2	0.0	0.1	0.1	0.0
Jcrea	3.9	4.8	-9.7	0.1	0.0	0.1	0.1	0.0
Jcrun	3.2	5.5	0.1	-9.6	0.0	0.1	0.1	0.0
SKCrea	3.9	4.8	0.1	0.1	-9.8	0.1	0.1	0.0
SKCrun	3.2	5.5	0.1	0.2	0.0	-9.7	0.1	0.0
STCrea	3.9	4.8	0.1	0.1	0.0	0.1	-6.3	0.0
STCrun	3.2	5.5	0.1	0.2	0.0	0.1	0.1	-6.4

A Generalized Mixture Regression Model (GLIMMIX)

Whereas the major thrust of development and application of mixture regression models has been in marketing and business research, there is potential for substantive applications in virtually all physical and social sciences. The applications of generalized linear models, which include as special cases linear regression, logit and probit models, log-linear and multinomial models, inverse polynomial models, and some models used for survival data, have been numerous (cf. McCullagh and Nelder 1989). Generalized linear models are regression models, where the dependent variable is specified to be distributed according to one of the members of the exponential family (see preceding chapter). Accordingly, those models deal with dependent variables that can be either continuous with normal, gamma or exponential distributions or discrete with binomial, multinomial, Poisson or negative binomial distributions. The expectation of the dependent variable is modeled as a function of a set of explanatory variables, as in standard multiple regression models (which are a special case of generalized linear models). The historical development of generalized linear models can be traced to the pioneering work of Gauss, Legendre, and Fisher (cf. Stigler 1986), but the term "generalized linear model" was coined by Nelder and Wedderburn (1972). The number of applications of generalized linear models in marketing research and other social sciences is vast. However, as argued above and demonstrated in the preceding examples, the estimation of a single set of regression coefficients across all observations may be inadequate and potentially misleading if the observations arise from a number of unknown groups in which the coefficients differ.

The model detailed in this section, proposed by Wedel and DeSarbo (1995), is a general extension of the various mixture regressions developed in the past, formulated within the exponential family. It is called the GLIMMIX model, for Generalized LInear Model MIXture. As indicated in the preceding chapter, the exponential family includes the most commonly used distributions such as the normal, binomial, Poisson, negative binomial, inverse Gaussian, exponential and gamma, as well as several compound distributions. Hence, it allows a wide range of marketing data to be analyzed. The exponential family is a very useful class of distributions and the common properties of the distributions in the class enable them to be studied simultaneously rather than as a collection of unrelated cases. The family thus affords a general framework that encompasses most of the previous developments in this area as special cases.

Assume the vector of observations on object n, $\mathbf{y_n}$, arises from a population that is a mixture of S segments in proportions $\pi_1,...,\pi_S$, where we do not know in advance the segment from which a particular vector of observations arises. The probabilities π_s are positive and sum to one as in equation 5.1. We assume that the distribution of $\mathbf{y_n}$, given that $\mathbf{y_n}$ comes from segment s, $f_s(y_{nk}|\mathbf{\theta}_s)$, is one of the distributions in the exponential family or the multivariate exponential family (see Table 6.2 in chapter 6). Conditional on segment s, the $\mathbf{y_n}$ are independent. If the y_{nk} cannot be assumed independent across k (i.e., repeated measurements on each subject or firm), a distribution within the multivariate exponential family, such as the multivariate normal or the multinomial, is appropriate. The distribution $f_s(y_{nk}|\theta_s)$ is characterized by parameters θ_{sk}. The means of the distribution in segment s (or expectations) are denoted by μ_{sk}. Some of the distributions, such as the normal, also have an associated dispersion parameter λ_s that characterizes the variance of the observations within each segment.

The development of the model is very similar to that of the mixture models described in the preceding chapter. A major difference, however, is that we want to predict the means of the observations in each segment by using a set of explanatory variables, as in the clusterwise regression models. To that end, we specify a linear predictor η_{nsk}, which is produced by P explanatory variables $X_1,...,X_P$ ($X_p = (X_{nkp})$; p = 1,...,P) and parameter vectors $\beta_s = (\beta_{sp})$ in segment s:

$$\eta_{nks} = \sum_{p=1}^{P} X_{nkp} \beta_{sp}. \tag{7.1}$$

The linear predictor is thus a linear combination of the explanatory variables, and a set of coefficients that are to be estimated. The linear predictor is in turn related to the mean of the distribution, μ_{sk}, through a link function g(.) such that in segment s:

$$\eta_{nsk} = g(\mu_{nsk}). \tag{7.2}$$

Thus, for each segment, a generalized linear model is formulated with a specification of the distribution of the variable (within the exponential family), a linear predictor η_{nsk} and a function g(.) that links the linear predictor to the expectation of the distribution. For each distribution there are preferred links, called canonical links. The canonical links for the normal, Poisson, binomial, gamma and inverse Gaussian distributions are the identity, log, logit, inverse and squared inverse functions, respectively (see Table 6.2). For example, for the normal distribution the identity link involves $\eta_{nsk} = \mu_{sk}$, so that by combining equations 7.1 and 7.2 the standard linear regression model within segments arises.

The unconditional probability density function of an observation vector y_n can now, analogous to chapter 6, be expressed in the finite mixture form:

$$f(y_n|\phi) = \sum_{s=1}^{S} \pi_s f_s(y_n|\theta_s), \tag{7.3}$$

where the parameter vector $\phi = (\pi_s, \theta_s)$, $\theta_s = (\beta_s, \lambda_s)$. Note that the difference with equation 6.3 is that here the mean vector of the observations (given each segment) is re-parameterized in terms of regression coefficients relating the means to a set of explanatory variables.

EM Estimation

The purpose is to estimate the parameter vector ϕ. To do so, we maximize the likelihood equation with respect to ϕ. The problem can again be solved using the EM algorithm as described in the preceding chapter. Estimates of the posterior probability, p_{ns}, that observation n comes from segment s can be calculated for each observation vector y_n, as shown in equation 6.4.

The EM algorithm is very similar to that described for ordinary mixture models, except for the estimation of the within-class generalized linear model. Once an estimate of ϕ has been obtained, the posterior membership probabilities p_{ns} are calculated in the E-step, as shown in equation 6.4. In the M-step, the expectation of ln L_c is maximized with respect to ϕ to obtain new provisional estimates. Further details about the EM estimation of GLIMMIX models are given in Appendix A1 at the end of this chapter.

EM Example

We now provide a simple example of the method discussed in this chapter. Here we demonstrate the EM algorithm in the case of a two-segment normal mixture regression model. For 20 hypothetical subjects, we generated S = 2 mixture regression data. There is only one observation for each of the 20 subjects (i.e. K = 1). Subjects 1 through 10 are members of segment 1 subjects 11 through 20 are members of segment 2. Data were generated in each segment with a regression model with one x-variate. The regression model in segment 1 was $y_n = 1 + x_n + \varepsilon_{n1}$; that in Segment 2 was $y_n = 2 + 2x_n + \varepsilon_{n2}$. The standard error of the (normally distributed) disturbances was set to 0.2 for the first and 0.3 for the second segment. The data are displayed in the second and third columns of Table 7.3. The EM algorithm was applied to this small hypothetical data set to illustrate the estimation algorithm; an extensive investigation of the performance of the method on synthetic data is presented in the next section. The EM algorithm was started with random values for the posterior probabilities.

Note that given the posteriors, the M-step of the algorithm constitutes a simple weighted least squares regression in which the regression parameters for each segment are obtained from the well-known closed-form expressions (i.e., $\beta_s = (X'P_s X)^{-1} X'P_s Y$, with P_s a diagonal matrix with the posteriors of segment s, p_{ns}). The posteriors in the second step are calculated on the basis of the estimates of the regression coefficients in step 1. New regression estimates are obtained in step 3 on the basis of the posteriors in step 2, and so on.

In this example, because the segments are very well separated, the structure in the data is recovered after only a few iterations. The EM algorithm converged after 11 iterations (the increase in the likelihood was less than 0.000001). The posteriors in the final step show that the segment memberships as well as the regression coefficients in each of the two segments are well recovered. In spite of the small number of observations, the EM algorithm appears to work very well in this example.

Standard Errors and Residuals

As in the case of unconditional mixtures, the approximate asymptotic covariance matrix of the estimates of $ß_s$, conditional on segment s, are calculated from the inverse of the observed Fisher information matrix, or, if required, calculated according to the mispecification consistent-formulas provided in the preceding chapter (equation 6.8).

For members of the exponential family, several types of residuals are defined that can be used to explore the adequacy of fit of the model with respect to the choice of the link function, the specification of the predictor and the presence of anomalous values. We use the deviance residuals, which can be defined for each segment according to McCullagh and Nelder (1989, pp. 391-418). Those authors extensively describe procedures for model checking that carry over to the models described in this and subsequent chapters. Their book provides a detailed discussion of those procedures.

Identification

Note that the above approach for mixture models is conceptually similar to that for the unconditional mixture models described in chapter 6. Therefore, the remarks on identification made in that chapter apply here. However, the mixture regression model has an additional identification problem: identification related to the condition of the predictor matrix **X**. As in linear models, collinearity among predictors may lead to problems of identification, resulting in unstable estimates of the regression coefficients and large standard errors. In mixture regression models that situation is compounded by the fact that there are fewer observations for estimating the regression model within each segment than there are at the aggregate level. Therefore identification related to the condition of the **X** matrix is an important issue in applications of mixture regression models.

Comments made in the preceding chapter on local maxima and possible remedies apply here as well, as do the comments about the slow convergence of the algorithm. Close starting estimates for the parameters β_s in the M-step of the algorithm are found by applying the link function to the data y_{nk}, and regressing the transformed dependent variable on the independent variables.

Part 2 Segmentation Methodology

Monté Carlo Study of the GLIMMIX Algorithm

Study Design

To assess the performance of the EM algorithm for mixture regression models, Wedel and DeSarbo (1995) conducted an extensive Monté Carlo study. Synthetic datasets were generated according to the following six factors, which were hypothesized to affect the performance of the algorithm.

1. Distribution of the dependent variable: normal, Poisson, binomial or gamma
2. Number of segments: $S = 2$ or $S = 4$
3. Number of x-variables: $P = 2$ or $P = 5$
4. Number of replicated measures per subject: $K = 1$ or $K = 5$
5. Cluster separation: low or high
6. Starting values: random or K-means

The factors and their levels were chosen to reflect variation in conditions representative of practical applications. Synthetic data were generated for 500 hypothetical subjects. The coefficients were generated in the (-2, 2) interval. The absolute difference of the coefficients between successive segments was set to 0.2 in the low cluster separation condition and 0.4 in the high cluster separation condition. The design used in the study was a full factorial in which datasets for all possible combinations of the levels of the factors were produced (192 synthetic data sets). To investigate the occurrence of local optima, each data set was analyzed with two different sets of random starting values, one generated from a uniform distribution in the interval (0,1) and scaled to satisfy the sum constraint in equation 6.1, and one from a partition obtained from a K-means clustering of the dependent variable, yielding 0/1 initial values.

Three measures of algorithm performance were calculated, assessing computational effort, and parameter recovery: the number of iterations required (ITER), the root mean squared error of the memberships (RMSE(p)), and the root mean squared error of the segment-specific regression parameters (RMSE(β)).

Table 7.3: Application of the EM Algorithm to Synthetic S = 2 Mixture Regression Data

			1		2		3		4		12	
n	y_n	x_n	p_{1n}^1	p_{2n}^1	p_{1n}^2	p_{2n}^2	p_{1n}^3	p_{2n}^3	p_{1n}^4	p_{2n}^4	p_{1n}^{11}	p_{2n}^{11}
1	0.76	0	0.324	0.676	0.524	0.476	0.525	0.475	0.542	0.458	0.000	1.000
2	3.81	3	0.759	0.241	0.531	0.469	0.502	0.498	0.480	0.520	0.000	1.000
3	1.82	1	0.536	0.464	0.536	0.464	0.532	0.468	0.550	0.450	0.000	1.000
4	3.03	2	0.684	0.316	0.537	0.463	0.525	0.475	0.532	0.468	0.000	1.000
5	9.07	8	0.342	0.658	0.340	0.660	0.221	0.779	0.059	0.941	0.000	1.000
6	7.66	7	0.684	0.316	0.383	0.617	0.271	0.729	0.100	0.900	0.000	1.000
7	4.85	4	0.583	0.417	0.512	0.488	0.466	0.534	0.402	0.598	0.000	1.000
8	9.27	8	0.135	0.865	0.348	0.652	0.231	0.769	0.067	0.933	0.000	1.000
9	4.01	3	0.718	0.282	0.531	0.469	0.505	0.495	0.487	0.513	0.000	1.000
10	5.03	4	0.552	0.448	0.514	0.486	0.471	0.529	0.412	0.588	0.000	1.000
11	7.68	3	0.342	0.658	0.498	0.502	0.523	0.477	0.555	0.445	1.000	0.000
12	3.75	6	0.076	0.924	0.539	0.461	0.601	0.399	0.707	0.293	1.000	0.000
13	6.23	2	0.930	0.070	0.481	0.519	0.503	0.497	0.517	0.483	1.000	0.000
14	7.99	3	0.532	0.468	0.493	0.507	0.522	0.478	0.556	0.444	1.000	0.000
15	5.58	2	0.023	0.977	0.495	0.505	0.511	0.489	0.527	0.473	1.000	0.000
16	19.8	9	0.561	0.439	0.617	0.383	0.720	0.280	0.878	0.122	1.000	0.000
17	13.9	6	0.680	0.320	0.539	0.461	0.605	0.395	0.715	0.285	1.000	0.000
18	17.9	8	0.431	0.569	0.589	0.411	0.681	0.319	0.832	0.168	1.000	0.000
19	18.1	8	0.620	0.380	0.592	0.408	0.690	0.310	0.844	0.156	1.000	0.000
20	4.27	1	0.571	0.429	0.475	0.525	0.490	0.510	0.488	0.512	1.000	0.000
π			0.504	0.496	0.504	0.496	0.504	0.496	0.505	0.495	0.500	0.500
μ			0.360	1.410	0.490	1.330	0.310	1.560	0.046	2.030	1.993	0.822
β			1.795	1.551	1.832	1.497	1.957	1.358	2.133	1.118	1.986	1.027
Log-likelihood			−31.72		−31.40		−30.29		−26.99		9.93	

Results

The 192 observations for each performance measure were submitted to an analysis of variance. Table 7.4 gives the means of the four dependent measures according to type of distribution and each of the five other factors. The conclusions of the study are that computational performance of the GLIMMIX algorithm improves:

- when the number of parameters estimated (number of x-variables, number of segments) decreases,
- when more replications are collected per subject, and
- when clusters are well separated

Parameter recovery is better:

- when a smaller number of parameters are estimated,
- when the clusters are well separated relative to the within-cluster variance (note that for the binomial, Poisson and gamma distributions, within-cluster variance increases when cluster separation increases), and
- when a larger number of repeated measures are available per subject (due to a stronger posterior update in the E-step).

Deterioration of algorithm performance when many parameters are to be estimated and clusters are not well separated may be largely due to convergence to local optima. The type of local optimum observed most often was a solution in which classes occurred more than once. Such solutions are easily recognizable in practical applications. Local optima of that type occurred in 6% of the analyses for the normal distribution, 10 % for the binomial distribution, 25% for the Poisson distribution and 31% for the gamma distribution. They also occurred more frequently for five than for two predictor variables (27% versus 9%) and for four than for two segments (33% versus 3%). Those findings show the necessity of using several random starts in applications to identify locally optimum solutions under such conditions. The use of a rational start using the K-means procedure did not improve algorithm performance. In fact, results from K-means starting points were sometimes poorer than those from random starts.

Marketing Applications of Mixture Regression Models

Here we review applications of mixture regression models to marketing problems. Table 7.5 is an overview of the applications. Most of the applications provide special instances of the general model outlined above, but not all. (Note that some of the distributions used do not provide special instances of the exponential family, and some models involve extensions that have not been discussed.) The applications can be classified according to the type of dependent variable used, and more specifically according to the distribution that is assumed to describe the dependent variable. The most important ones are within the exponential family: *normal data, binomial data, multinomial data, and count data.* A final and important category of applications that has received much attention is conjoint analysis. The applications are described briefly below, but are discussed more extensively in chapter 18.

Normal Data

One of the first mixture regression models was proposed by DeSarbo and Cron (1988), as described above. They assumed a dependent variable that is normally distributed, so that the M-step in their EM algorithm involves a standard multiple regression of a dependent on a set of independent variates. A similar model was applied by Helsen, Jedidi and DeSarbo (1993) to a country segmentation problem in international marketing. The purpose was to identify segments of countries with similar patterns of diffusion of durable goods. Ramaswamy, et al. (1993) applied the mixture regression approach to cross-sectional time-series data. Here the purpose was to pool the time series for effective estimation of the effects of marketing mix variables on the market shares of brands. The model assumes a multivariate normal distribution of the dependent variable (market shares), and allows the time-series data within the pools (classes) to have an arbitrary covariance structure.

Binary Data

In pick-any experiments, subjects are required to pick any number of objects/brands from sets of size K. If one assumes that each decision to pick (or not) an object is independent from all others, the data are binary. The binary decisions are to be explained from the attributes of the brands. Subjects may be heterogeneous in the extent to which they consider the attributes important. De Soete and DeSarbo (1991) developed a mixture of binomial distributions (with a non-canonical probit link) for the analysis of such data. Using that model, they identified segments in the sample, and simultaneously estimated the importances of the brand attributes. The model was applied to explain consumers' pick-any/K experimental choice data for a set of communication devices.

Table 7.4: Results of the Monté Carlo Study on GLIMMIX Performance (adapted from Wedel and DeSarbo 1995)[a]

		Iter				RMS (β)				RMS (p)			
		N	B	P	O	N	B	P	G	N	B	P	G
REPLICATION	1	58.7ª*	44.2ª	42.6ª	115.4ᵇ	0.018ª	0.077ᵃᵇ	0.103ᵇ	0.300ᶜ	0.041	0.048	0.066	0.063
	5	25.8ᶜ	22.4ᶜ	23.3ᶜ	81.7ᵈ	0.032ª	0.037ª	0.093ᵈ	0.143ᵈ	0.024	0.027ª	0.044	0.061
X-VARIATES	2	53.8ª	36.6ª	38.0ᵇ	76.7ᵇ	0.013ª	0.046ᵃᵇ	0.078ᵇ	0.057ᵃᵇ	0.034ª	0.027ª	0.075ᵇ	0.056ª
	5	30.6ᵇ	30.0ᵃᵇ	27.9ᵃᵇ	120.4ᶜ	0.036ª	0.068ᵃᵇ	0.118ᵇ	0.386ᶜ	0.031ª	0.047ª	0.036ª	0.068ª
SEGMENTS	2	29.8ª	11.0ª	21.2ª	38.1ᵇ	0.008ª	0.010ª	0.027ª	0.174ᶜ	0.031ª	0.006ª	0.045ᵇ	0.027ª
	4	54.7ᵇ	55.7ᵇ	44.7ᵇ	159.0ᶜ	0.041ª	0.104ª	0.169ᶜ	0.269ᵈ	0.034ᵃᵇ	0.068ᵇ	0.066ᵇ	0.097ᵇᶜ
SEPARATION	low	56.8ª	30.3ᵇ	37.0ᵇ	111.6ᶜ	0.019ª	0.022ª	0.050ª	0.212ᶜ	0.021ª	0.016ª	0.068ᵇ	0.066ᵇ
	high	27.7ᵇ	36.3ᵇ	28.9ᵇ	85.5ᵈ	0.030ª	0.092ᵇ	0.146ᵇ	0.231ᶜ	0.044ᵃᵇ	0.058ᵇ	0.043ᵃᵇ	0.057ᵇ
START	random	40.4ª	33.2ª	34.2ª	98.6ᵇ	0.010ª	0.042ᵃᵇ	0.100ᵇ	0.225ᶜ	0.025	0.028	0.054	0.060
	K-means	45.9ª	33.5ª	30.4ª	98.6ᵇ	0.034ª	0.086ª	0.094ᵃᵇ	0.214ᶜ	0.047	0.056	0.057	0.064

[a] N = normal, B = binomial, P = Poisson, G = gamma distributions. Means sharing a superscript are not significantly different at $p < 0.01$

A similar model was developed by Wedel and DeSarbo (1993). It is a mixture of binomial distributions with a logit-link (instead of De Soete and DeSarbo's probit-link) function, applied to paired comparisons data. In paired comparisons, subjects are offered stimuli in all possible pairs and are asked to choose one within each pair. Such data are also commonly modeled by a binomial distribution. The binary model was applied to data on risk perceptions in the choice of automobiles, to identify segments of subjects using different attributes in making their pairwise choices. Dillon and Kumar (1994) used a similar approach for paired comparisons data.

Multichotomous Choice Data

One of the first mixture regression models applied in marketing was the one developed by Kamakura and Russell (1989). They proposed a mixture of multinomial logits to identify market segments that are homogeneous in brand preferences and in how they respond to price promotions. Their model starts with the same assumptions of random utility maximization as the multinomial logit model for each consumer. However, the model allows consumers to differ in their brand preferences and response to price promotions. One of the major limitations of the multinomial logit model is that it considers every consumer as a replication of the "typical" consumer, implying not only that all consumers have the same preferences for the brands available, but also that they respond equally to the marketing mix, thus leading to a competitive structure in which a brand's price promotion has the same impact (i.e., cross-elasticities) on all competitors. Because of its finite-mixture formulation, the Kamakura and Russell model produces more realistic patterns of brand competition at the market level, in which each brand draws differentially from every competitor, depending on how closely they compete at the market-segment level. The model is discussed in more detail in chapter 17. Bucklin and Gupta (1992) extended the approach by using a sequential combination of mixture multinomial logit models within segments, which allows for heterogeneity in purchase incidence (the probability that the product category is bought) and brand choice (conditional on a purchase of the category). Gupta and Chintagunta (1994) extended the Kamakura and Russell approach and simultaneously estimated the relationship between segment membership probabilities and consumer characteristics, using a concomitant variables mixture model (Dayton and MacReady 1988). They also applied their model to scanner panel data, demonstrating how the concomitant model can be used to determine the profile of each market segment (discussed in more detail in chapter 9).

Russell and Kamakura (1994) present an approach to link store- (macro-) and household- (micro-) level scanner data, using a multinomial mixture model estimated at the micro level and a Dirichlet regression estimated at the macro (market) level. The macro-level analysis estimates a market response model at the market level that takes into account the segment preference structure identified at the micro level. In that analysis certain parameters are replaced by their estimates from the micro-level model. The resulting model assumes that the coefficients of the marketing mix are constant across segments, but allows elasticities of the marketing mix to be segment-specific.

Count Data

Wedel, et al. (1993) developed a mixture of Poisson distributions (with log link) that they used to analyze the frequency with which consumers purchase brands in response to direct mail offerings. The frequency of purchases was assumed to be described by a Poisson distribution, and the model identified the effects of mailing and consumer characteristics on response within each segment. The model was applied to the analysis of a customer database from a direct marketing company. Such customer lists contain accumulated data on received mailings, orders and payments of the customers of the company, and are an essential instrument in direct marketing. As an additional feature, the posterior probabilities of the Poisson mixture model can be used to select consumers who have optimal probabilities of response for future mailings. A similar model was applied to scanner panel data by Wedel and DeSarbo (1995) to investigate heterogeneity in the effects of determinants of the frequency of coupon usage.

Bucklin, Gupta and Siddarth (1991) used a related approach, but assumed that the Poisson distribution is truncated at zero. Ramaswamy, Anderson and DeSarbo (1993) extended these models, and allowed for within-segment heterogeneity. They assumed that the parameters of the Poisson model within segments follow a gamma distribution and used the resulting compound mixture of negative binomial distributions (with a log link) to analyze purchase frequencies derived from scanner data. Deb and Trivedi (1997) applied a similar model to medical care for the elderly.

Choice and Count Data.

A general framework for the analysis of choice and purchase frequency data has been proposed by Böckenholt (1993a). The framework is a direct extension of the mixture framework for brand choice data described in the preceding chapter. Böckenholt extended that class of models to a mixture regression context. As explained in chapter 6, he assumed an (N×K) data set of brand choice frequencies. For those data, a hierarchy of distributional assumptions was used, with a component for the overall purchase frequency and another for the brand choice probabilities given purchase frequencies. The most general of his models entails the negative binomial for the overall purchase frequency, obtained by compounding Poisson frequencies with gamma heterogeneity, and the Dirichlet-multinomial for the brand choice probabilities, modeled as a multinomial process with Dirichlet heterogeneity (the DG model, see Table 6.5). The gamma and Dirichlet distributions account for within-segment heterogeneity in purchase rates and brand preferences, respectively. The hierarchy of models is subsumed under the mixture of generalized linear models presented in this chapter.

Within the general framework, Böckenholt addresses two types of explanatory variables, those related to the consumers and those relating brands to consumers. The first set of explanatory variables is included in the linear predictor of the means of the negative binomial distributions. The brand-specific variables are included in the linear predictor of the Dirichlet-multinomial distributions. If within-segment heterogeneity is ignored, the model reduces to a Poisson and/or multinomial mixture regression for purchase frequency and choices, respectively. Böckenholt uses a log link for the purchase frequency models and a logit link for the brand choice models. Using synthetic data, he shows that in cases of within-segment heterogeneity, a larger number of segments is needed to represent the data with the simpler homogeneous-segments formulation than is needed for the general heterogeneous model.

This class of models was demonstrated empirically on choices among four brands of canned tuna for 424 households. The household-specific variables were household size, college education (yes/ no), and retirement status. Brand-specific variables were price, display and feature. Starting with the simplest Poisson-multinomial mixture model, the CAIC statistic revealed an $S = 2$ segment solution. Both of the regression parts (consumer and brand specific) were tested by likelihood-ratio tests. The consumer-specific variables appeared not to contribute significantly. Testing the simple PM model against models that included gamma heterogeneity of purchase intensity and Dirichlet heterogeneity in brand preferences, Böckenholt concluded that for this particular data set it was not necessary to address within-segment heterogeneity through either the gamma or the Dirichlet models, and that the covariates in the mixture regression model captured within-segment heterogeneity very well.

Response-Time Data

Rosbergen, Pieters and Wedel (1997) developed a mixture of gamma regressions (with log-link functions) to describe consumers' gaze duration (registered by eye-movement devices) when looking at print advertisements. The results revealed three segments with quite distinct patterns of eye movements, related differentially to advertising recall.

Conjoint Analysis

An important field of application for GLIMMIX models that has attracted substantial interest is conjoint analysis. Segmentation research on the basis of conjoint analysis is treated in more detail in a subsequent chapter; here we present an overview of applications.

Table 7.5: GLIMMIX Applications in Marketing

Authors	Year	Application	Mixture Distribution[a]
DeSarbo and Cron	1988	Trade show performance	N
Ramaswamy et al.	1993	Marketing mix effects	N
Helsen, Jedidi and DeSarbo	1993	Country segmentation	N
Wedel and DeSarbo	1994	Conjoint analysis, SERVQUAL	N
Wedel and DeSarbo	1995	Customer satisfaction	N
Kamakura and Russell	1989	Brand choice analysis	M
Kamakura and Mazzon	1991	Value segmentation	M
Bucklin and Gupta	1992	Purchase incidence and brand choice	M
Gupta and Chintagunta	1994	Brand choice and segment description	M
Russell and Kamakura	1994	Linking micro- and macro-level data	M
Kamakura and Russell	1992	Measuring brand equity	M
DeSarbo, Wedel et al.	1992	Conjoint analysis	M
Kamakura, Wedel and Agrawal	1994	Conjoint analysis and segment description	M
Kamakura, Kim and Lee.	1996	Mixture of nested logits	NM
De Soete and DeSarbo	1991	Pick-any choices	B
Wedel and DeSarbo	1993	Paired comparison choices	B
Dillon and Kumar	1994	Paired comparison choices and segment description	B
Wedel et al.	1993	Direct mail	P
van Duÿn and Böckenholt	1995	Repeated count data	P
Wedel and DeSarbo	1995	Coupon usage	P
Deb and Trivedi	1997	Elderly's health care	NB
Bucklin, Gupta and Siddarth	1991	Purchase frequency analysis	TP
Ramaswamy et al.	1993	Purchase frequency analysis	NB
Rosbergen et al	1997	Eye movements, print ads	G
Böckenholt	1993	Brand choice	DG, MG, DP, MP

[a] N = normal, M = multinomial, B = binomial, P = Poisson, G = gamma TP = truncated Poisson, NB = negative binomial, DM = Dirichlet-multinomial, DG = Dirichlet-gamma, MG = multinomial-gamma, DP = Dirichlet-Poisson, MP = multinomial Poisson, NM = nested multinomial.

Mixture regression models have been applied to both traditional and choice-based conjoint data. DeSarbo, Wedel et al. (1992) developed a conjoint segmentation model for metric data. The preference values of a subject were assumed to follow a multivariate normal distribution, allowing for possible covariances among preference judgments for the profiles. The authors applied their model to a conjoint experiment on remote controls for cars, simultaneously identified segments, and estimated the part-worth and covariance structure within those segments. Similar applications of a mixture regression model to conjoint experiments are provided by Wedel and DeSarbo (1994, 1995), who analyzed conjoint data on the measurement of service quality and on customer satisfaction.

A multinomial mixture model for both conjoint choice and rank-order experiments was developed by Kamakura, Wedel and Agrawal (1994). As an additional feature, those authors provided a simultaneous profiling of the segments with consumer descriptor variables (using a concomitant variables mixture model). The model was applied to the analysis of a conjoint experiment on banking services. A similar model for conjoint choice experiments was proposed by DeSarbo, Ramaswamy and Cohen (1995).

DeSarbo, Ramaswamy and Chatterjee (1995) proposed a model for constant-sum data collected in conjoint analysis, where consumers are asked to allocate a fixed number of points across the alternative profiles. They used a mixture of Dirichlet distributions to describe the data (a log-link function was used). Their model, like the preceding ones, simultaneously estimates segments and identifies the part-worth of the conjoint attributes within the segments. It was applied to a conjoint study on industrial purchasing, where profiles were constructed on the basis of supplier selection criteria.

Conclusion

The mixture regression model has had numerous applications in marketing, where it has revealed consumer heterogeneity in various markets and various types of behavior. The approach has been developed predominantly in marketing research, where interest centers on consumer heterogeneity, as opposed to other disciplines (microeconomics, econometrics) where unobserved heterogeneity has been considered mainly a nuisance. The mixtures of generalized linear models presented in this chapter provide a response-based normative approach to market segmentation that subsumes most of the work done in that area, including compound models addressing within-segment heterogeneity.

APPENDIX A1
The EM Algorithm for the GLIMMIX Model

In this appendix we describe the EM algorithm for the generalized linear mixture model (GLIMMIX), discussed in this chapter. Note that the EM algorithm for the unconditional mixture model, as described in chapter 6, is a special case of the algorithm described here with the matrix **X** consisting of a set of dummies indicating each of the K replications.

We explicitly provide the formulation of the exponential family here. We assume that the conditional probability density function of y_{nk}, given that y_{nk} comes from segment s, is in the exponential family, taking the form:

$$f_s(y_{nk}|\xi_{nks},\lambda_s) = \exp\left[(y_{nk}\xi_{nks} - b(\xi_{nks}))/a(\lambda_s) + c(y_{nk}\lambda_s)\right], \quad (A.1.1)$$

for specific functions a(·), b(·) and c(·), where conditional on segment s the y_{nk} are i.i.d. with canonical parameters ξ_{nsk} and means μ_{nsk}. The parameter λ_s is called the dispersion parameter, and is assumed to be constant over observations in segment s, while $a(\lambda_s) > 0$. See McCullagh and Nelder (1989) for further details.

The EM Algorithm

To derive the EM algorithm, we introduce unobserved data, z_{ns}, indicating whether observation vector $y_n = (y_{nk})$ from subject n belongs to segment s: $z_{ns} = 1$ if n comes from segment s and $z_{ns} = 0$ otherwise. The z_{ns} are assumed to be i.i.d. multinomial:

$$f(z_n|\pi) = \prod_{n=1}^{N} \pi_n^{z_{ns}}, \quad (A.1.2)$$

where the vector $z_n = (z_{n1},...,z_{nS})'$, and we denote the matrix $(z_1,...,z_n)$ by **Z**. With z_{ns} considered as missing data, the complete log-likelihood function can be formed:

$$\ln L_c(\phi|y,Z) = \sum_{n=1}^{N}\sum_{s=1}^{S} z_{ns} \ln f_{n|s}(y_n|\beta_s,\lambda_s) + \sum_{n=1}^{N}\sum_{s=1}^{S} z_{ns} \ln \pi_s. \quad (A.1.3)$$

The complete log-likelihood is maximized by using the iterative EM algorithm. Dempster et al. (1977) prove that the EM algorithm provides monotone increasing values of ln L_c. Under moderate conditions, ln L_c is bounded from above, in this case by zero, and convergence to at least a local optimum can be established by using a limiting sums argument (as well as Jensen's inequality) (cf. Titterington, Smith and Makov 1985).

The E-Step

In the E-step, the expectation of ln L_c is calculated with respect to the conditional distribution of the unobserved data **Z**, given the observed data y and provisional estimates of ϕ. It can be easily seen that $E(\ln L_c(\phi|\mathbf{y},\mathbf{Z}))$ is obtained by replacing z_{ns} in equation A.1.3 by their current expected values, $E(z_{ns}|\mathbf{y},\phi)$. To obtain that expectation, we first calculate the conditional distribution of y_n, given z_{ns}, which is:

$$f(y_n|Z,\phi) = \prod_{s=1}^{S} f_s(y_n|\beta_s,\lambda_s)^{z_{ns}}. \tag{A.1.4}$$

Using Bayes' rule, we can now derive the conditional distribution of z_{ns} given y_n from equations A.1.4 and A.1.2, which is in turn used to calculate the required conditional expectation:

$$E(z_{ns}|y_n,\phi) = \frac{\pi_s f_s(y_n|\beta_s,\lambda_s)}{\sum_{s'=1}^{S} \pi_{s'} f_{s'}(y_n|\beta_{s'},\lambda_{s'})}, \tag{A.1.5}$$

which is easily seen to be equal to the posterior probability p_{ns}.

The M-Step

To maximize the expectation of ln L_c with respect to ϕ, in the M-step, the unobserved data **Z** in equation A.1.3 are replaced by their current expectations:

$$E(\ln L_c(\phi|y,Z)) = \sum_{s=1}^{S}\sum_{n=1}^{N} p_{ns} \ln f_s(y_n|\beta_s,\lambda_s) + \sum_{s=1}^{S}\sum_{n=1}^{N} p_{ns} \ln \pi_s. \tag{A.1.6}$$

As the two terms on the right have zero cross-derivatives, they can be maximized separately. The maximum of equation A.1.6 with respect to π, subject to the constraints on those parameters, is obtained by maximizing the augmented function:

$$\sum_{s=1}^{S}\sum_{n=1}^{N} p_{ns} \ln \pi_s - \mu(\sum_{s=1}^{S} \pi_s - 1), \tag{A.1.7}$$

where μ is a Lagrangian multiplier. Setting the derivative of equation A.1.7 with respect to π_s equal to zero and solving for π_s yields:

$$\hat{\pi}_s = \sum_{n=1}^{N} \hat{p}_{ns} / N. \tag{A.1.8}$$

Maximizing equation A.1.4 with respect to ß and λ is equivalent to independently maximizing each of the S expressions:

$$L_s^* = \sum_{n=1}^{N} p_{ns} \ln f_s(y_n | \beta_s, \lambda_s). \tag{A.1.9}$$

The maximization of L_s^* is equivalent to the maximization problem of the generalized linear model for the complete data, except that each observation y_n contributes to the log-likelihood for each segments with a known weight p_{ns}, which is obtained in the preceding E-step. The stationary equations are obtained by equating the first-order partial derivatives of equation A.1.9 to zero:

$$\frac{\partial L_s^*}{\partial \beta_{sp}} = \sum_{n=1}^{N} p_{ns} \frac{\partial \ln f_s(y_n | \beta_s, \lambda_s)}{\partial \beta_{sp}} = 0. \tag{A.1.10}$$

Introducing the canonical parameter ξ_{nsk} and using the chain rule yields:

$$\frac{\partial L_s^*}{\partial \beta_{sp}} = \sum_{n=1}^{N} \sum_{k=1}^{K} p_{ns} \frac{\partial \ln f_s(y_n | \xi_{nsk}, \lambda_s)}{\partial \xi_{nsk}} \frac{d \xi_{nsk}}{d \mu_{nsk}} \frac{d \mu_{nsk}}{d \eta_{nsk}} \frac{d \eta_{nsk}}{d \beta_{sp}}, \tag{A.1.11}$$

where we have assumed independence of the K components of y_n. McCullagh and Nelder (1989) show that:

$$\frac{d b(\xi_{nsk})}{d \xi_{nsk}} = \mu_{nsk}, \quad \frac{d \mu_{nsk}}{d \xi_{nsk}} = \frac{d^2 b(\xi_{nsk})}{d \xi_{nsk}^2} = V_{nsk}, \quad \frac{\partial \eta_{nsk}}{\partial \beta_{sp}} = x_{nkp},$$

where V_{ns} is the variance function of $f_s(y_n|\xi_{nsk},\lambda_s)$. Therefore:

$$\frac{\partial \ln f_s(y_n | \beta_s, \lambda_s)}{\partial \beta_{sp}} = \frac{r_{nsk}}{a(\lambda_s)} \frac{d \eta_{nsk}}{d \mu_{nsk}} (y_{nk} - \mu_{nsk}) x_{nkp}, \tag{A.1.12}$$

where

$$r_{nsk} = \left(\frac{d\mu_{nsk}}{d\eta_{nsk}}\right)^2 V_{nsk}. \tag{A.1.13}$$

Substituting equation A.1.12 in A.1.11 and assuming $a(\lambda_s) = \lambda_s$ (McCullagh and Nelder 1989) yields:

$$\frac{\partial L_s^*}{\partial \beta_{sp}} = \sum_{n=1}^{N} p_{ns}^{(0)} r_{nsk} (y_{nk} - \mu_{nsk}) x_{nkp} \frac{d\eta_{nsk}}{d\mu_{nsk}} = 0. \tag{A.1.14}$$

We can see that equation A.1.14 is the ordinary stationary equation of the generalized linear model fitted across all observations, where observation n contributes to the estimating equations with fixed weight p_{ns}. Therefore, for each segment, L_s^* can be maximized by the iterative re-weighed least squares procedure, proposed by Nelder and Wedderburn (1972) for ML estimation of generalized linear models, with each observation y_n weighted additionally with p_{ns}.

Given some provisional estimates of ß$_{sp}$, this procedure involves forming an adjusted dependent variable for each segments:

$$\hat{\mu}_{nsk} = \hat{\eta}_{nsk} + (y_{nk} - \hat{\mu}_{nsk}) \frac{d\eta_{nsk}}{d\mu_{nsk}}, \tag{A.1.15}$$

where η_{nsk} is the current estimate of the linear predictor and μ_{nsk} is the corresponding fitted value derived from the link function. The derivative of the link function in character A.1.15 is evaluated at the current estimate of μ_{nsk}. The adjusted dependent variable $\hat{\mu}_{nsk}$ is regressed on x_{skp} ($p = 1,...,P$), with weight $\hat{w}_{nsk}$ for each segment s, to obtain new estimates of ß$_{sp}$, where:

$$\hat{w}_{nsk} = \left(\frac{d\mu_{nsk}}{d\eta_{nsk}}\right)^2 V_{nsk}^{-1} \hat{p}_{ns}. \tag{A.1.16}$$

The derivative of μ_{nsk} in equation A.1.16 is evaluated at the current estimate of η_{nsk}. On the basis of those revised estimates of ß$_{sp}$, new estimates of η_{nsk} and μ_{nsk} are calculated. These values are input for a new weighted regression with the dependent variable calculated according to equation A.1.15, and weights calculated according to equation A.1.16. That procedure is repeated until changes in the log-likelihood are sufficiently small.

Estimates of λ_s are obtained by setting the derivative of equation A.1.9 with respect to λ_s equal to zero and solving for λ_s:

Part 2 Segmentation methodology

$$\sum_{n=1}^{N} P_{ns} \left(\frac{y_{nk} \xi_{ns} - b(\xi_{nsk})}{a(\lambda_s)^2} \cdot \frac{d\, a(\lambda_s)}{d\, \lambda_s} + \frac{d\, c(y_{nk}, \lambda_s)}{d\, \lambda_s} \right) = 0. \qquad (A.1.17)$$

The iterative weighted least squares procedure maximizes the likelihood equation (A.1.9) according to Fisher's scoring method, and is equivalent to a Newton-Raphson procedure for canonical link functions (McCullagh and Nelder 1989).

8
MIXTURE UNFOLDING MODELS

We describe a general mixture unfolding approach that allows simultaneously for a probabilistic classification of observations into segments (similar to the GLIMMIX models described in the preceding chapter) and the estimation of an internal unfolding model within each segment. This multidimensional scaling (MDS) based methodology is formulated in the framework of the exponential family of distributions, whereby a wide range of data types can be analyzed. Possible re-parameterizations of stimulus coordinates by stimulus characteristics, as well as of probabilities of segment membership by subject background variables, are permitted. We also review previous applications of the approach to market segmentation problems.

Multidimensional scaling (MDS) is a class of statistical methods developed in the behavioral sciences to represent subjects' perceptions of stimuli (brands or products) as multidimensional spatial structures. For the purposes of MDS analyses, two types of data can be collected: proximity data (indices defined for each pair of stimuli, quantifying the degree of similarity/affinity in the pair) or dominance data (indices defined over sets of stimuli that quantify their relative degrees of superiority). In this chapter, we focus on dominance data, typically preference or choice data, contained in a persons-by-brands two-way data matrix, $\mathbf{Y} = (y_{nk})$ as before. See, for example, Davison (1983) for a comprehensive introduction to the MDS literature.

The major types of MDS models for dominance data are *unfolding* and *vector* models. In those models, the brands are represented by their coordinates on a number of dimensions hypothesized to underlie the subjects' judgments. In unfolding MDS models, ideal-point coordinates on the underlying dimensions are estimated for each subject, which specify the ideal combination of brand characteristics for the subject. According to the simple ideal-point model, a subject's preference for a stimulus object decreases as a function of the distance of the object from her/his ideal. An extension of that model is the weighted ideal-point model, which allows for differential weighting of the dimensions by each subject. The most general unfolding model allows for differential weighting and rotation by each subject. Alternatively, a vector model specifies preference vectors for each subject, indicating the direction in which her/his liking for the brands increases along the attribute space.

MDS preference/choice maps can be obtained through either *internal* or *external* analyses. *Internal analyses* estimate the stimulus coordinates, whereas in *external analyses* the stimulus coordinates are assumed given or known, for example from previous analyses. External preference/choice models are usually estimated by using generalized linear models. Therefore, in the presence of assumed heterogeneity, the external models can be estimated by the GLIMMIX models in the preceding chapter. Here we focus on internal analysis, conceptually extending the class of models presented in chapter 6. Note that in internal analyses, brand coordinates can be re-parameterized later by using information on the stimuli (e.g., features) to help the interpretation of the derived dimensions (this is commonly called property fitting).

A main limitation of several previous unfolding procedures is that they are deterministic, thus not providing any significance tests for the estimated parameters or the number of dimensions. Moreover, to account for heterogeneity, an excessive number of individual-level parameters must be estimated, reducing the degrees of freedom available for estimation. Major progress has been made in MDS research with the development of stochastic unfolding (STUN) MDS preference models using either vector- or ideal-point representations for dominance data. Such models can be used for statistical inferences (e.g., Böckenholt and Gaul 1986; DeSarbo and Hoffman 1987; DeSarbo and Cho 1989). Assumptions on the form of the statistical distribution of the dominance data make possible the estimation of STUN models by the method of maximum likelihood and, consequently, statistical tests identifying the appropriate models, and the computation of confidence intervals for the estimates. The development of STUN models has essentially moved MDS preference modeling from a descriptive tool to an inferential tool.

A common limitation of both deterministic and stochastic MDS preference models is that they are not suited to provide a joint-space representation for the large samples of subjects frequently encountered in marketing research. The ability of the models to portray the structure in such volumes of data is in fact limited. The number of parameters to be estimated is excessive and the derived joint spaces become saturated with subjects' vectors or ideal points, rendering interpretation very difficult. Often, the individual-level parameters are clustered in a second step of the analysis to aid interpretation. The interest then shifts to the derived groups of subjects, or segments.

Those problems have stimulated the development of stochastic unfolding mixture models (STUNMIX), for which the number of parameters is significantly reduced relative to the STUN models. STUNMIX models are estimated by the method of maximum likelihood. This chapter describes a fairly general model that builds on prior research, integrating previously developed models into a general framework. This family of models also encompasses a wide range of new options, including different distributions of the data and a variety of functional forms. The model was recently developed by Wedel and DeSarbo (1996). Similar to the GLIMMIX model developed in the preceding chapter, the STUNMIX model is formulated in the framework of the exponential family of distributions, including the most commonly used distributions such as the normal, binomial, Poisson, negative binomial, inverse Gaussian, exponential and the gamma, as well as several compound distributions. It allows for the spatial representation of a wide range of data, such as ratings, pick-any data, choices, times and frequencies. In describing the STUNMIX models, we draw from the work by DeSarbo, Manrai and Manrai (1994) and Wedel and DeSarbo (1996). First, however we provide examples of STUNMIX applications.

Examples of Stochastic Mixture Unfolding Models

Example 1: Television Viewing

We describe an application of a stochastic mixture vector model, recently developed by DeSarbo, Ramaswamy and Lenk (1993). DeSarbo and his co-authors used a sample of 100 households included in a panel that is used by a large US market research company to record TV viewing behavior. For each household, the data consist of the proportion of TV-viewing time (in minutes) allocated across five alternatives (ABC, CBS, NBC, Fox and cable) over a one-year period.

The two-way data were analyzed with a STUNMIX model that simultaneously derives a spatial map of the five alternatives, as well as the preference vectors of segments of consumers. As the data take the form of proportions that sum to one for each subject, DeSarbo and colleagues assumed a Dirichlet distribution for them. The Dirichlet is a multivariate extension of the beta distribution (in the exponential family), which is suited for the modeling of proportions because it satisfies the sum constraint. Using the mixture Dirichlet unfolding model, the authors investigated several numbers of dimensions ($T = 1$ to $T = 4$) and segments ($S = 1$ to $S = 4$), and found that the two-dimensional, two-segment solution provided the minimum value of the CAIC. That solution was selected. Figure 8.1 depicts the results. The two segments of the solution were well separated, yielding an entropy value of 0.992. The relative locations of the stimuli in the spatial map suggest a horizontal dimension defining cable versus traditional broadcast networks and a vertical dimension defining the overall popularity of the alternatives. The authors name segment 2 (94%) *traditional network viewers*, because they devote a substantial amount of time to watching traditional programming, especially NBC and CBS. Segment 1 (6%) is defined as *cable viewers*, who watch more cable TV, presumably for special programs. The relatively new network Fox appears to have a rather weak position with respect to both segments (note that the linear projection of a network onto a segment vector provides the relative preference of that segment for that network). The authors justify the relatively small size of the cable viewers segment by noting that cable TV had just reached 50% penetration in the test market and probably lacked the promotional push of the traditional network services. Socioeconomic and demographic variables did not explain the differences between the segments very well in an analysis of the posterior probabilities of segment membership.

Part 2 Segmentation methodology

Figure 8.1: TV-Viewing Solution (adapted from DeSarbo, Ramaswamy and Lenk 1993)

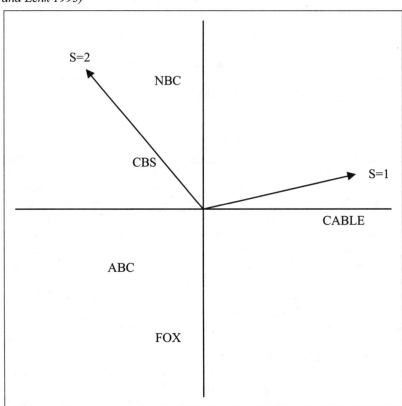

Example 2: Mobile Telephone Judgments

This application was provided by Wedel and DeSarbo (1996). The data for the application were collected by a major telecommunications firm in the United States. The study aimed at understanding consumers' intentions to buy portable telephones. A sample of 279 respondents were asked to mark, of 12 actual products exhibited, the ones they would consider buying within the next six months. From those "pick-any/K" data, the purpose was to uncover the underlying dimensions of portable telephones by using a binomial distribution to describe the zero-one pick-any data. Seven portable telephone features were available.

1. Calling options: CALL (1 = send and receive; 0 = receive only)
2. Number of telephone memories: MEMO
3. Speaker phone option: SPEAK (1 = yes; 0 = no)
4. PRICE in dollars

5. RANGE in feet
6. SETUP (1 = stand-up, 0 = lie-down)
7. STYLE (1 = walkie-talkie; 0 = cradle)

The names of the brands were not provided, and the products were identified only by the letters A through L. Several alternative model specifications were considered, including a simultaneous re-parameterization of the dimensions with the above seven product features. On the basis of the ICOMP criterion, the unrestricted $S = 4$, $T = 2$ simple ideal-point model appeared to represent the data best. That solution is depicted in Figure 8.2. Wedel and DeSarbo calculated the 95% confidence regions of the estimates, which are added to Figure 8.2. The product-feature vectors fitted to the unrestricted solution by standard procedures also are depicted, as they may be somewhat useful in interpretation of the dimensions.

The figure shows that the two recovered dimensions separate the choice alternatives well, because on both dimensions the confidence intervals of the alternatives do not overlap. The first dimension separates alternatives A, B and H from the others. Price, style, setup and send-receive options load positively on dimension 1. The second dimension separates the brands E and C from H, I, J, K and L. The brands B, G, F and D take an intermediate position, where the location of G differs significantly from that of the brands in the two aforementioned groups (note that because of identification restrictions the confidence ellipsoids of K and L were not estimated). Memory, speakerphone, range and price load negatively on this dimension, and style positively. The authors labeled the two dimensions *quality* and *function*, respectively.

Figure 8.2 also shows that the ideal points of the four segments are very well separated. The ideal point of segment 2 (20.3% of the sample) is close to the products A, B and H. The binomial selection probabilities of those alternatives (not shown here) are much higher than the corresponding probabilities in the other segments. The ideal-point of Segment 4 (10.5%) is on the opposite end of the first dimension, among the remaining alternatives C - G, and I - L. The selection probabilities for those alternatives in segment 4 are much higher than the corresponding probabilities in the other segments. The ideal points of segment 1 (34.6%) and segment 3 (34.7%) take opposite positions on the second dimension, indicating a preference for brands that load positively and negatively on that dimension, respectively. Segment 1 has high selection probabilities for brands I, J and L in comparison with segments 2 and 3, but note that the probabilities for those brands in segment 4 are higher. Segment 3 has high probabilities for brands C, E, F and G, in comparison with segments 1 and 2, but segment 4 has higher probabilities of selecting those brands. Segment 2 appears to prefer low quality, low price brands, whereas segment 4 prefers higher quality brands. Segment 1 prefers alternatives with additional options and segment 3 prefers brands with a walkie-talkie style. Thus, the analysis gives the firm strategic insights on how to target specific segments with portable telephones with specific sets of features.

Figure 8.2: Spatial Map of the Telephone Brands (adapted from Wedel and DeSarbo, 1996)

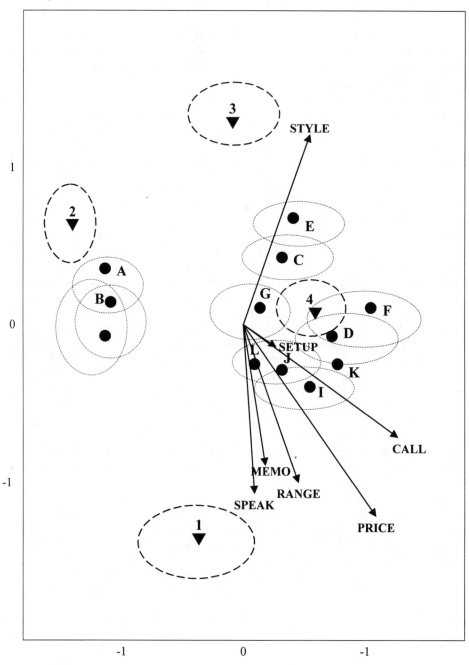

A General Family of Stochastic Mixture Unfolding Models

Here we describe a general stochastic mixture unfolding model that includes most of the previously published models as special cases. The framework for this STUNMIX model is similar to that for the GLIMMIX model in the preceding chapter: the exponential family. The model described in this chapter is from Wedel and DeSarbo (1996). We add the following notation to that introduced before.

Notation

Let: b_{kt} denote the position of brand k on latent dimension t,
a_{st} denote the ideal point/vector terminus of segment s on latent dimension t,
w_{st} denote the weight of latent dimension t in segment s, and
A_{nl} denote the l-th descriptor variable for subject n.

Assume, as in the preceding two chapters, that the n subjects in the sample come from a population that is a mixture of S segments in proportions $\pi_1,....,\pi_S$. The membership of the subjects in the S segments is not known in advance. Again we assume that the probability distribution of y_{nk}, given that y_{nk} comes from segment s, is a distribution in the exponential family. Conditional on segment s, the y_{nk} are independently distributed with means μ_{ks}. Note that those expectations differ across the unobserved segments, s, and the brands, k. Again, the parameter λ_s is a dispersion parameter and is assumed to be constant within segment s. We specify a predictor η_{sk} and a link function $g(\cdot)$ for each unobserved segment s (see the preceding chapter), such that:

$$\eta_{ks} = g(\mu_{ks}). \tag{8.1}$$

Thus, the predictor is a function of the mean of the observations. The specific link function chosen depends on the specific distribution thought to be appropriate for the data, for example the identity, log, logit, inverse and squared inverse functions for the normal, Poisson, binomial, gamma and inverse Gaussian distributions, respectively (see Table 6.1).

The predictor is assumed to be produced by the locations of the brands on T unobserved dimensions, b_{kt}, and by segment-specific preference vectors with termini, a_{st}, according to a vector model:

$$\eta_{ks} = c_{ks} + \sum_{t=1}^{T} b_{kt}\, a_{st}, \tag{8.2}$$

or by ideal points, a_{st}, and (optional) weights (w_{st}):

Part 2 Segmentation methodology

$$\eta_{ks} = c_{ks} - \sum_{t=1}^{T} w_{st} (b_{kt} - a_{st})^2. \tag{8.3}$$

Thus, conditional on segment s, a nonlinear model is formulated consisting of a specification of the distribution of the random variable y_{nk}, as one of the exponential family, a predictor produced by a vector or ideal-point unfolding model in each unobserved segment s, and a function g(.) linking the random and systematic components. Note that the formulation of the STUNMIX model is similar to the formulation of the GLIMMIX model in the preceding chapter. Here, however, the predictor is not a linear function of exogenous variables, but a function of latent variables that must be estimated.

Equations 8.2 and 8.3 present the well-known vector and ideal-point models which, contrary to standard MDS models, are not formulated directly for the data, but for some transformation of the expectations of the data. Also, the ideal points and vectors are not specific to individual subjects as in standard MDS models, but are specific to segments, which are assumed to share a common vector or ideal point. Those segments are to be identified from the data simultaneously with the segment structure. If the linear predictors and the segments were observed, one could estimate a vector or unfolding model directly on that S × K matrix.

The following methodological options are accommodated in the STUNMIX procedure:

- (M1) **Distributions from the exponential family.** The y_{nk} are allowed to be distributed according to any distribution in the exponential family. As explained before, that family includes the normal (e.g., for rating scale measures), the Poisson (e.g., for purchase frequencies), the binomial (e.g., for choice or pick-any data), and the gamma and inverse Gaussian (e.g., for duration-type data) distributions.

- (M2) **Different types of preference models**. Vector, simple and weighted ideal-point unfolding models are all accommodated. For the ideal-point model, constraining $w_{st} = 1$, for all s and t yields the simple unfolding model. A weighted unfolding model is obtained by not imposing any constraints on w_{st}.

- (M3) **Internal versus external analyses.** One can estimate b_{kt} and a_{st} from the data or, if a configuration b_{kt}^* of the stimuli is known, one can constrain $b_{kt}^* = b_{kt}$ and estimate a_{st} only. In fact, this class of models can be estimated by the GLIMMIX mixture regression models described in the preceding chapter.

- (M4) **Constraints on the intercepts.** The model that includes all c_{ks} as well as the MDS unfolding component is not uniquely identified. One should impose one of three types of constraints to identify the model. An intercept can be estimated either for each segment, $c_{ks} = c_s$, or for each brand, $c_{ks} = c_k$, or both. Thus, overall differences in the preference or choice measure between segments or brands can be removed from the scaling model.

- (M5) **Simultaneous re-parameterization or property fitting**. One can constrain the b_{kt} to be a linear function of some set of preselected brand descriptor variables contained in the $(N \times M)$ matrix $\mathbf{B} = ((B_{km}))$:

$$b_{kt} = \sum_{m=1}^{M} B_{km} \beta_{mt}. \tag{8.4}$$

$\boldsymbol{\beta}$ is an $(M \times T)$ matrix of coefficients $((\beta_{mt}))$. The constraints can aid in the interpretation of the derived dimensions. We formulate the unconditional distribution as in the preceding chapter (equation 7.3), except for the segment-specific parameters, which now are defined by $\theta_s = (\mathbf{a}_s, \mathbf{b}_s, \boldsymbol{\beta}_s, \lambda_s)$.

EM Estimation

The EM algorithm proceeds as in the preceding mixture models. In the E-step, posterior segment membership probabilities (p_{ns}) are computed as shown in equation 6.4. The M-step updates the estimates of the segment sizes (π_s) as described in equation 6.6 and of the other parameters within each segment. Details of the EM estimation for the STUNMIX model are provided in Appendix A2.

Some Limitations

The limitations of the EM algorithm noted in the preceding chapter also apply to this class of models: its convergence may be rather slow, and convergence to a global maximum is not ensured. In the case of the STUNMIX models, those problems reportedly are minimized by having a "close" start at the first step of the algorithm (Böckenholt and Böckenholt 1991). A close start can obtained by the following procedure. An unrestricted mixture model is estimated in which an intercept is specified for each brand. There is no spatial-structure component in the model. That procedure provides close starting values for the posterior membership probabilities of subjects, p_{ns}. The starting procedure itself should be applied with several sets of random starting values for p_{ns}. Close starting values for the vector or ideal-point models are obtained by computing the segment-by-stimulus matrix of totals of the dependent variable, transformed by the appropriate link function, $g(\Sigma_n p_{ns} y_{nk})$, and performing a singular value decomposition of that matrix (see Kamakura and Novak 1992 for an example). Starting values for $\beta_{mt}^{(0)}$ are obtained from a regression of $b_{kt}^{(0)}$ on $\mathbf{B}$. That procedure for obtaining starting values has been reported to greatly improve the properties of numerical algorithms for estimating STUNMIX models. Under typical regularity conditions, the ML estimates are asymptotically normal, and the standard errors can be obtained as discussed in the preceding chapter.

Issues in Identification

It is important to note the particular indeterminancies involved in the STUNMIX models (see Wedel and DeSarbo 1996). Table 8.1 is a summary of the effective numbers of parameters estimated for the various STUNMIX models and options. The unfolding models require the estimation of S - 1 class size or mixing parameters, S or J constants, TJ brand coordinate parameters, ST ideal-point coordinates or vector termini and possibly ST dimensional weights. The effective number of parameters for the various models is affected by the respective centering, scaling and rotational invariance to which all MDS models are subject. Essentially, the fact that an MDS solution can be translated, expanded or rotated without any effect on the solution obtained leads to the indeterminancies. The derived spaces for the weighted unfolding model are invariant under joint centering of a_{st} and b_{kt} and scale indeterminancies exist between w_{st} and (a_{st}, b_{kt}). Consequently, 2T are subtracted from the number of parameters estimated. For the simple unfolding model, because the solutions are invariant under centering and under rotation through an orthogonal matrix, $T(T + 1)/2$ is subtracted.

For the vector model, the solution is invariant under translation of an arbitrary T-dimensional vector and transformations of a non-singular $(T \times T)$ matrix, whereby $T(T + 1)$ is subtracted from the number of parameters estimated. Re-parameterization of the coordinates removes translational indeterminancies in the various models, which alleviates T indeterminancies and adds T to the number of effective parameters for the object coordinates. The models are identified only for $S \geq T$. According to the authors' experience in applications of the models, additional constraints may need to be imposed to identify the models. In general, searching for the appropriate identifying restrictions is not straightforward.

Model Selection

When the STUNMIX models are applied to real data, the actual numbers of both dimensions, T, and segments, S, are unknown and must be inferred from the data. Ideally, one would want to test the null hypothesis of S segments against the alternative hypothesis of S + 1 segments, or the null hypothesis of T dimensions against the alternative of T + 1 dimensions. As noted in chapter 6, the standard likelihood ratio statistics for the tests of the number of segments are not asymptotically distributed as chi-square, and information criteria such as AIC, CAIC or ICOMP must be used to determine the appropriate number of segments. Much the same holds for the number of underlying dimensions T. As the likelihood ratio tests are invalid, information criteria must be used to determine the appropriate value of T (cf. Anderson 1980), in deciding between the non-nested vector and simple ideal-point models. To determine both S and T, we consider a three-step procedure (cf. De Soete and Winsberg 1993).

Table 8.1. Numbers of Parameters for STUNMIX Models

Model/constant	Non-re-parameterized brand coordinates	Re-parameterized brand coordinates
Vector model		
Segment	$S + (S-1)(L + 1) + T(J + S) - T(T + 1)$	$S + (S + 1)(L + 1) + T(K + S) + T^2$
Brand	$J + (S - 1)(L + 1) + T(J + S) - T(T + 1)$	$J + (S - 1)(L + 1) + T(K + S) - T^2$
Simple unfolding		
Segment	$S + (S - 1)(L + 1) + T(J + S) - T(T + 1)/2$	$S + (S - 1)(L + 1) + T(K + S) - T(T - 1)/2$
Brand	$J + (S - 1)(L + 1) + T(J + S) - T(T + 1)/2$	$J + (S - 1)(L + 1) + T(K + S) - T(T - 1)/2$
Weighted unfolding		
Segment	$S + (S - 1)(L + 1) + T(J + 2S) - 2T$	$S + (S - 1)(L + 1) + T(K + 2S) - T$
Brand	$J + (S - 1)(L + 1) + T(J + 2S) - 2T$	$J + (S - 1)(L + 1) + T(K + 2S) - T$

1. The appropriate number of segments, S^*, is decided by comparing the AIC, CAIC and/or ICOMP statistics of saturated models that have different numbers of segments and maximum dimensionality. Such a saturated model with maximum dimensionality is identical to a simple mixture model, where no constraints are imposed on the constants, c_{ks}, and where the spatial MDS structure is absent.

2. Once the appropriate number of segments is determined, the appropriate dimensionality, T^* ($T^* \leq S^*$), is identified by comparing the information statistics of S^*-segment models that have different numbers of dimensions.

3. Given S^* and T^*, several tests of nested models can be performed. The appropriateness of constraints on the intercepts (M4) and the stimulus coordinates (M3, M5) can be tested by using likelihood ratio tests. The information criteria can be used to decide on the appropriate model form (M2).

The deviance residuals for generalized linear mixture models described in the preceding chapter can be applied to explore the adequacy of the fit STUNMIX models and the presence of outliers.

Synthetic Data Analysis

Several authors have tested STUNMIX models on synthetic data to investigate their performance. Here we summarize and extend the results of Wedel and DeSarbo (1996), who analyzed various synthetic datasets generated from the normal, binomial and Poisson distributions. To generate the synthetic data, eight stimuli were located on a 3×3 equally spaced grid on [-1,1]. The synthetic data were generated for three segments, where each segment contained eight subjects. Ideal points of the three segments were located in the interior of the grid (see Figure 8.3). Three datasets, representing the binomial, Poisson and normal, were constructed by using the simple ideal-point model.

A second set of three datasets was generated to investigate the robustness of the models to within-segment heterogeneity of the stimulus coordinates and ideal points. Here, random numbers were added to the coordinates of the stimuli and ideal points for each subject to simulate within-segment heterogeneity. Six datasets were generated for the simple ideal point model. A similar set of six datasets was generated according to the vector model. The datasets were analyzed by using the STUNMIX model with the proper distributional assumptions and the appropriate vector model.

The root mean squared error of the true and estimated segment ideal points and stimulus coordinates, RMSE(b) and RMSE(a), were calculated to evaluate the solutions. The results are reported in Table 8.2.

For those analyses, the EM algorithm converged quickly to the final solutions. In each analysis all subjects were classified correctly into the true segments. The posterior segment membership probabilities p_{is} equaled zero or one up to four decimal places. The small values of RMSE(a) and RMSE(b) show that the actual spatial structure is recovered very well by the STUNMIX procedure (Table 8.2). Under the conditions of within-segment heterogeneity, the RMSEs are up to three times as large. (Note that additional constraints had to be imposed beyond the number indicated in Table 8.1 to identify vector model 2.)

Table 8.2: Results of STUNMIX Synthetic Data Analyses (adapted from Wedel and DeSarbo 1996)

Distribution	Ideal-Point RMSE(b)	Model RMSE(a)	Vector RMSE(b)	Model RMSE(a)
Normal	0.0220	0.0152	0.0163	0.0344
H normal[a]	0.0359	0.0985	0.0553	0.0505
Binomial	0.0707	0.0707	0.1287	0.1173
H binomial	0.0715	0.0885	0.1097	0.1157
Poisson	0.0336	0.0648	0.0353	0.0378
H Poisson	0.0989	0.0566	0.0470	0.0396

[a] H denotes within-segment heterogeneity.

Figure 8.3 depicts the true configuration and the configurations recovered by the STUNMIX methodology (after target rotation) for the normal, binomial and Poisson ideal-point data; the respective solutions for the data with within-segment heterogeneity have been superimposed. The 95% confidence ellipsoids of the coordinates are included in each of the plots. Figure 8.3 illustrates the previous findings.

Additionally, the authors investigated the robustness of STUNMIX to mispecification of the distribution. For that purpose they generated, as a benchmark, three datasets with varying error levels, the error being drawn from the normal distribution with different standard errors. Second, three additional datasets were generated with the error drawn from the log-normal distribution, with the same standard deviations. The six datasets were analyzed with the normal STUNMIX model. The dependent measures are reported in Table 8.3.

For the normal and log-normal data with $\sigma = 0.1$ and $\sigma = 1$, recovery was relatively good, as evidenced by the RMSE(a) and RMSE(b) values. All subjects were correctly classified, with the p_{is} close to zero or 1 up to four decimal places. Interestingly, the classification of subjects under mispecification is still relatively good: 23 subjects were classified into the correct class with a posterior of 0.95 or higher and only one subject was misclassified. We conclude that the performance of the STUNMIX methodology appears satisfactory. Parameter recovery is good, and the models appear to be relatively robust to mispccification of the functional and distributional forms.

Table 8.3: STUNMIX Results for Normal and Log-Normal Synthetic Data (adapted from Wedel and DeSarbo 1996)

Distribution	RMSE(y)	RMSE(x)
Normal, $\sigma = 0.1$	0.0059	0.0059
Log-normal, $\sigma = 0.1$[a]	0.0058	0.0044
Normal, $\sigma = 1$	0.0302	0.0411
Log-Normal, $\sigma = 1$[a]	0.1658	0.1025
Normal, $\sigma = 1.25$	0.0747	0.0604
Log-Normal, $\sigma = 1.25$[a]	0.6622	0.3606

[a] Mispecified model.

Marketing Applications

Several stochastic mixture unfolding models have been developed and applied in the marketing and psychometrics literature. A review of that literature has been provided by DeSarbo, Manrai and Manrai (1994). Models for various types of data (normal, binomial, Poisson, multinomial and Dirichlet) have been developed. The DeSarbo, Manrai and Manrai review shows that the literature on mixture unfolding models consists of a collection of independent special cases, in which various different model specifications and estimation procedures have been used. The work by Wedel and DeSarbo described above integrates the previous developments into a common framework. Here we discuss the applications by distribution type.

Normal Data

The first published STUNMIX model is that of DeSarbo, Howard and Jedidi (1990), who developed a STUNMIX vector model for normally distributed data. The model was applied to consumers' evaluations of microcomputers. DeSarbo et al. (1991) extended the model to a weighted ideal-point model, which was applied to the identification of strategic groups of businesses on the basis of the PIMS data. De Soete and Winsberg (1993) and De Soete and Heiser (1993) extended the model by incorporating linear restrictions on the brand coordinates. In those models, the brand locations were profiled simultaneously with brand characteristics, as in property fitting. Their applications involved consumers' preferences for houses and consumer sympathy for political parties, among other issues. The derived latent dimensions were described simultaneously by exogenous variables describing the stimuli.

Figure 8.3: Synthetic Data and Results of Their Analyses with STUNMIX (adapted from Wedel and DeSarbo 1996)

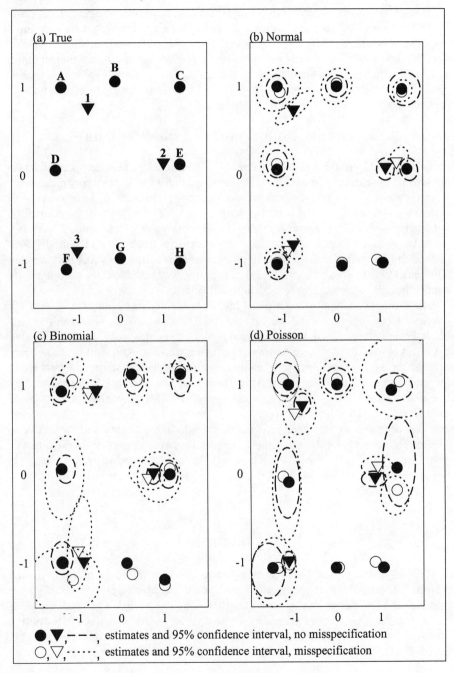

Binomial Data

Böckenholt and Böckenholt (1991) applied vector, simple and weighted ideal-point models to binary data. The models were used to derive simultaneously segments' ideal points/vectors and the latent dimensions underlying the preferences of magazines. We have already described Wedel and DeSarbo's (1996) application of the binomial STUNMIX model to the analysis of pick-any data on mobile telephones. A related binomial STUNMIX model was applied by DeSarbo et al. (1995) to the analysis of pick-any data on new bundles of car features.

Poisson, Multinomial and Dirichlet Data

Wedel (1995) used a STUNMIX Poisson model to analyze scanner panel data on consumers' purchases of soup flavors. He attempted to derive latent sensory dimensions from consumers' purchases of soups over one year, and identified segments with different preference vectors for the soup flavors. The segments were profiled simultaneously with consumer characteristics as concomitant variables. A problem related to such applications is that the data are non-experimental and the model does not account for the marketing efforts of producers and retailers. DeSarbo, Ramaswamy and Lenk (1993) developed a vector mixture model for Dirichlet-distributed data. Its application was presented previously in this chapter.

Another application to scanner panel data was provided by Chintagunta (1994). He developed a multinomial STUNMIX model that he used to analyze consumers' choice behavior for detergents. An important additional feature of his model is that it included the effect of marketing mix variables such as price, display and featuring. The effects of those marketing variables are taken into account separately, so that the perceptual map derived is adjusted for them. An extension of that approach was proposed by Wedel et al. (1998). They applied the multinomial STUNMIX model to conjoint choice experiments. A perceptual map of brands was estimated simultaneously with the ideal points of segments of consumers, and the part worths of the attributes of the cars were included in the conjoint design. Simultaneously, they also accounted for the effect of brand name (un)familiarity. A more thorough discussion of the latter two models is beyond the scope of this chapter. Table 8.4 summarizes the literature on applications of STUNMIX models.

Conclusion

This chapter extends the preceding one to mixture unfolding models. In the STUNMIX methodology described here, latent variables and latent segments underlying a dependent variable are identified simultaneously. The resulting MDS model has the same level of generality as the mixture models described in preceding chapters. Its results may provide useful insights for strategic decision making, especially for product (re)positioning within targeted market segments. So far, the

STUNMIX models have allowed for heterogeneity in preferences (i.e., ideal points or vectors) across and within segments, but assumed perfect homogeneity in perceptions (i.e., location of the stimuli). Further research is needed to explore the possible heterogeneity in perceptions across consumers.

Table 8.4: STUNMIX Marketing Applications

Authors	Year	Application	M^a	R^b	f^c
DeSarbo, Howard and Jedidi	1990	Microcomputer evaluation	V	-	N
DeSarbo et al.	1991	Identification of strategic groups	V	-	N
Böckenholt and Böckenholt	1991	Magazine preferences	I,V	-	B
De Soete and Heiser	1993	Political party sympathy	I	C	N
De Soete and Winsberg	1993	1. Housing preferences	V	C	N
		2. Student success		C	N
DeSarbo, Ramaswamy and Lenk	1993	TV-channel viewing times	V	-	D
Wedel	1995	Soup flavor purchase frequency	V		P
Wedel and DeSarbo	1996	Mobile telephone choices	I	C	B
DeSarbo and Jedidi	1995	Consideration set formation	B	-	-
Chintagunta	1994	Detergent choices	V	X	M
DeSarbo et al.	1996	Bandle Composition	I	X	B
Wedel et al.	1997	Conjoint choices on cars	I	X	M

[a] Model type: I = ideal point, V = vector.
[b] C = restrictions on stimulus coordinates, X = exogenous variables included.
[c] Mixture distributions: N = normal, B = binomial, D = Dirichlet, P = poisson, M = multinomial.

APPENDIX A2
The EM Algorithm for the STUNMIX Model

To estimate the parameter vector ϕ in the STUNMIX model, we maximize the likelihood with respect to ϕ subject to the restrictions on the priors, similar to the way that was done in Appendix A1. To derive the EM algorithm, we introduce unobserved data, z_{ns}, indicating whether consumer n belongs to segment s: $z_{ns} = 1$ if n comes from segment s and $z_{ns} = 0$ otherwise. With z_{ns} considered as missing data, the complete log-likelihood function is:

$$\ln L_c(\phi \mid y, Z) = \sum_{n=1}^{N} \sum_{k=1}^{K} \sum_{s=1}^{S} z_{ns}(\ln f_s(y_{nk} \mid \phi) + \ln \pi_s). \quad (A.2.1)$$

Note that if, say, R independent repeated measurements are taken for each subject, the equation for the complete log-likelihood involves an additional summation over r = 1,...,R. We drop that summation for convenience of notation. The distribution $f_s(y_{nk}|\phi)$ is assumed to be one of the exponential family, as defined in equation A.1.1 in Appendix A1.

The E-step

In the E-step the expectation of $\ln L_c$, equation A.2.1, is calculated with respect to the conditional distribution of the unobserved data $Z = ((z_{ns}))$, given the observed data y and provisional estimates of ϕ. We can easily see that $E(\ln L_c(\phi|y,Z))$ is obtained by replacing z_{ns} in equation A.2.1 by its current expected value $E(z_{ns}|y, \phi) = p_{is}$, as before (see equation A.1.5), conditional on provisional estimates of the STUNMIX model parameters within any iteration.

The M-step

To maximize the expectation of $\ln L_c$ with respect to ϕ, in the M-step, the unobserved data Z in equation A.2.2 are replaced by their current expectations p_{is}:

$$E(\ln L_c(\phi \mid y, Z)) = \sum_{s=1}^{S} \sum_{n=1}^{N} \sum_{k=1}^{K} \hat{p}_{ns}(\ln f_s(y_{nk} \mid \phi) + \ln \pi_s). \quad (A.2.2)$$

As the two terms on the right have zero cross-derivatives, they be maximized separately. The maximum of equation A.2.2 with respect to π_s, subject to the constraints

Appendix 2 The EM Algorithm for the STUNMIX Model

on those parameters, is obtained by maximizing the augmented function:

$$\sum_{s=1}^{S}\sum_{i=1}^{I} \hat{p}_{ns} \ln \pi_s - \mu \left(\sum_{s=1}^{S} \pi_s - 1\right), \tag{A.2.3}$$

where μ is a Lagrangian multiplier. Setting the derivative of equation A.2.3 with respect to π_s equal to zero and solving for π_s yields:

$$\hat{\pi}_s = \sum_{n=1}^{N} \hat{p}_{ns} / I. \tag{A.2.4}$$

Maximizing equation A.2.2 with respect to **C**, **W**, **X**, or **ß**, **Y** and λ, is equivalent to maximizing:

$$L^* = \sum_{s=1}^{S}\sum_{n=1}^{N}\sum_{k=1}^{K} \hat{p}_{ns} \ln f_s(y_{nk}|\phi). \tag{A.2.5}$$

The stationary equations are obtained by equating the first-order partial derivatives of equation A.2.5 to zero. Introducing the canonical parameter ξ_{ks} (see Appendix A1), using the chain rule and defining $\varsigma = c_{ks}, w_{st}, a_{st}, b_{kt}$ or β_{mt} yields:

$$\frac{\partial L^*}{\partial \varsigma} = \sum_{n=1}^{N}\sum_{k=1}^{K}\sum_{s=1}^{S} \hat{p}_{ns} \frac{\partial \ln f_s(y_{nk}|\phi)}{\partial \xi_{ks}} \frac{d \xi_{ks}}{d \mu_{ks}} \frac{d \mu_{ks}}{d \eta_{ks}} \frac{\partial \eta_{ks}}{\partial \varsigma}. \tag{A.2.6}$$

McCullagh and Nelder (1989) show that:

$$\frac{d b(\xi_{ks})}{d \xi_{ks}} = \mu_{ks}, \quad \frac{d \mu_{ks}}{d \xi_{ks}} = \frac{d^2 b(\xi_{ks})}{d \xi_{ks}^2} = V_{ks}, \tag{A.2.7}$$

where V_{ks} denotes the variance function of $f_s(y_{nk}|\phi)$. We now elaborate on the last term in equation A.2.7 involving the partial derivative of η_{ks} with respect to ς. The partial derivatives with respect to c_s or c_j equal 1. For the vector model, the partial derivatives with respect to a_{st} and b_{kt} equal b_{kt} and a_{st}, respectively. For the ideal-point models, the derivatives with respect to w_{st}, a_{st}, b_{kt} and β_{mt} are:

$$\frac{\partial \eta_{ks}}{\partial w_{st}} = (b_{kt} - a_{st})^2, \tag{A.2.8}$$

$$\frac{\partial \eta_{ks}}{\partial a_{st}} = -2 w_{st}(a_{st} - b_{jt}), \tag{A.2.9}$$

$$\frac{\partial \eta_{ks}}{\partial b_{kt}} = 2 w_{st}(b_{kt} - a_{st}), \qquad (A.2.10)$$

$$\frac{\partial \eta_{ks}}{\partial \beta_{mt}} = 2 w_{st}(b_{kt} - a_{st}) B_{km}. \qquad (A.2.11)$$

We assume $a(\lambda_s) = \lambda_s$. Estimates of λ_s are obtained by setting the derivative of equation A.2.5 with respect to λ_s equal to zero and solving for λ_s:

$$\sum_{n=1}^{N}\sum_{k=1}^{K} \hat{p}_{ns} \left(\frac{y_{nk}\xi_{ks} - b(\xi_{ks})}{a(\lambda_s)^2} \cdot \frac{d\,a(\lambda_s)}{d\,\lambda_s} + \frac{d\,c(y_{nk},\lambda_s)}{d\,\lambda_s} \right) = 0. \qquad (A.2.12)$$

In each E-step, the estimates are obtained by numerically maximizing expression A.2.5 with respect to the parameters, using numerical methods such as conjugate gradients or quasi-Newton methods based on the derivatives in equations A.2.6 through A.2.11. For a discussion of those numerical procedures, we refer to standard texts such as Scales' (1985).

9
PROFILING SEGMENTS

In this chapter we further expand on the mixture models provided in the preceding three chapters. We discuss procedures that simultaneously profile segments. Those procedures, called concomitant variable mixture approaches, apply to any of the MIX, GLIMMIX and STUNMIX model frameworks described in the preceding chapters. They allow for a simultaneous profiling of the segments derived with external "concomitant variables" and alleviate some disadvantages of previously used two-step procedures.

Profiling Segments with Demographic Variables

Segments derived by mixture models, whether standard mixture models, mixture regression models or mixture unfolding models, cannot always be targeted directly with new products, promotions and so on. Segments must first be profiled with consumer (firm) descriptors to fulfill the accessibility requirement for effective market segmentation (see chapter 1). For profiling purposes, demographic and socioeconomic variables are often used. Although we have shown that those variables are not the most effective in developing segments, they are used in segmentation studies to profile segments and thus enhance identifiability and accessibility. In that way, segments can be targeted, as media profiles and market areas are often described along the demographic variables. Also, describing the segments with demographic variables enables the researcher to investigate the face validity of the segments.

Segment profiling is related to the hybrid approaches to segmentation. In segment profiling, segments are described post-hoc with passive (e.g., demographic) variables. In contrast, hybrid approaches first use demographic and socioeconomic variables to define a-priori segments. Post-hoc methods (clustering or mixture models) of segmentation are then used to further segment the sample, given the previously defined segments.

In mixture models, post-hoc profiling of segments is typically performed by using the posterior segment membership probabilities (see equation 6.4, Table 6.3). Those membership probabilities, p_{ns}, provide the probability that a particular subject belongs to each of the derived segments. The segments derived from mixture models have been profiled by most authors in a second step of the analyses: taking a logit transform of the probabilities and regressing $\log(p_{ns}/(1 - p_{ns}))$ on the descriptor variables of interest (see, e.g., DeSarbo, Ramaswamy and Chatterjee 1995, DeSarbo, Wedel et al.1992).

However, that procedure has several disadvantages. First, this second step is performed independently from the estimation of the mixture model, and optimizes a different criterion. Therefore, segments identified in the first step might not have an "optimal" structure with respect to their demographic profile. Second, the logit model used to profile the segments in the second step does not take into account the estimation error of the posterior probabilities.

Several authors in marketing have proposed models that simultaneously profile the derived segments with descriptor variables (e.g., Dillon, Kumar and Smith de Borrero 1993; Gupta and Chintagunta 1994; Kamakura, Wedel and Agrawal 1994). Their work was based on the concomitant variable latent class model proposed by Dayton and MacReady (1988). They coined the terminology "concomitant variable" mixture models, and proposed the term "sub-model" for the model relating the segments to the concomitant (i.e., profiling) variables. Formann (1992) used the term "linear logistic latent class analysis" for similar techniques. Before discussing concomitant variable models and linking them to the mixture, mixture regression and mixture unfolding models described in the preceding chapters, we report two illustrative applications of the models.

Examples of Concomitant Variable Mixture Models

Example 1: Paired Comparisons of Food Preferences

Dillon, Kumar and Smith de Borrero (1993) analyzed data from 550 women who were responsible for food shopping. In the survey, respondents were asked to think about their primary concerns in purchasing food products when shown 10 pairs of the following five food-related concerns (the stimuli).

1. Nutritional value
2. Convenience of preparation
3. Calories
4. Ingredients
5. Price

For each pair of concerns, respondents were asked to judge which one was more important when purchasing food products. The model used to analyze the paired comparisons was based on the Bradley-Terry-Luce (BTL) model. That model re-parameterizes the probabilities of choosing one of each pair as a function of unobserved scale values of the concerns included in the pair. The scale values determine the locations of the stimuli on a unidimensional underlying scale. The authors assumed that the underlying scale may differ across unobserved consumer segments and extended the BTL model to a mixture context. Thus, they estimated a set of scale values for each segment and, moreover, simultaneously estimate the relationships of those segments to consumer descriptors. The following three concomitant variables were used.

1. Frequency of exercising per week
2. Number of dependent children living at home
3. Employment status (1 = work outside home, 0 = otherwise)

Monté Carlo testing procedures revealed that a four-segment solution of the extended BTL model provided an adequate fit to the paired comparison data. The authors tested various forms of the sub-model (for S=4) and found that a main-effects sub-model including the three consumer descriptor main effects fitted the data best. Table 9.1 provides the results.

According to the authors, the four-segment solution provides a fairly clear and interesting set of segment typologies. Segment 1 (34.6% of the sample) is primarily concerned with convenience of preparation, and to a lesser extent with price as evidenced by the high scale values for these concerns. Note that the sub-model coefficients for this segment were not estimated due to identification constraints. As a result, the profiles of the other segments are presented relative to those of Segment 1.
From the coefficients for the other segments one may conclude that employment outside the house increases the probability of belonging to this segment, which has a high concern for convenience of preparation. Segment 2 (21.1%) is concerned with nutritional value, and to a lesser extent with price. A higher frequency of weekly exercising, and employment outside the home increases the probability of membership to this segment, as compared to Segment 1. Segment 3, the smallest segment (11.2%), has rather distinctive concerns, primarily related to ingredients, nutritional value and calories. A higher frequency of weekly exercise increases the probability of belonging to this segment, relative to Segment 1, as evidenced by the sub-model coefficients. This segment appears to be a health-conscious segment. Segment 4 (comprising 33.1% of the sample), appears to be the most price-sensitive segment. Consistent with this, the authors found from the sub-model coefficients that the probability of belonging to this segment increases if the female has no work outside the house, and if the number of children living at home increases.

Dillon and his co-authors investigated the predictive validity of the sub-model, that is, its ability to classify subjects in a holdout sample into the four segments. They found substantial agreement, with (intra-class) correlation coefficients ranging from 0.76 to 0.89. Note that the application by Dillon and Kumar (1994; see chapter 6) was a similar application to paired comparison data of product forms, where geographic location was used as a concomitant variable.

Example 2: Consumer Choice Behavior with Respect to Ketchup

Gupta and Chintagunta (1994) analyzed A.C. Nielsen scanner panel data on purchases of ketchup for a sample of 709 households. They used the six largest brand-size combinations for the analysis: Heinz 28 oz, 32 oz, 40 oz and 64 oz, Hunts 32 oz, and

Table 9.1: Concomitant Variable Mixture Model results for Food Concern Data (adapted from Dillon, Kumar and Smith de Borrero 1993)[a]

Variable	Segment 1	Segment 2	Segment 3	Segment 4
Nutritional Value	0.167* (14%)	1.310* (45%)	0.998* (32%)	-0.691* (7%)
Convenience	1.398* (49%)	-1.401* (3%)	-1.731* (2%)	-0.127* (11%)
Calories	-1.222* (4%)	-0.941* (5%)	0.871* (28%)	-0.347 (9%)
Ingredients	-1.298 (3%)	0.186* (14%)	1.087* (35%)	-0.417* (9%)
Price	0.911* (30%)	1.011* (33%)	-1.217* (3%)	1.591* (64%)
Sub-model				
Constant	-	-3.377*	-6.327*	-3.247*
Exercise	-	0.798*	1.121*	-0.701*
Children	-	0.101	-0.301	1.341*
Work	-	-0.201*	-0.298	-0.412*

[a]Relative importances implied by the BTL model are in parentheses.
* $p < 0.05$.

Del Monte 32 oz Those brand-size combinations accounted for 85% of the market. The authors attempted to assess the impact of price, display and featuring, simultaneously deriving segments in which the impact of the marketing mix differed. The segments were simultaneously described by a set of three concomitant variables.

1. Income (1 = < $5000; 2 = $5000-$10,000; 3 = $10,000-$15,000, etc.)
2. Household size (number of subjects)
3. Average age of head of households (years)

The model used by Gupta and Chintagunta is a direct extension of the model by Kamakura and Russell (1989) to include concomitant variables (the latter model is discussed extensively in chapter 16). Gupta and Chintagunta assume a multinomial distribution for the observed choices, where the marketing mix variables enter the utilities giving rise to the multinomial choice probabilities. They identified three segments (using the BIC criterion). They used a likelihood ratio test to investigate the contribution of the demographic variables, which was significant. Table 9.2 reports the parameter estimates for the S = 3 solution.

We can see from the brand constants in Table 9.2 that segment 1 (62.1% of the sample) has the highest preference for Heinz 28 oz (the constant for Del Monte was not estimated in order to identify the model). Segment 1 also appears to be most price and promotion sensitive. The significant negative income coefficient in the sub-model indicates that lower income results in a higher probability of belonging to segment 1 than to segment 3. That finding supports the face validity of the model. Note that the concomitant-variable coefficients for segment 3 were not estimated because of identification constraints, and that the coefficients for the other segments are relative to segment 3.

Segment 2 (13.5%) prefers Heinz 64 oz, but the preference for Heinz 40 oz is also larger than it is for the other segments. This segment is characterized by the lowest levels of price and display coefficients. The coefficient of household size in the sub-model indicates that larger households have a higher probability of belonging to this segment than to segment 3. The preference for large packs among larger households makes sense intuitively. Segment 3 (24.4%) appears to have a high preference for Heinz 28 oz, and is the segment least sensitive to featuring. From the sub-model coefficients, we conclude that this segment consists of higher income, smaller households. Here, the concomitant variable approach has provided managerially useful and consistent findings. Gupta and Chintagunta note, however, that the improvement in fit obtained from the concomitant variables was very small (an increase of 0.003). They compare the concomitant variable approach to a two-stage procedure in which regression models were used to explain the posterior membership probabilities in a second step. They demonstrate the superiority of their approach over the two-step procedure, which yielded counter intuitive results.

Part 2 Segmentation methodology

Table 9.2: Concomitant Variable Mixture Model Results for Ketchup Choice Data (adapted from Gupta and Chintagunta 1994)

Variable	Segment 1	Segment 2	Segment 3
Heinz 64 oz	-0.394	6.522*	0.759
Heinz 40 oz	1.861*	5.830*	4.386*
Heinz 32 oz	2.377*	4.558*	2.647*
Heinz 28 oz	2.714*	4.924*	5.363*
Hunt's 32 oz	0.957*	2.659*	2.256*
Price	-1.899*	-1.043*	-1.231*
Display	1.068*	0.843*	1.043*
Feature	1.165*	0.607*	0.434*
Sub-model			
Constant	0.886	-1.070	-
Income	-0.168*	-0.082	-
Household size	0.167	0.356*	-
Age	0.014	-0.016	-

* $p < 0.05$.

The Concomitant Variable Mixture Model

To distinguish the concomitant descriptor variables from the segmentation bases used in the main part of the mixture model, we need to introduce the following additional notation.

Notation		
l	=	1,...,L indicate concomitant variables
z_{nl}	=	value of the tth concomitant variable for subject n
Z	=	$((z_{nl}))$

We start from the basic model formulations provided for standard mixture (MIX), mixture regression (GLIMMIX) and mixture unfolding (STUNMIX) models provided in chapters 6, 7 and 8. The data are assumed to be distributed according to some member of the exponential family (the binomial and multinomial distributions in the two examples above), conditional on the unobserved segments. We now directly formulate the unconditional distribution:

$$f(y_n|\phi) = \sum_{s=1}^{S} \pi_{s|z} f_s(y_n|\theta_s), \qquad (9.1)$$

where θ_s denotes the set of parameters to be estimated in the respective model, that is, the μ_s in the mixture model, the β_s in the mixture regression model and the a_s, b_s, c_s in the mixture unfolding model, and the dispersion parameters λ_s if applicable. Note that the prior probabilities π_s in the preceding chapter have been replaced by $\pi_{s|z}$. This is the core of the concomitant variable models: the prior probabilities of segments membership are re-parameterized in terms of the concomitant variables. For that purpose, most frequently a logistic formulation is used in the sub-model (although other formulations are possible; cf. Dayton and MacReady 1988):

$$\pi_{s|Z} = \frac{\exp(\sum_{l=1}^{L} \gamma_{ls} z_{nl})}{\sum_{s=1}^{S} \exp(\sum_{l=1}^{L} \gamma_{ls} z_{nl})}. \qquad (9.2)$$

To include an intercept for each segment, $Z_{1s} = 1$ is specified for all s. The sub-model (9.2) relates the prior probabilities to the concomitant variables. The parameter γ_{ls} denotes the impact of the lth consumer characteristic on the prior probability for class s. For reasons of identification, usually $\gamma_{lS} = 0$. The parameters of the logit sub-model are specific to each descriptor variable and market segment: a positive parameter value γ_{ls} implies that a higher value of variable l increases the prior probability that consumer n belongs to segment s. Note that the posterior probabilities are also affected by the concomitant variables, as they are calculated by updating the priors according to Bayes' rule:

$$\alpha_{ns} = \frac{\pi_{s|Z} f_s(y_n|\theta_s)}{\sum_{s=1}^{S} \pi_{s|Z} f_s(y_n|\theta_s)}. \qquad (9.3)$$

Another way of looking at the concomitant variable model is to note that it enables the prior probabilities, or equivalently the segment sizes, to vary across demographic and other variables. The concomitant variable model determines the segment sizes on the basis of a set of descriptors. Thus, in the application of Dillon and his co-authors above, the latent segment 4 is 34% ($e^{-0.412}$) smaller among working women than segment 1.

Estimation

Although the concomitant variable models can be estimated by a full Newton-type search over the entire parameter space to maximize the likelihood (cf. Kamakura, Wedel and Agrawal 1994), here we use the convenient EM algorithm to estimate the models (cf. Dillon,. Kumar and Smith de Borrero 1993; Wedel and DeSarbo 1995). The EM algorithms for estimating the MIX, GLIMMIX and STUNMIX models in chapters 6, 7 and 8, can be easily modified to accommodate concomitant variables. The reason is that the likelihood can be maximized separately with respect to the parameters of the main model and the sub-model in the M-step. The EM - algorithm for concomitant variable models can be shown to consist of one additional estimation step in the M-step (step 2 of the algorithm provided in chapter 6): here the multinomial model (9.2) is fitted directly to the estimates of the posterior probabilities, obtained in the previous M-step. That can be accomplished by gradient-search-type algorithms, such as Newton and quasi-Newton algorithms (Scales 1985).

Model Selection and Identification

For the CV-MIX, CV-GLIMMIX and CV-STUNMIX models the same identification constraints hold as for the corresponding models without concomitant variables. In addition, however, in the sub-model the coefficients γ_{lS} for the last segment S in the sub-model are set to zero for identification.

For model selection, again the information criteria AIC, CAIC, ICOMP etc. can be used. In addition, because the models with concomitant variables is nested on the mixture model (by setting all $\gamma_{ls} = 0$, $l = 2,..,L$), likelihood ratio tests can be performed to test for the contribution of all (or subsets of) concomitant variables.

Monté Carlo Study

Kamakura, Wedel and Agrawal (1994) performed a small Monté Carlo study on the performance of a concomitant variable model. They started from a GLIMMIX multinomial regression model, in which concomitant variables were included. The model was applied to two synthetic datasets to illustrate its performance in recovering true model and sub-model parameters. The data generated in the small simulation study were multinomial choices by subjects in three segments, the prior memberships being defined by concomitant variables. To create the concomitant variable and choice data, the researchers used a choice set of four alternatives, and two explanatory variables affecting choice probabilities. The prior probability of membership of a consumer in a segment was defined by equation 9.2, where (extreme-value) errors were added to represent omitted variables. The synthetic data were generated by first specifying true values for the coefficients β_{ks} in the main model and for the coefficients γ_{ls} in the sub-model (see Table 9.3). A sample of 500 hypothetical consumers was then created with values of the explanatory variables **X** and concomitant variables **Z**. Each consumer was assigned to the segment with the highest probability of membership. Ten choices were generated for each consumer, based on the coefficients for the segment to which the consumer was assigned, according to a multinomial model with extreme-value errors. That procedure was used to generate data for two sets of known true values for the model and sub-model parameters under two different scenarios.

In the first scenario, segment allocations and discrete choices were generated under highly noisy conditions; that is, the deterministic components of the utility functions and latent class sub-models were dominated by their stochastic components. The second scenario reflected a more deterministic choice behavior and a more accurate definition of prior membership probabilities based on demographics. Table 9.3 compares the true values with the estimates obtained with the ML estimation for each of the two scenarios.

The results in Table 9.3 clearly show the high degree of randomness in the choice data under the "high noise" condition, as the R^2 equals 4.7%. The log-likelihood of the model is -6128.8. Even under such severe condition, the parameter estimates are reasonably close to the actual values used to generate the data. One cannot reject the hypothesis that the estimated parameters equal the actual values used to generate the data ($p < 0.05$), and none of the parameters differ significantly from the true values, although some of the estimates of the sub-model for segment 3 are relatively far off. The R^2 under the second scenario is 21.9% and the log-likelihood of the model under that scenario was -5024.0, reflecting the low noise condition. Once again, one cannot reject the hypothesis that the estimated parameters equal the actual values used to generate the data ($p < 0.05$). The analysis of synthetic data by Kamakura and his co-authors thus illustrates the ability of the concomitant variable mixture models to recover true parameters.

Alternative Mixture Models with Concomitant Variables

An alternative way to accommodate concomitant variables is simply to include them among the y-variables in the mixture model, see equation (6.2). This approach is fairly common in latent class modelling and is usually taken if there are no theoretical reasons to separate the variables into the two types that are used to identify, respectively discriminate groups (for example, Langeheine and Rost 1988). If conditional independence of the y and z-variables given s is assumed, the distribution of the core (y) and concomitant (z) variables can be formulated as:

$$f(y_n, z_n | \phi) = \sum_s \pi_s f_s(y_n | \theta_s) f_s(z_n | \theta_s). \tag{9.4}$$

The posterior probabilities of membership are:

Part 2 Segmentation methodology

$$\alpha_{ns} = \frac{\pi_s f_s(y_n | \theta_s) f_s(z_n | \theta_s)}{\sum_s \pi_s f_s(y_n | \theta_s) f_s(z_n | \theta_s)}, \tag{9.5}$$

The question arises, which of the two formulations, (9.1) versus (9.4) is to be preferred. Wedel (1999) shows the relations between the concomitant variable (9.1) and standard (9.4) mixture models, by representing them as directed graphs. Figure 9.1 displays the directed graph for the *standard mixture model*. It shows that that applying this model one assumes conditional independence of the y- and z- variables given the segments (s).

Figure 9.1: Directed Graph for the Standard Mixture

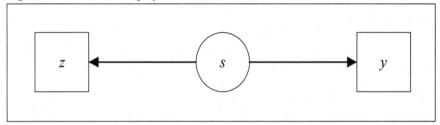

Figure 9.2 contains the directed graph for the *concomitant variable mixture model*. It shows that here too the x- and z- variables are conditionally independent, given the segments, s.

Figure 9.2: Directed Graph for the Concomitant Variable Mixture

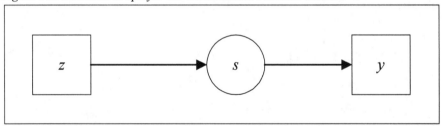

From Figures 9.1 and 9.2 it can be seen that the concomitant variable model specifies the conditional distribution of y given z, whereas the standard mixture model formulates their joint distribution. Thereby inference for the concomitant variable model is conditional upon the fixed observed values of z in the sample, which makes it more relevant to the data at hand. But, the concomitant variable model does not allow for inference under assumptions of repeated sampling, which does apply to the case of the standard mixture model (Wedel 1999). To show the relations between the posterior membership probabilities of the two models, let us start from the posterior for the standard mixture model provided in (9.5). Dividing numerator and denominator by the marginal distribution $f(z_n | \phi)$ one obtains:

$$\alpha_{ns} = \frac{f_s(y_n|\theta_s)(\pi_s f_s(z_n|\theta_s))/f(z_n|\phi)}{f_s(y_n|\theta_s)(\pi_s f_s(z_n|\theta_s))/f(z_n|\phi)} = \frac{\pi_s f_s(z_n|\theta_s)}{\sum_s \pi_s f_s(z_n|\theta_s)}, \tag{9.6}$$

which equals the posterior probability of membership for the concomitant variable mixture model, so that (9.3) and (9.5) are the same.

The standard and concomitant variable approaches, however, differ in the rules they provide for classifying new data. Assume that for a new consumer q data are available only on the concomitant variables. On the basis of the incomplete data the new subject needs to be classified into one of the S segments, using the estimates obtained from fitting the model to previously obtained complete data. Taking a decision theoretic framework (McLachlan 1992, p.7), allocation rules can be derived by minimizing the expected risk of misclassification is minimized (assuming 0/1 loss). It can be shown (Wedel, 1999) that for the standard mixture model classification of a new consumer, q, is provided by assigning the subject to the segment with the largest marginal posterior probability:

$$\alpha_{ns}^x = \frac{\hat{\pi}_s f_s\left(z_q|\hat{\theta}_s\right)}{\sum_s \hat{\pi}_s f_s\left(z_q|\hat{\theta}_s\right)}, \tag{9.7}$$

which is the posterior probability in (6.4) computed on the basis of the observed concomitant variables z_q only. This rule is identical to the *discriminant rule* for classification in supervised pattern recognition (McLachlan, 1992, p. 60). For the concomitant variable mixture model, the optimal classification of a new observation is provided by assigning the subject, q, to the segment s, with the largest posterior probability:

$$\hat{\pi}_{s|Z} = \frac{\exp\left(\sum_l \hat{\gamma}_{ls} z_{ql}\right)}{\sum_s \exp\left(\sum_l \hat{\gamma}_{ls} z_{ql}\right)}, \tag{9.8}$$

which is the prior probability in (9.2), evaluated for the observations on the concomitant variables, z_{ql}. This classification rule is the same as the *logistic discrimination rule* in supervised pattern recognition (McLachlan 1992, p. 255). Thus, the standard mixture model identifies groups and at the same time discriminates them, using the discriminant rule based on the concomitant variables, while the concomitant variable mixture discriminates them using logistic discrimination. Wedel (1999) shows that the information carried by the concomitant variables on group membership is the same in the standard and concomitant variable mixture models. However, the standard mixture model may be preferable if assumptions on the distribution of the concomitant variables can be justified; the concomitant variable model is more robust under such distribution assumptions. The standard mixture model is more appropriate if, on subjects that need to be classified, missing observations occur for both types of variables, which is more easily accommodated in this model.

Marketing Applications

The marketing applications of concomitant variable mixture models are relatively limited in number. Most of the applications pertain to mixture or mixture regression models, and are mentioned above or will be discussed in subsequent chapters. The STUNMIX procedure of Wedel and DeSarbo (1996) also includes options to re-parameterize prior mixture probabilities. They do not illustrate that option, but an illustration is provided by Wedel (1996). For completeness we summarize the applications in marketing in Table 9.4.

Conclusions

A question that comes to mind when working with concomitant variable models is: What is the difference in this procedure when the concomitant variables are included directly in the main part of the model as additional predictor variables? Aside from the fact that such an approach may be cumbersome for some models (e.g., the mixture unfolding models), the difference is probably more important conceptually than empirically. The difference is equivalent to the difference between a *clustering approach* to segmentation where, for example, both attitudinal and demographic (concomitant) variables are included as clustering variables, and a *hybrid approach* where segments are first delineated on the basis of demographics, and subsequently the consumers are clustered according to attitudinal data within each demographic group. It is the underlying theory of the substantive application that should drive the choice of approach.

Despite the limited improvements in fit found in previous empirical applications, the addition of the concomitant model to the GLIMMIX model produces an integrated approach that satisfies several of the requirements for effective market segmentation, because it produces a profile of each segment along the concomitant variables. For example, when combined with the multinomial GLIMMIX (e.g., Gupta and Chintagunta 1994; Kamakura, Wedel and Agarwal 1994), it can lead to segments that are *identifiable, accessible, responsive* and *actionable*. Stability of the segments in composition and profiles is implicitly assumed in all the models discussed so far. The possibility that segments can shift over time in terms of their underlying characteristics and composition is discussed in the following chapter. A criticism sometimes raised about the concomitant variable mixture approach is that when there is voluminous information on each individual (i.e., a large number of replications per subject), that

Table 9.3: Estimates from Synthetic Datasets Under Two Conditions (adapted from Kamakura, Wedel and Agrawal 1994).

Coefficient	Highly Stochastic Condition			Less Stochastic Condition		
	Segment 1	Segment 2	Segment 3	Segment 1	Segment 2	Segment 3
β_{1s}	-0.5[a] -0.479[b]	0.5 0.552	-0.2 -0.210	-1.0 -1.052	1.0 1.070	-0.5 -0.531
β_{2s}	0.2 0.181	-0.2 -0.244	-0.2 -0.295	0.5 0.511	-0.5 -0.539	-0.5 -0.531
γ_{0s}	0.0 -	0.0 -0.164	0.0 -0.441	0.0 -	0.0 -0.073	0.0 -0.057
γ_{1s}	0.0 -	0.2 0.152	-0.2 -0.065	0.0 -	0.5 0.535	-0.5 -0.329
γ_{2s}	0.0 -	-0.5 -0.456	0.5 0.680	0.0 -	-1.0 -0.959	1.0 0.952

[a] "True" value used to generate the choice data.
[b] Parameter estimate.

Part 2 Segmentation methodology

information dominates the prior information in the Bayesian update of the prior membership probabilities, which might reduce the differences in the posteriors across demographic groups. As experience with the empirical application of concomitant-variable mixture models accumulates, we will be able to delineate the conditions under which it affords the greatest advantages.

Table 9.4: Marketing Applications of Concomitant Variable Models

Authors	Year	Application	Model
Dillon, Kumar and Smith de Borrero	1993	Consumer segmentation on the basis of paired comparisons of concerns in food consumption	Binomial MIX
Dillon and Kumar	1994	Segmentation of consumers on the basis of product-form paired comparisons	Binomial MIX
Kamakura, Wedel and Agrawal	1994	Conjoint segmentation for bank services	Multinomial GLIMMIX
Gupta and Chintagunta	1994	Consumer brand choice segmentation using scanner data	Multinomial GLIMMIX
Wedel	1996	Consumer food choice segmentation and perceptual mapping	Poisson STUNMIX
Wang, Cockburn and Puterman	1998	Analysis of relationships of patents and R&D	Poisson GLIMMIX

10
DYNAMIC SEGMENTATION

In the segmentation methods discussed so far, we make the implicit assumption that the segments are stationary in structure and characteristics. Segment change is an important but currently under-researched area. In this chapter we describe approaches to track segment changes over time. Three types of models are identified and discussed. Two of the dynamic segmentation approaches build directly on the concomitant variable methods described in the preceding chapter.

The segmentation methods discussed in the preceding chapters are based on the assumption that the composition and profile of the uncovered segments are stable over time or, at the very least, during the sampling period. Aside from the *stability* and *substantiality* requirements for effective segmentation, from a managerial perspective, violation of the assumption of *stability* or stationarity may invalidate model estimation when data collection spans a long time period, such as in tracking studies. For example, Kamakura, Kim and Lee (1996) considered the possibility that the decision process and brand preferences could change within the two-year period of their scanner data. They tested two versions of their mixture of nested - logits. One of the extensions allowed households to switch among stable (in preferences and decision process) segments over time; the other version specified segments of stable composition, but changing brand preferences and responses to marketing variables over time. In their particular application, Kamakura and his co-authors find that the stationary model could not be rejected in favor of the non-stationary extensions.

Calantone and Sawyer stated in 1978 that the extent to which segments are stable over time is a neglected aspect of segmentation studies. If a segment identified at one point in time changes in terms of desired benefits or size, the marketing effort targeted at that segment cannot be maintained. Current segment sizes may provide insufficient justification to warrant a target marketing strategy for specific segments if those segments appear to be decreasing in size or changing in terms of demographic and socioeconomic characteristics. Market studies are needed that assess the stability of segments over relatively long periods of time (Wind 1978). On the basis of the work by Calantone and Sawyer (1978) and Böckenholt and Langeheine (1996), we formulate two major sources of segment instability.

1. Manifest change: The segment membership is stable, but changes may occur in the preference or choice structure of customers (or firms) in a segment over time.

Part 2 Segmentation Methodology

2. *Latent change*: The preference structure of segments is stable, but changes may occur in segment size and/ or the segment membership of consumers (or firms) over time.

Accordingly, we can distinguish two types of models that address the *stability criterion* for effective segmentation. One is the mixture model approach for manifest change. It is an extension of the GLIMMIX approach and includes structural change in the regression part of the model (i.e., time series, Markov, hazard and other approaches that address time - dependence within segments). The other is the mixture approach for latent change which addresses changing segment membership and size in basically two ways: through a simple extension of the concomitant variable approach presented in chapter 5 or through Markov - type transition probabilities among segments over time.

We provide examples of both approaches, but first introduce the following additional notation.

Notation
r = 1, ... , R indicates time periods
u = 1, ..., U indicates segments

Models for Manifest Change

It is difficult to provide a general formulation for models of manifest changes because of the great variety of possible formulations. Nevertheless, the formulations presented here may contain many others (though certainly not all) as special cases. We provide two empirical examples to illustrate other possible formulations and approaches. Models for manifest change extend the GLIMMIX and STUNMIX models by including structural time change in the linear function. For example, the GLIMMIX model can be modified to:

$$\eta_{ksr} = \sum_{p=1}^{P} X_{nkp} \beta_{sp} + \sum_{q=1}^{Q} G_q(r) \xi_{qs} \qquad (10.1)$$

or to

$$\eta_{ksr} = \sum_{p=1}^{P} X_{nkp} \beta_{spr}. \qquad (10.2)$$

The STUNMIX equation (8.2) can be modified to:

Chapter 10 Dynamic segmentation

$$\eta_{ksr} = c_{ks} \sum_{t=1}^{T} b_{kt} q_{st} + \sum_{q=1}^{Q} G_q(r) \xi_{qs} \tag{10.3}$$

or to

$$\eta_{ksr} = c_{ks} + \sum_{t=1}^{T} b_{kt} a_{str}. \tag{10.4}$$

Here, $G_q(r)$ denote Q different functions of time (r). For Q = 3, for example, $G_1(r) = r$, $G_2(r) = r^2$, and $G_3(r) = \ln r$. The ξ_{qs} are parameters capturing time dependence, to be estimated.

The conceptual difference between formulations 10.3 and 10.4 is that 10.3 includes a time trend in the mean of the expectation of the dependent variable, whereas 10.4 specifies the preference vector terminus of each segment, a_{str}, to vary over time. Note that other formulations are possible, for example by specifying β_{spr} or by constraining the $\beta_{spr} = \beta_{sp}(r)$ to be specific functions of time in 10.2 or by constraining a_{str} to be some specific function of time, $a_{str} = a_{st}(r)$, in 10.3. That can be accomplished through the re-parameterization option (M5) for STUNMIX models provided in chapter 4. Estimation of those models proceeds by appropriate modifications of the EM algorithm. We provide two examples taken from the literature.

Example 1: The Mixed Markov Model for Brand Switching

Poulsen (1990) modeled a data set provided by Aaker from the MRCA panel on a frequently purchased consumer good. The data represented purchases by the new triers of a certain brand, and subsequent purchases of that and other brands by those triers in five subsequent waves. The purpose was to model the sequence of purchases after the initial trial of a new brand, for which Poulsen used (among others) a mixed Markov model. The model postulates that given a segment s, a sequence of buy/not buy decisions follows a first-order Markov process. Suppose there are two possible outcomes, buy (1) or not buy (0), at two consecutive occasions, indexed by k and l. The probability of buying or not buying at the two occasions is denoted by $\psi_{kl|s}$, conditional on segment s as throughout the preceding chapters. Now, according to the Markov - model, that probability is written as:

$$\psi_{kl|s} = \psi_{k|s} \psi_{l|ks}. \tag{10.5}$$

In words, the probability of choosing alternative k at occasion 1 and alternative l at occasion 2, respectively (given segment s), is equal to the probability of choosing k times the probability of choosing l given that k has been chosen before. The former probabilities are the initial choice probabilities. The latter probabilities form the

161

Part 2 Segmentation Methodology

transition matrix describing the probabilities of switching among the brands. The unconditional probability can be written in the well-known form as:

$$\psi_{kl} = \sum_{s=1}^{S} \pi_s \psi_{k|s} \psi_{l|ks}. \qquad (10.6)$$

The π_s are the prior segment probabilities.

Fitting that model to the MRCA data, Poulsen recovered the two segments shown in Table 10.1 apparently segment 1 (19%) has a high initial probability of buying the brand (68%) and a probability of 50% of switching into the brand at succeeding purchase occasions. Segment 2 (81%) has a much lower probability of buying the brand (18%) and a very low probability of switching into it, given that no purchase has been made yet. Also, when the brand has been bought previously, the probability of a repeat is much lower in segment 2 than in segment 1. The model fits the data quite well, but tends to over predict the choice probabilities in a five-wave holdout period (Figure 10.1).

Table 10.1: Mixed Markov Results for MRCA Data (adapted from Poulsen 1990)

Segment		s = 1			s = 2	
		no buy	buy		no buy	buy
$\psi_{k\|s}$		0.32	0.68		0.87	0.13
$\psi_{l\|ks}$		no buy	buy		no buy	buy
	no buy	0.50	0.50	no buy	0.94	0.06
	buy	0.29	0.71	buy	0.79	0.21

We further illustrate this class of approaches to change in segment preference structure over time by the mixture hazard model approach developed by Wedel, et al. (1995).

Example 2: Mixture Hazard Model for Segment Change

Wedel et al. (1995) used the following hypothetical example to set the scene for their study. A consumer visits a large retail chain once a week. Visiting the supermarket last week, s/he had noticed that Hunts was reduced in price, but had decided not to buy. Leaving home for this week's shopping trip, s/he notices that the 32 oz bottle of Heinz ketchup purchased four weeks ago is almost empty. S/he purchases two 32 oz bottles of ketchup of her regular brand (Heinz), which is under promotion this week. The issues demonstrated by this example are that (1) the timing of purchases can be seen as a discrete process, being determined by the timing of (weekly) supermarket visits; (2) as time since the last purchase in one product category increases and inventory depletes, consumers' sensitivity to the marketing mix may change, and (3) those effects may vary in magnitude across different segments of consumers. The second point is of special relevance here, as it pertains to changes in segment structure over time.

Figure 10.1: Observed and Predicted Shares for the Mixed Markov Model (adapted from Poulsen 1990)

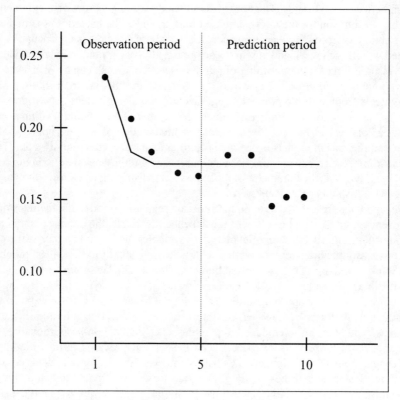

Wedel and his co-authors present a duration model for the analysis of brand switching behavior that accommodates the three issues. Duration or hazard models deal with time-dependent occurrences of events, brand purchases in this case. The model builds on the previous work of Jain and Vilcassim (1991) in the area of duration modeling for brand-switching behavior. However, it differs from their approach because it is formulated in discrete-time and it accounts for unobserved heterogeneity across consumers by using a mixture model formulation.

The discrete-time model was applied to the A.C. Nielsen scanner panel data on ketchup purchases. The three major national brands, Heinz, Hunt's and Del Monte, were used and the other brands were aggregated into one category. For those four brands, all brand sizes were included in the analyses, coded as integer-equivalent units of 14 oz. A random sample of 200 subjects was used. Data were obtained for a period of 37 weeks. The purchase histories of all households were censored at the final observation week. The purchase quantities (the dependent variable) and information on price registered in each week in which a store visit was made by a household were used in the analyses.

The authors set out to model the hazard of making either a repeat purchase or a switch between two brands. The hazard is modeled by using a piecewise exponential distribution function for the inter-purchase times, in which the hazard is assumed constant within each week. As a consequence, the inter-purchase time is exponentially distributed within each week and purchases are assumed to occur randomly within weeks. (The assumption of random timing of purchases within a week can be justified from a behavioral point of view, because most consumers plan shopping trips within a week to buy items from multiple product categories, not just a single item.) If within-week purchase times are exponentially distributed, the number of units bought within a week follows a Poisson distribution. Piecewise exponential models of duration are equivalent to Poisson regression models of the quantity of units purchased within a week. They are called piecewise exponential because purchases within weeks are exponentially distributed, whereas the parameter of the exponential distribution is allowed to vary in a piecewise fashion across weeks.

The model is formulated as a mixture Poisson regression model that includes time-dependent variables. The log-mean of the dependent variable, the purchase frequency, is modeled as a linear function of explanatory variables that include price (x, the only marketing variable included), as well as a variable indicating time since last purchase (r, in weeks), and log of time since last purchase (ln(r)). The latter two variables allow for a flexible representation of the time dependence of repeat buying and switching (by presenting a piecewise approximation to a variety of hazards, such as the exponential, Weibull and Gompertz). Moreover, the interactive effect of price and time (linear, in weeks) is included. This interactive effect represents potential non-proportionality, that is, a change in the effects of price over time. Most previous hazard models in marketing assumed proportionality, i.e. that the effects of the variables remains constant over time. The effects of those variables on transitions from brand k to brand l on two consecutive purchase occasions are estimated. In this model, l equaled one if the same brand (repeat purchase), or zero if another brand (switch) was purchased on the last occasion. Thus, in effect the model is a relatively simple semi-Markov mixture Poisson model. The structural part of the Poisson regression model, given segment s, is:

$$\ln(\lambda_{klr|s}) = \alpha_{kls}^0 + \alpha_{ks}^1 r + \alpha_{ks}^2 \ln(r) + \beta_{ks} x_{nr} + \gamma_{ks} x_{nr} r \ . \tag{10.7}$$

In fitting the model to the ketchup data, the CAIC statistic indicted that the S = 2 model provided an appropriate representation. Several alternative forms of the brand switching specification were investigated, but the results reported here provide the most parsimonious representation of the data.

The results of the S = 2 model selected are reported in Table 10.2. The entropy statistic, $E_s = 0.835$, indicates that the two segments are well separated. Figure 10.2 shows the estimated hazard functions for repeating and switching for each of the four brands in both segments. Note that the hazard is a step function, because of the discrete-time formulation of the model.

Table 10.2: Segment Level S=2 Nonproportional Model Estimates (adapted from Wedel et al. 1995)

	s = 1		s = 2	
	Repeat	Switch	Repeat	Switch
Week	-0.008	-0.065*	0.034*	0.045*
Log(week)	1.303*	1.457*	0.559*	0.585*
Price	-0.756*	-1.020*	-1.163*	-0.921*
Price*week	0.044*	0.022	-0.059*	0.038*
Heinz	-1.043*	-1.169*	-2.322*	-2.536*
Delmonte	-1.452*	-3.987*	-2.645*	-2.999*
Hunts	-4.582*	-4.220*	-2.605*	-3.622*
Others	-6.630*	-5.177*	-2.562*	-3.697*

* $p < 0.01$

Segment 1 constitutes 45.6% of the sample. Heinz has a very large market share (86%), and almost 80% of purchases are repeat purchases. Consequently, this segment appears to consist of predominantly of loyal Heinz buyers. The average inter-purchase time is about five weeks. Both repeat purchases and switches are significantly affected by price, but the effect of price changes on switches is about one-and-a-half times as great. Apparently, price discounts are less effective in attaining last-brand-loyal behavior for Heinz than in inducing switching behavior away from Heinz. The repeat buying hazard shows a positive nonproportional price effect (PRICE*WEEK), indicating that the effect of a price discount decreases as time since last purchase elapses. In other words, the longer a consumer resists repeat-purchasing a brand, the less likely he/she is to repeat purchase in response to a price-promotion. A possible explanation may be, that as time elapses and stocks deplete, the need for the product rises, and these loyal consumers buy their favorite brand, whether it is being promoted or not.

Segment 2 constitutes 54.4% of the sample. The average inter-purchase time (seven weeks) is about 30% longer than among consumers in segment 1. Consequently, there appears to be considerable heterogeneity with respect to duration dependence across the two segments. In segment 2, the market share of Heinz is half that in segment 1, whereas the shares of Del Monte, Hunt's and the other brands are three to four times as high. Almost 60% of all purchases in segment 2 consist of switches. Segment 2 can be designated a segment of brand-switchers. It accounts for almost a 20% smaller purchase volume than segment 1. The hazards of both switching and repeat-buying increase as a brand offers a price discount (PRICE). Here, the effects of price changes on switches and repeats are about equal. The repeat-purchase hazard for this segment shows a significant negative nonproportional effect of price (PRICE*WEEK), which indicates that the impact of a price discount on repeat purchases increases as the time since last purchase elapses. Nonproportional price effects for switches, on the other hand, are positive, indicating that the impact of price discounts on switches decreases over time. These consumers apparently wait until their favorite brand is being promoted, so the effect of price promotions for that brand increases as time since last purchase elapses and stock deplete.

Figure 10.2: Stepwise Hazard Functions in Two Segments (adapted from Wedel et al. 1995)

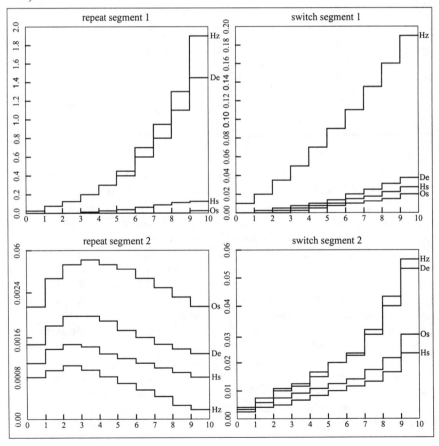

The authors compare the predictive validity of their model to the hazard model for brand switching of Vilcassim and Jain (1991) and the mixture multinomial logit model of choice of Kamakura and Russell (1989) on a hold-out sample. The discrete-time hazard model predicted choices substantially better than either of those models, and predicted switch durations more than 20% better than the model of Vilcassim and Jain (1991).

Models for Latent Change

Here we describe models in which segment sizes and memberships change over time. There are two main approaches in this class of models, based on different re-parameterizations of the priors for the mixture components. Therefore, these models are directly related to the concomitant-variable mixture models described in the preceding chapter.

Dynamic Concomitant Variable Mixture Regression Models

This particular model for latent change is a simple extension of the concomitant variable approach o accommodate changing segment size and membership over time. We assume the sizes of the segments, π_s may vary across time intervals. The segment structure, defined by the parameter vector θ_s of the distribution function $f_s(y_n|\theta_s)$ is assumed to remain stable over the entire time horizon considered. Remember that θ_s may represent the μ_s in the mixture model, the β_s in the mixture regression model, and the a_s, b_s, c_s in the mixture unfolding model, and the dispersion parameters λ_s if applicable (see chapters 6, 7 and 8). Note that this assumption is in line with the findings of Calantone and Sawyer (1978) for benefit segmentation. They found that although segment size and membership of a number of benefit segments changed over a two-year period, segment preferences remained stable.

The above assumptions imply that the prior probabilities or equivalently the segments sizes change over time intervals. It is further assumed that the prior probabilities may be functions of household descriptor variables, Z, as in the preceding chapter. This implies that depending on those characteristics, some households have a higher prior probability of belonging to a given segment than others.

The sub-model for the prior probabilities is formulated as:

$$\pi_{sr|Z} = \frac{\exp[\sum_{l=1}^{L} \gamma_{ls} Z_{nl} + \sum_{q=1}^{Q} \xi_{qs} G_q(r)]}{\sum_{s=1}^{S} \exp[\sum_{l=1}^{L} \gamma_{ls} Z_{nl} + \sum_{q=1}^{Q} \xi_{qs} G_q(r)]} \tag{10.8}$$

Here, $G_q(r)$ denote Q different functions of time r, for example; for $Q = 2$: $G_1(r) = r$ and $G_2(r) = r^2$. The parameter γ_{ls} denotes the impact of the lth household characteristic on the prior probability as before, and the ξ_{qs} are parameters that capture time dependence. (For reasons of identification, $\xi_{qS} = 0$ and $\gamma_{lS} = 0$.)

We directly formulate the unconditional distribution:

$$f(y_n|\phi) = \sum_{s=1}^{S} \pi_{sr|Z} f_s(y_n|\theta_s). \qquad (10.9)$$

Note that the prior probabilities $\pi_{s|Z}$ of the preceding chapter have been replaced by $\pi_{sr|Z}$, because they depend on time (r) due to the inclusion of the G(r) functions in equation 10.8. With the formulation above, the segment sizes may change over time. For example, if only a linear function of time is included in equation 10.8, and we have a two-segment model, with ξ_{11} negative and ξ_{12} positive, the size of segment 1 decreases over time at the cost of segment 2. Note that the posterior probabilities are also affected and some consumers may switch segments over time, as they are calculated by updating the priors according to Bayes rule:

$$\alpha_{nrs} = \frac{\pi_{sr|Z} f_s(y_n|\theta_s)}{\sum_{s=1}^{S} \pi_{sr|Z} f_s(y_n|\theta_s)}. \qquad (10.10)$$

This dynamic mixture model can be estimated by using the EM algorithm described in the preceding chapter. To date, no applications of this model in marketing are known, other than the simplified version by Kamakura, Kim and Lee (1996), described at the beginning of this chapter.

Latent Markov Mixture Regression Models

For ease of exposition, two time periods indicated by $r = 1$ and $r = 2$ are assumed. The segments at time $r = 1$ are denoted by $s = 1,...,S$, and those at time $r = 2$ by $u = 1,...,U$. The dependent variable for subject n is $y_n = ((y_{nkt}))$. The extension to more than two periods is straightforward.

The expected values of the dependent variable within segments is assumed to be provided by the mixture, mixture regression or mixture unfolding models described in chapters 5, 6, and 7. In the latent Markov model, subjects are assigned a simultaneous prior probability of belonging to segment s at time $r = 1$ and to segment u at time $r = 2$: π_{su}. This simultaneous probability of segments s and u is specified as the product of the marginal probability of being in segment s at time r, and the conditional probability of being in segment u at time $r = 2$, given segment s at time $r = 1$:

$$\pi_{su} = \pi_s \pi_{u|s}. \qquad (10.11)$$

Equation 10.11 presents the sub-model of the latent Markov mixture regression model. The unconditional distribution of the data is:

$$f(y_n|\phi) = \sum_{s=1}^{S} \sum_{u=1}^{U} \pi_s f_s(y_n|\theta_s) \pi_{u|s} f_u(y_n|\theta_u). \quad (10.12)$$

The simultaneous posterior probability that subject n is in segment s at r = 1 and in segment u at r = 2 is calculated using Bayes' rule:

$$p_{nsu} = \frac{\pi_s \pi_{u|s} f_s(y_n|\theta_s) f_u(y_n|\theta_u)}{\sum_{s=1}^{S} \sum_{u=1}^{U} \pi_s \pi_{u|s} f_s(y_n|\theta_s) f_u(y_n|\theta_u)}. \quad (10.13)$$

The marginal probability of being in s at time r = 1 is obtained by summing p_{nsu} across u, and the marginal probability of being in u at r = 2 is obtained by summing p_{nsu} across s. In this way, the latent Markov mixture model can be used, for example, to investigate the effects of a set of independent variables **X** on a dependent variable **y** in a number of unobserved segments, and at the same time the transitions of subjects among those segments over time.

Estimation

We briefly describe how the EM algorithm provided in appendices A1 (chapter 7) and A2 (chapter 8) can be modified to estimate the latent Markov mixture regression model. The likelihood (obtained by a product of the density in 10.9 over n) can be maximized separately with respect to the parameters of the main model, θ_s and θ_u, and the sub-model (equation 10.11) in the M-step. The M step of the EM-algorithm consists of two independent sub-steps. The first sub-step estimates the parameters θ_s and θ_u and the second sub-step fits the model (10.10) to the estimates of the posterior probabilities (10.13) obtained in the E-step, which involves averaging the posteriors in the appropriate manner. The use of the EM algorithm to estimate latent Markov models has been described in detail by Van de Pol and De Leeuw (1986). Ramaswamy (1997) illustrates the performance of the latent Markov approach on synthetic data. His results illustrate the ability of the latent Markov framework to accommodate and detect different segment evolution scenarios, to recover the true model parameters, and to accurately segment the market in two time periods.

Similar identification constraints hold for the latent Markov mixture regression models as for the mixture regression models, for each of the two sets of prior probabilities: $\sum_u \pi_{u|s} = \sum \pi_s = \sum_u \sum_s \pi_{us} = 1$. The same information criteria (i.e., AIC, CAIC, and ICOMP) can be used for selecting the numbers of segments S and U. Several restrictions on the model can be tested by using Likelihood ratio tests. See Ramaswamy (1997), Poulsen (1990) and Böckenholt and Langeheine (1996) for details.

Examples of the Latent Change Approach

Example 1: The Latent Markov Model for Brand Switching

Poulsen (1990) combined the stochastic description of individual choices with changes in the segment structure of the population. The idea is to define a set models for the behavior of subjects within a segment. Poulsen used first-order (binomial) choice models for that purpose. The shifts among segments are modeled by using a stationary transition matrix. The model we describe differs from the Markov model application by Poulsen (1990) in that here the purpose is to model changes in the latent segment structure, whereas in the earlier model the purpose was to model changes in preferences within each segment.

We assume two purchase occasions, on which the segments are indicated by s and u, respectively, and the choice options by k and l. The model postulates that given a segment s, a purchase sequence of buy/not buy decisions follows a zero- order process. There are two possible outcomes of the purchase, buy (k,l = 1) or not buy (k,l = 0). The probability of buying at the two occasions is denoted by $\psi_{k|s}$ and $\psi_{l|u}$, conditional upon segment s on occasion r = 1 and segment u on occasion r = 2, respectively. On the first occasion the prior probabilities of segment membership are π_s. Given the segment s on the first occasion, the prior of segment u on the second occasion is $\pi_{u|s}$. Now, the simultaneous probability of choosing brand k and being in segment s on occasion 1 and choosing brand l and being in segment u on occasion 2 is:

$$\psi_{klsu} = \pi_s \psi_{k|s} \pi_{u|s} \psi_{l|u} \qquad (10.14)$$

In words, the probability of choosing k on occasion 1 and alternative l on occasion 2, respectively (given segments s and u), is equal to the probability of choosing k given segment s, times the probability of segment s, times the probability of choosing l given segment u, times the probability of segment u given segment s on the previous occasion. The unconditional probability can be written:

$$\psi_{kl} = \sum_{s=1}^{S} \sum_{u=1}^{U} \psi_{klsu}. \qquad (10.15)$$

Poulsen estimated his latent Markov model on Aaker's data on triers of a new brand and their succeeding purchase pattern. He estimated a two-segment model, with segment sizes of 0.75 and 0.25, respectively (Poulsen restricted S to equal U). The results of this model are presented in Table 10.3.

Table 10.3: Latent Markov Results for MRCA Data (adapted from Poulsen 1990)

Segment	s = 1		s = 2	
π_s	0.75		0.25	
	no buy	buy	no buy	buy
$\psi_{k\|s}$	0.95	0.05	0.23	0.77
	u = 1		u = 2	
$\pi_{u\|s}$ s = 1	0.98		0.02	
s = 2	0.19		0.81	

The results in Table 10.3 show that the two latent segments differ in their probability of choosing the new brand. Segment 1 has no or only minor interest in the brand, whereas segment 2 is characterized by a fairly high choice probability. Initially, after the first trial purchase, 75% of the sample is in segment 1 and 25% is in Segment 2. Then, at the new purchase, the probability of switching out of segment 2 is more than eight times as large as the probability of switching into that segment. Segment 2 represents a potential market of the new brand, so one concludes that this state of affairs is adverse to the brand. The proportion of buyers of the brand will decrease over time as is shown in the path towards the steady state distribution in Figure 10.3. When Poulsen compared the mixed and latent Markov models, he concluded that both give a satisfactory fit and that a clear choice between them could not be made.

Example 2: Evolutionary Segmentation of Brand Switching

Ramaswamy (1997) extended the Poulsen (1990) latent Markov model. His model extracts preference segments in two periods, before and after the introduction of a new product, and for tracking the evolution among those segments. The model is calibrated on panel survey data of brand choice. It enables one to test whether the newly introduced brands have:

1. Penetrated one or more existing segments, without affecting the segment sizes (the *Stability model*)
2. increased the size of one or more existing segments in the process of penetration (the *Expansion model*),
3. created entirely new segments, while possibly also penetrating existing segments (the *Evolution model*).

Figure 10.3: Observed and Predicted Shares for the Latent Markov Model (adapted from Poulsen 1990)

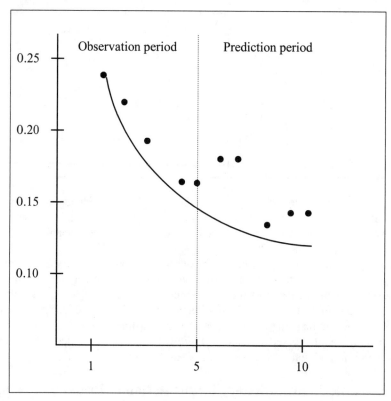

The model formulation is conceptually similar to that of Grover and Srinivasan's (1987) model, when only one period is considered. The objective is to model the switching among brands in the two periods. Here we use the following additional notation.

Notation	
i, j	= 1, ..., I indicate brands on occasion r = 1
k, l	= 1,....,K indicate brands on occasion r = 2

Conditional upon segment s in the first time period, the probability of switching among brands is assumed as a zero-order process, so that:

$$\psi_{ijs} = \pi_s \psi_{i|s} \psi_{j|s}. \tag{10.16}$$

As in Poulsen's model, given the segment s on the first occasion, the prior of segment u on the second occasion is $\pi_{u|s}$. Therefore the simultaneous probability of being in u

Chapter 10 Dynamic segmentation

and switching from brand k to brand l on occasion 2 is (assuming again a zero order process given segment u):

$$\psi_{klu} = \pi_{u|s} \psi_{k|u} \psi_{l|u}. \tag{10.17}$$

Now, the simultaneous probability of switching from i to j at period 1 and from k to l at period 2 is provided by:

$$\psi_{ijkl} = \sum_{s=}^{S} \sum_{u=1}^{U} \pi_s \pi_{u|s} \psi_{i|s} \psi_{j|s} \psi_{k|u} \psi_{l|u} \tag{10.18}$$

In addition, it is assumed that among the S segments on occasion 1, there are I brand-loyal segments (where i and j indicate the same brand), and (S-I) brand switching segments. Similarly, at occasion 2 there are K brand-loyal segments (where k and l indicate the same brand), and (U-K) brand switching segments. Note that neither S and U nor I and K are assumed to be equal. Ramaswamy discusses several restrictions that can be placed on the model parameters, yielding the three respective model forms above.

The model was applied to scanner panel data on heavy-duty laundry detergents to illustrate the approach. In the period the data encompass, two new brands were introduced into the particular geographic market: liquid Cheer and liquid Surf. Two periods of six months before and after introduction were used to calibrate the model. The data sample consisted of 1450 households, each with two purchases in each of those two periods before and after introduction. The data pertained to nine national brands that constituted 80% of the market in the first period.

By a modified AIC criterion (here the value of d = 3 in equation 6.9), s = 3 and u = 4 were decided upon. Also, the full first-order *evolution* model dominated the *stability*, and *expansion* models. Table 10.4 reports the transition probabilities between the switching segments in the two periods, as well as the marginal (prior) probabilities of segment membership.

We can see from the table that there is a certain level of segment stability. s=2 in the first period corresponds to u = 2 in the second period (62%), and s = 3 in the first period corresponds to u = 4 in the second period (67%). Further, consumers in s = 1 in the first period tend either to switch to u = 4 in the second period after the new product introduction or to remain in u = 1.

Table 10.5 reports the estimated choice probabilities within segments. These probabilities are useful in interpreting the choice behavior within segments. For each period, the Table lists the sizes of the brand loyal segments (e.g., π_i) and the zero-order choice probabilities in the brand-switching segments (e.g., ψ_{is}). (Parameter estimates that equal 0.00 in the table were constrained to that value). Parameters in boldface type were used to interpret the results.

In period 1, segment s = 1 appears to be a liquid segment, led by Wisk. Segment s = 2 is a powder segment, led by Tide. Whereas these first two segments are form primary (powder and liquid), the third segment is a hybrid. Tide powder is the largest brand, followed by Wisk and Tide liquid.

Table 10.4 Conditional Segment Transition Matrix Between Periods 1 and 2 (adapted from Ramaswamy 1997)

Period 2		u = 1	u = 2	u = 3	u = 4
Period 1		0.09	0.21	0.07	0.19
s = 1	0.17	0.35	0.10	0.00	0.34
s = 2	0.24	0.04	0.62	0.04	0.00
s = 3	0.17	0.00	0.00	0.07	0.67

Table 10.5 Preference Structure in Three and Four Switching Segments in Two Periods (adapted from Ramaswamy 1997)

Brand	\multicolumn{4}{c}{Period 1}			\multicolumn{4}{c}{Period 2}			
	π_i	ψ_{i1}	ψ_{i2}	ψ_{i3}	π_k	ψ_{k1}	ψ_{k2}	ψ_{k3}	ψ_{k4}
Bold-P	0.05	0.04	0.09	0.02	0.01	0.02	0.05	0.00	0.00
Cheer-P	0.12	0.08	**0.25**	0.02	0.03	0.04	**0.15**	0.02	0.01
Oxydol-P	0.11	0.02	**0.22**	0.05	0.04	0.00	**0.26**	0.05	0.04
Tide-P	0.24	0.00	**0.33**	**0.29**	0.11	0.01	**0.36**	**0.57**	0.14
Bold-L	0.15	**0.27**	0.03	**0.25**	0.04	0.12	0.06	0.02	**0.33**
Wisk-L	0.04	0.14	0.00	0.04	0.01	**0.16**	0.00	0.00	0.01
Era-L	0.11	**0.19**	0.04	0.11	0.05	0.09	0.04	0.01	0.09
Solo-L	0.07	0.17	0.03	0.00	0.01	**0.21**	0.00	0.00	0.00
Tide-L	0.11	0.12	0.02	**0.23**	0.05	0.12	0.01	**0.23**	**0.18**
Cheer-L	-	-	-	-	0.01	0.09	0.02	0.01	0.08
Surf-L	-	-	-	-	0.02	0.13	0.06	0.10	0.11

In period 2, the relative proportions of switchers and loyals are about the same as in period 1. The largest switching segment is Segment 2. It is a powder segment, dominated by Tide powder and Oxydol. Segment 4 is also a relatively large liquid switching segment, led by Wisk and Tide. Segment u = 1 is a small liquid switching segment, where Bold and Solo dominate. Segment 3 is a brand-primary segment, led by Tide (powder and liquid). The two new brands, liquid Cheer and liquid Surf, appear to positioned in both liquid segments, u = 1 and u = 4. However, Surf also seems to have penetrated the powder segment u = 2, and the Tide primary segment, u = 3. Going back to Table 10.4, we can see that powder buyers (s = 2) continue to favor the powder after the new brand introductions (u = 2). However, whereas a portion of the liquid buyers appears to retain their preference for liquids, a substantial part of the liquid buyers in period 2 (u = 4) has shifted preference from the hybrid segment in period 1 (s = 3).

The model by Ramaswamy (1997) is related to the model by Ramaswamy, Chatterjee and Cohen (1996). For a discussion of identification issues of these models, see Walker and Damien (1999) and the reply by Ramaswamy, Chatterjee and Cohen (1999).

Example 3: Latent Change in Recurrent Choice

The model for latent change described in this chapter was extended by Böckenholt and Langeheine (1996) to a dynamic model of purchase timing, quantity, and brand choice. Those authors assume that quantity purchased in a product category by consumers in a latent class during a fixed time period (e.g., a month) follows a Poisson process, whereas the multichotomous brand choices in the same product category follow a multinomial process. With those assumptions, Böckenholt and Langeheine develop a Poisson-multinomial model for the purchase quantities for each brand. This product of a Poisson for category volume and a multinomial for brand choices is conceptually appealing for modeling purchase volume at the brand level because it can incorporate income effects on the demand for the product category, independently of price effects on brand preferences.

To account for non-stationarity in purchase rates and brand preferences, those authors consider several constrained versions of the latent-change model: a) *first-order markov* ($\pi_{abc} = \pi_a \pi_{b|a} \pi_{c|a}$), b) *random change* ($\pi_{abc} = \pi_a \pi_b \pi_c$) and *latent symmetry* ($\pi_{ab} = \pi_{ba}$), where π_{abc} is the prior membership probabilities for classes a, b and c in the first, second, and third periods, respectively. Their general formulation also allows the choice probabilities and category buying rate to vary over time. The authors use purchase incidence data for two brands of tuna from a sample of 512 households to test several constrained versions of their general model. They find that although consumers switched classes from month to month, the purchase rates, choice probabilities and segment sizes remained stationary during the three months analyzed.

Marketing Applications

The examples presented in this chapter comprise all of the applications of dynamic segmentation that have appeared in the marketing literature to date. For completeness, we summarize them in Table 10.6. It is clear from the table that the issue of non-stationarity in market segmentation has received limited attention. In fact, STUNMIX models have not yet been tested for dynamic segmentation purposes.

Table 10.6: Marketing Applications of Dynamic Segmentation

Authors	Year	Application	Change Model[1]	Distri-bution[2]	Mixture Model
Poulsen	1990	Analyses of new brand triers	MC	B	MIX
Poulsen	1990	Analyses of new brand triers	LC	B	MIX
Wedel et al	1995	Hazard model of brand switching	MC	P	GLIMMIX
Ramaswamy	1997	Brand switching after new brand introduction	LC	M	MIX
Kamakura et al.	1996	Nested logit model of brand choice	MC	NM	GLIMMIX
Böckenholt and Langeheine	1996	Purchase quantity and brand choice	LC	P-M	GLIMMIX

[1] MC = manifest change LC = latent change
[2] B = binomial P = Poisson M = multinomial NM = nested multinomial

Conclusion

Because dynamic segmentation models have been developed only recently, relatively little is known about their performance. For example, which factors determine the choice between models of manifest and latent change? The only satisfactory answer at this point in time is that primarily substantive arguments should be used to guide such choices. If one can assume that segment sizes and memberships are stable, the manifest-change approach may be more appropriate. If not, obviously, a latent change model is needed. In addition, empirical arguments, such as the relative fit of the two types of models, may be used. However, limited experience in this field (Poulsen 1990) shows that the two types of approach may yield comparable fit measures. A third solution may be to include both latent and manifest changes in the same model, and test for the statistical significance of the two components. Several researchers have attempted that approach, with little success because of problems of collinearity, identification, and instability of the estimates.

With this chapter we conclude our discussion of the market segmentation methodology currently available to the marketing researcher. We started with a review of the traditional methodology for marketing segmentation: cluster analysis (chapter 5). We then followed with descriptions of general formulations of finite-mixture models for simple latent-class analysis (chapter 6), latent-class regression (chapter 7), and latent-class unfolding (chapter 8). Those general models form a complete "toolbox" for descriptive and predictive market segmentation. The STUNMIX models for latent-class unfolding are not only useful for segmentation purposes, but may also produce valuable insights for product positioning. The concomitant-variables extension discussed in chapter 9 complement the finite-mixture models by allowing for the simultaneous identification and profiling of market segments. The dynamic extensions discussed in the last chapter relax the common assumptions of stationarity in composition and/or profiles in the segments. The stationarity assumptions can be potentially restrictive, especially when the segmentation models are applied to new product categories where new brands are introduced and preferences are still being formed. Appendix A3 describes the computer software currently available for application of these mixture models.

APPENDIX A3
Computer Software for Mixture models

Cluster analysis, both hierarchical and non-hierarchical, is imp lemented in all statistical packages, including SPSS, BMDP, SAS, GENSTAT, SYSTAT and S-PLUS. In addition CLUSTAN is a set of dedicated programs that includes a large variety of clustering procedures. However, most of the standard statistical software packages do not include programs for mixture models. Many of the mixture models described in this book can be programmed in flexible packages that include programming options such as GAUSS, GENSTAT, GLIM, MATLAB etc. But several dedicated programs are available that implement part of the analyses described in chapters 6 through 9. Specifically, mixture models are implemented in the programs MLSSA (Clogg 1977), LCAG (Hagenaars and Luijkx 1990), PANMARK (Van de Pol, Langeheine and de Jong 1991), C.A.MAN (Böhning, Schlattmann and Lindsay 1992), LEM (Vermunt 1993) and GLIMMIX. Below, we provide a brief description of the three recent and general programs: PANMARK, LEM and GLIMMIX. The three programs are complementary to a large extent. PANMARK and LEM subsume most of the analyses that are implemented in LCAG and MLSSA. The first two programs are specifically developed for discrete data and the analysis of contingency tables. GLIMMIX handles both discrete and continuous data using the mixture of exponential family framework as described in part 2 of this book.

PANMARK

Authors:	van de Pol, Langeheine and de Jong (1991)
Manual:	Frank van de Pol, Rolf Langeheine and Wil de Jong (1991), PANMARK User's Manual. Report number 8493-89-M1-2 of the Netherlands Central Bureau of Statistics, Department of Statistical Methods, P.O. Box 959, 2270 AZ Voorburg, The Netherlands
Obtainable from:	Netherlands Central Bureau of Statistics, Department of Statistical Methods, P.O. Box 959, 2270 AZ Voorburg, The Netherlands.

General description

PANMARK is tailored to the analysis of discrete data, that is, frequency data from a multi way contingency table, classified by a number of discrete variables. It implements the traditional latent class models, that is, mixture models for discrete data. Moreover, it is a program for mixed Markov latent class models as described in Chapter

9. The general class of Mixed Markov model includes special cases such as the latent Markov chain, the mixed Markov model, the mover-stayer model, simultaneous latent structure analysis in several groups, and the classical latent class model. Applications such as those by Grover and Srinivasan (1989), Poulsen (1990) and Ramaswamy (1997) can be implemented with this program.

Some Program Details

The program is written in Turbo Pascal, and implements maximum likelihood estimation through the EM algorithm. It runs under MS-DOS. A user-friendly interactive questionnaire specifies the data, the model and several options. Parameters may be fixed or set equal, standard errors of the parameters may be requested (approximations are obtained through finite differences), a search of the best of a set of random starting values may be performed to avoid local maxima, and identification tests may be requested.

The program comes in two versions, PM.EXE and PMUNACC.EXE. The former gives more precise results, whereas the latter runs two times faster on computers without a coprocessor. At least 300 Kb of memory is required, the maximum number of non-zero cells in the observed table (the number of observations) is 3119, the maximum number of categories of a variable is 20, and the maximum number of free non-boundary parameters is 114. The data should be prepared in free format and should contain only frequencies or "indexed" frequencies (where the indexes of the discrete variables are included to identify the frequencies). In the first case the program generates the indices for the categorical variables.

The program starts by reading a batch-type setup job from the files SETUP.DAT and SETUP.JOB (other file names can be provided when starting the program). If these files are not present, PANMARK works interactively using menus (Main, Data-Definition, Model Definition, and Output, among others). The Data Definition menu requires one to indicate the number of groups present, the number of indicators for latent variables, the numbers of categories of indicators and their labels. The Model Definition menu requires one to choose between classical latent class, latent Markov chain and latent mixed Markov models. The output options include covariance, identification tests, restrictions, residuals, posterior probabilities, and others. Estimation results, standard errors and printer output are produced in separate output files (respectively NAME.EST, NAME.MAT and NAME.LST).

LEM

Authors: Vermunt (1993)
Obtainable from: J. K.Vermunt, Tilburg University, Methodology Department, P.O. Box 90153 LE Tilburg, The Netherlands.
Manual: Jeroen K. Vermunt (1993). ALEM: Log-Linear and Event History Analysis with Missing Data Using the EM "lgorithm" Working Paper 93.09015/7, Tilburg University,

Methodology Department, Tilburg, The Netherlands.

General Description

LEM is a program to obtain maximum likelihood estimates of parameters of log-linear path models with latent variables and log-linear hazard models. The log-linear path models are extensions of log-linear models for the analysis of categorical data. The path models are used to specify log-linear models with latent variables, analogous to the LISREL approach. As in LISREL, the models combine a measurement part for the latent constructs with a structural part specifying the relations between the latent variables. All variables are categorical. The structural model is a system of simultaneous logit models, the measurement model consists of a latent class model in which relations between the latent constructs and their indicators are specified. The program also handles non-response, making use of dummy response indicators. LEM allows the user to specify a variety of linear constraints on the parameters.

LEM can also be used to estimate log linear hazard models (see Chapter 10). It estimates a general class of hazard models with missing information on the explanatory variables. Specifically, it handles piecewise exponential survival models with categorical covariates, time-varying covariates, multiple spells and multiple risks. Models with misclassified covariates, models with partially observed covariates, and models with non-parametric unobserved heterogeneity can be estimated.

Some Program Details

LEM implements the EM algorithm for estimation of the model parameters. A variant of EM, GEM, is used where the likelihood is not maximized, but only increased in each M-step: the M-step consist of a single iteration. LEM is written in Turbo Pascal, uses a co-processor if available and runs under MS-DOS. One starts the program by typing LEM, specifying the input and output files as parameters, if possible; if not, a help screen appears. The input is read in free format.

The commands to specify the log-linear path model include the numbers of response indicators, latent variables, manifest variables and the number of categories of those variables and their labels. Further, the log-linear effects for the structural and measurement models need to be specified, as well as the marginal tables from which they are estimated. This is done in the usual log-linear model notation. The commands to specify the discrete time hazard model include the numbers of response indicators, latent variables, manifest variables and the number of categories of those variables and their labels, as before. Further, it requires specification of the number of time points, the number of states and the transitions or events to be analyzed, and the hazard model, in the usual log-linear model notation.

Several extra input settings may be specified, including the starting values of the parameters, weights, scale transformations, the number of iterations and the convergence criterion. Outputs that may be requested include the parameter estimates, estimated and expected frequencies, residuals, hazard rates and survival probabilities, the increase in the log-likelihood during iterations, Pearson and LR chi-square

Appendix A3 Computer software for mixture models

measures, and AIC and BIC information criteria. Output can be displayed on the screen or saved on files specified by the user.

GLIMMIX

Authors:	Wedel (1997).
Obtainable from:	PROGAMMA, P.O. Box 841 9700 AV Groningen, The Netherlands.
Manual:	Michel Wedel (1997), GLIMMIX User Manual. Programma, Groningen, The Netherlands.

General Description

GLIMMIX implements the mixture and mixture regression models described in chapters 6. and 7. It simultaneously estimates a number of unobserved segments (latent classes), as well as the regression parameters within each segment. The program allows both dependent and independent variables to be either discrete or continuous, and distributed according to a number of members of the exponential family: normal, binomial, Poisson and gamma. It accommodates a variety of link functions, including the identity, log, logistic, probit and inverse, leading to nine different combinations of distributions and link functions. GLIMMIX includes many latent class models and hazard models for discrete data as special cases. Unconditional mixtures (chapter 6) can be estimated by specifying a regression model with intercepts only. Most applications listed in Tables 6.5 and 7.5 can be estimated with this program

Some Program Details

GLIMMIX runs under WINDOWS 3.11, 95 and NT. It is installed by double-clicking the INSTALL.EXE file on the floppy disk. After double-clicking the GLIMMIX ICON, the first screen is activated, showing a tool bar and a power bar. GLIMMIX operates in five major steps, which are activated from the five buttons on the toolbar.

The first three buttons on the toolbar are: *CLEAR* (clears the current project), *OPEN FILE* (opens a new datafile), and *SAVE FILE* (saves the current datafile). The specifications for the analysis are to be provided in five steps. There are five *STEP* buttons:

STEP 1: Define the variables in the data set.
STEP 2: Recode the variables in the data set.
STEP 3: Make a selection of cases from the data set.
STEP 4: Provide the specifications for the analysis.
STEP 5: View the results of the analysis.

Each step is activated automatically after the preceding one.

Part 2 Segmentation methodology

STEP 1: Define

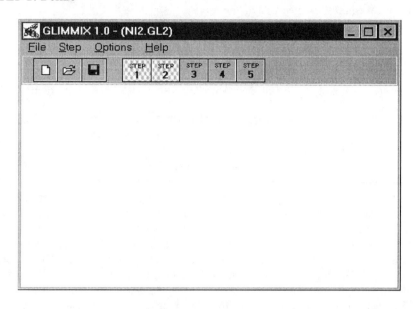

A data file (with the extension .DAT) needs to be opened. You can do this with the OPEN FILE button, which activates the OPEN FILE dialog box. The STEP 1 button activates the same dialog box if a file has not yet been opened. The datafile should contain a variable identifying the observations to be classified (indicated as "SUBJECTS"), a variable identifying replications for each subject (referred to as "BRANDS"), a dependent variable and independent variables. These variables may be in an arbitrary order. The DEFINE DATA dialog box enables you to provide the details of the variables. If the file has been analyzed previously, a dialog box appears asking you whether the previous specifications need to be loaded.

The VIEW button enables you to inspect the data set. GLIMMIX automatically determines the number of variables when reading the file in free format. A fixed format can be specified, analogous to FORTRAN formats, where for example 2X skips two positions, F4.0 reads an integer of four positions, 2F6.2 reads two reals with six positions and two decimals -- 2X,F4.0,3(1X,3F6.2).

To define a variable, choose that variable from the source variable list by clicking on it. Then click on the EDIT button. You may type a name for the variable and indicate its order number in the file. For a dependent or independent variable you should indicate whether it is continuous or discrete. Clicking the USE box forces the variable to be used in the analysis. The variable names can be saved in a file with the extension .VAR, which can be loaded in subsequent applications by using the LOAD button.

Appendix A3 Computer software for mixture models

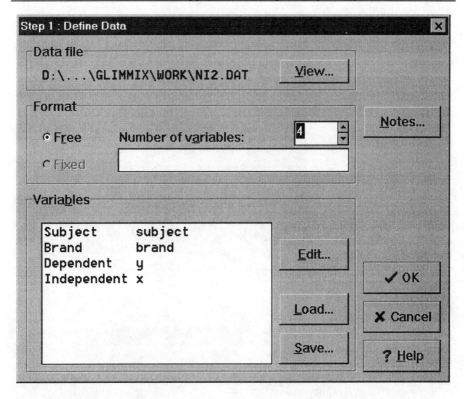

STEP 2: Recode

STEP 2 allows you to transform variables in the data before analysis. You select a variable from the variable source list, and the display box shows summary statistics for that variable.

The MISSING button allows you to indicate which value indicates missing observations, and whether these observations should be deleted or replaced with a certain value. The TRANSFORM button allows you to indicate whether a variable is to be centered, scaled, or multiplied by some constant. For discrete variables a dialog box is activated that allows you to replace each old value of a level by a new one.

Through the DUMMIES button and dialog box you can generate dummy variables from a discrete independent variable. For a variable with K levels, K-1 dummies are generated, one for each level of the variable that occurs in the data, with exception of the last level. You may use two types of codings: dummy coding, where a dummy equals 1 for a level of the original variable and 0 otherwise, or effects coding, where the dummy equals 1 for a certain level, -1 for the last level, and 0 otherwise.

Part 2 Segmentation methodology

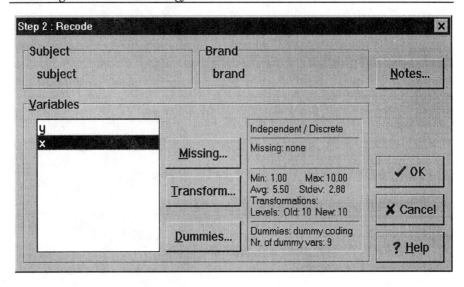

STEP 3: Select

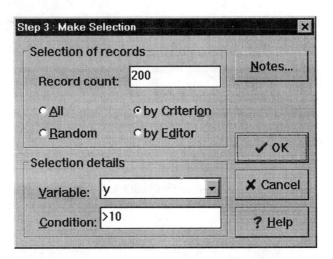

The MAKE SELECTION dialog box allows you to select a part of the records in the data for analysis.

This selection may be a random selection of subjects, where you indicate the percentage selected and the seed for the random number generator. You may also select on the basis of some criterion. Select a variable from the variable source list, and type the selection condition in symbolic notation, i.e., $= 0$ or < 3 or $<= 3$, $= 5$, $>= 6$. Manual selections can be made for small files by using the editor.

Appendix A3 Computer software for mixture models

STEP 4: Analyze

In STEP 4 you specify the details for the mixture model analysis.

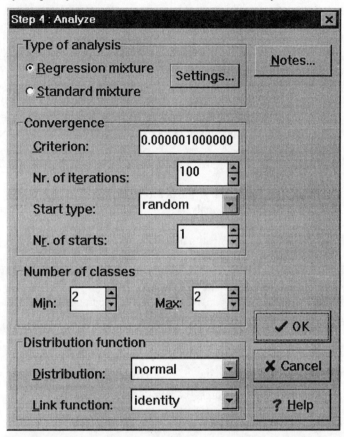

You may choose a standard mixture model or a regression mixture model. In the standard mixture model there are no independent variables. For each class, one mean parameter is estimated for all brands or one for each brand separately. You can choose between those two options by using the SETTINGS button. The same button allows you to include or exclude an intercept in the regression equation for each class in the mixture regression model and to select the independent variables.

The convergence criterion specifying the minimum change in the log-likelihood between EM steps can be specified, and the maximum number of MAJOR (EM) and MINOR (in the M-step) ITERATIONS can be specified. Setting the number of minor iterations to 1 results in the GEM algorithm, which may be more efficient in terms of computation. The M step is based on Bunch, Gay and Welsch (1993).

The START-TYPE for the EM algorithm can be chosen to be a RANDOM

Part 2 Segmentation methodology

selection of membership probabilities, or some PREFIXED set of starting values can be read from a file (with the extension .SET). Two spin-boxes allow you to specify the minimum and maximum numbers of classes for the analysis. The analysis is repeated for all intermediate numbers of classes.

The DISTRIBUTION box allows you to specify the statistical distribution of the dependent variable, where the options NORMAL and GAMMA for continuous variables and BINOMIAL and POISSON for discrete variables are available.

For each distribution, two LINK FUNCTIONS can be chosen. The preferred canonical links are: NORMAL-IDENTITY, POISSON-LOG, BINOMIAL-LOGIT, and GAMMA-INVERSE. For the BINOMIAL distribution after pressing the OK button, you will be prompted to specify the binomial total (if it is the same for all observations) or the variable in the file that contains the binomial total for each observation.

STEP 5: Results

In STEP 5 the results are inspected. A file is produced that contains:
- The specifications of the analysis
- For each number of classes:
    - The parameter estimates and standard errors.
    - The posterior probabilities of membership.
    - Selection criteria to select the number of classes, including the log-likelihood, AIC, CAIC, MAIC, BIC.
    - Entropy criterion to determine the separation of the classes.
    - Diagnostic error messages if errors are encountered.

This file can be EDITed, SAVEd or PRINTed. In addition, several graphs are produced:
- The change in the log-likelihood during EM iterations.
- Information statistics against the number of classes.
- Histograms of the coefficient values for the classes.
- Plots of residuals and fitted values for each class.

PART 3
SPECIAL TOPICS IN MARKET SEGMENTATION

Each chapter in this third section of the book describes an extension of the finite-mixture models discussed in part 2, developed with a specific purpose in mind. The joint segmentation model described in chapter 11 extends the basic mixture model (chapter 6) to develop a segmentation solution that simultaneously considers two or more segmentation bases. This model identifies segments along each segmentation base and at the same time determines the relationships among the various types of segments.

The tailored interviewing procedure discussed in chapter 12 uses a previous segmentation solution obtained from the basic mixture model to classify new customers into the segments, based only on a subset of the original variables, leading to substantial reductions in interviewing time. The mixture of structural equations discussed in chapter 13 estimates a structural equation model within latent classes of subjects, allowing for distinct structures among observed and latent variables within each class. This finite-mixture of structural equations allows researchers to verify whether structures observed cross-sectionally reflect relationships inferred within homogeneous groups of subjects. In chapter 14 we discuss mixture model analyses of consumer dissimilarity judgements. We focus on two important representations of those judgements: spaces and trees, and formulate mixture model versions of both of them.

11
JOINT SEGMENTATION

The mixture models discussed so far are applicable for segmentation along a single dimension or basis, or at most, for segmentation along one criterion while using another dimension for segment-membership predictions (e.g., concomitant-variable mixture models). In this chapter we describe an extension of the latent-class model for joint segmentation, which define segments along multiple segmentation bases.

Joint Segmentation

Here we discuss an application of the latent Markov approach (described in chapter 10) to a nondynamic segmentation problem. This application is due to Ramaswamy, Chatterjee and Cohen (1996), and addresses an important issue in segmentation: How do we identify segments when distinct types of segmentation bases are present (e.g., benefits and product usage)? The joint segmentation model explicitly considers segments for each type of basis, and the potential interdependence between these two types of segments. This line of research was initiated by Morwitz and Schmittlein (1992), who demonstrated the benefit of identifying joint segments in the context of conversion rates from intention to behavior. They demonstrated that demographic variables moderate the link between intentions and behavior, and their framework offers an approach for uncovering the relationship between segmentation bases that may be induced by unobserved consumer variables. In addition, the joint segmentation framework capitalizes on the notion that there is not one single best way to segment a market, but, depending on the purposes of the study and the bases used for segmentation, several sets of segments may be identified in the sample.

The Joint Segmentation Model

The joint segmentation model was originally formulated for polychotomous categorical variables, but here we formulate it for two sets of variables of any type. The distribution of those variables, assumed to be one of the exponential family, is denoted by f as before. We use the following notation:

Part 3 Special topics in market segmentation

Notation
k = 1 ,..., K indicates variables composing the first set of segmentation bases
l = 1 ,..., L indicates variables composing the second set of bases
s = 1, ... , S indicates segments on bases of type 1
u = 1 , ..., U indicates segments on bases of type 2

The joint segmentation model is formulated as follows, where the simultaneous distribution of the variables in the two sets is described from their conditional distributions as:

$$f_{kl}(y_k, y_l | \theta_{su}) = \sum_{s=1}^{S} \sum_{u=1}^{U} \pi_{su} f_{k|s}(y_k | \theta_s) f_{l|u}(y_l | \theta_u) \quad (11.1)$$

In equation 11.1, π_{su} denotes the joint probability of a consumer belonging to segment s on the first base and to segment u on the second base, where it holds that $\sum_{s=1}^{S} \sum_{u=1}^{U} \pi_{su} = 1$. The sizes of the two types of segments are calculated simply as the marginals of π_{su} by summing over s or over u, respectively. Equation 11.1 states that the simultaneous distribution of the variables k and l is described from the conditional distribution of k given s, and that of l given u, times the joint probability of the segments s and u. Issues of model identification and estimation are not discussed here, as they are quite similar to those issues as discussed for the latent Markov model above.

The model is estimated by maximizing the likelihood using the EM algorithm. One of the likelihood-based information criteria, BIC or CAIC, may be used to determine the optimal numbers of segments, S and U, in both types of bases. Posterior probabilities of membership to both types of segments are calculated as shown for the latent Markov approach.

Two alternative approaches to joint segmentation arise as special cases of the model in equation 11.1. In the *combined approach* no distinction is made between the two types of bases, and only one set of segment sizes or a-priori probabilities are estimated, say: π_s. This approach is in fact the simple mixture approach described in chapter 6., where all K + L variables are assumed to indicate the same S underlying groups. It arises as a special case of the joint segmentation model, where the joint probability matrix is a diagonal matrix (i.e. $\pi_{su} = \pi_s$ for s = u, and 0 otherwise). The disadvantage of the approach is that the amorphous combination of conceptually distinct bases can be less meaningful theoretically, and may therefore lead to suboptimal managerial action. In the example above, the approach would amount to considering segments formed from benefits and purchases, without distinguishing the two variables, even though benefits may be a more appropriate base for developing new products, and product usage may be a more appropriate base for targeting current products.

The second approach is the *separate approach,* in which both sets of bases are analyzed separately with a finite mixture model (see chapter 6). It yields two sets of segments and their prior probabilities π_s and π_u. These estimates are identical to those of the joint segmentation model under the assumption that the two sets of segments are independent (i.e., $\pi_{su} = \pi_s \pi_u$). The disadvantage of this approach is that because the two sets of segments are considered to be independent, one cannot assess the relationship between the segments in developing marketing strategy. In our example, the approach would amount to considering segments formed from benefits and segments formed from purchases, without the possibility of relating the two. If benefits are used to develop a new product for particular segments and product usage is used to target segments with current products, it is appropriate to investigate the relationships between the two sets of segments (e.g., to avoid cannibalization).

Synthetic Data Illustration

We report part of the synthetic data analyses of Ramaswamy, Chatterjee and Cohen (1996). Synthetic data were generated for 500 hypothetical subjects, with six binary variables measured in each of two sets (i.e. K = 6 and L = 6). For each of the two sets of variables, three segments were generated, with a certain pattern of association among them (S = 3, U = 3, the true π_{su} are shown in Table 11.1. In addition, within-segment heterogeneity was simulated by adding small random numbers to the response probabilities of the variables within segments.

Table 11.1: True and Estimated Joint Segment Probabilities (adapted from Ramaswamy et al. 1996)

Segments s, u	u = 1	u = 2	u = 3
s = 1 True	0.22	0.19	0.00
Joint	0.25	0.14	0.00
Separate	0.19	0.12	0.06
s = 2 True	0.00	0.15	0.09
Joint	0.00	0.17	0.11
Separate	0.08	0.11	0.07
s = 3 True	0.24	0.00	0.11
Joint	0.23	0.00	0.11
Separate	0.20	0.08	0.09

The joint (binary mixture) segmentation model was estimated for different values of S and U, and the optimal number was determined on the basis of the BIC-criterion. BIC reached a minimum value for S = 3 and U = 3, indicating the true numbers of segments on each of the bases. Table 11.1 shows the estimated simultaneous probabilities and the true probabilities.

The data were also analyzed by the separate approach, where each of the two sets of variables is analyzed with a separate binomial mixture model. In this case, the BIC criterion points to $S = 3$ for the first data set, and to $U = 3$ for the second, correctly indicating the number of segments in both cases. For comparison, a-posteriori joint segment sizes were calculated by cross classifying the a-posteriori membership values calculated for the two solutions. The results are shown in Table 11.1. The joint approach produces more accurate estimates of the joint segment sizes than the separate approach. Moreover, the joint approach better assigns consumers to their true segments based on the a-posteriori probabilities: the hit rate is 76%, whereas that for the separate model is 66%. In addition, the combined model was fitted ($S = 6$), yielding an even lower hit rate of 58% correct assignment of consumers to their true segments. Further, it appeared that separate analyses may produce biased estimates of the response probabilities within segments when the joint model holds (not shown here).

Banking Services

Here we reiterate the findings of Ramaswamy et al. (1996) in an application of the joint segmentation model to a banking survey. The goal of the survey was to segment customers on the basis of both investment goals (benefits) and banking services used (usages). Customers had to pick their financial investment goals from a list of 13, and the banking services they used, also from a list of 13. The final sample comprised 554 customers. The binomial mixture joint segmentation model was applied to the data and, on the basis of BIC values, an $S = 4$, $U = 4$ solution was selected. Table 11.2 reports the segment-level probabilities for the items from both sets.

From Table 11.2, we see that segment $s = 1$ is characterized by *starter goals*, such as buying a first house and saving for a car. Segment $s = 2$ favors *self-oriented goals* such as preserving assets, minimizing taxes and personal enjoyment. *Family oriented goals* appear to be more prevalent for segment $s = 3$, including saving for education, protection of the family and retirement. The last segment, $s=4$, is driven by *empty nest goals*, as they are primarily involved with leaving an estate to their heirs and preserving assets.

From the lower panel of Table 11.2, we see that segment $u = 1$ can be characterized as *infrequent investors*. They have little or no investment in any financial securities or brokerage accounts, and do not seek investment counseling. Segment $u = 2$ is called the *simple investors*, as they use retirement-related services. Their usage mirrors the sample average. Segment $u = 3$ seems to consist of *advanced investors*. They seek investment counseling, have brokerage accounts, stocks and bonds, and have a broadly diversified portfolio. Segment $u = 4$ consists of *conservative investors*, who are likely to use the safer financial services, such as certificates of deposit.

Table 11.2: Segmentation Structure of Binary Joint Segmentation Model for Banking Services

Financial goals	s = 1 (27%)	s = 2 (19%)	s = 3 (14%)	s = 4 (40%)
Save for retirement	0.66	0.78	0.93	0.32
Preserve assets I already have	0.14	0.85	0.43	0.59
Safety net for emergencies	0.39	0.63	0.34	0.42
Personal enjoyment	0.53	0.71	0.20	0.27
Save for education	0.24	0.35	0.80	0.11
Minimize taxes	0.05	0.73	0.32	0.20
Leave an estate for my heirs	0.00	0.39	0.22	0.34
Protect my family in case of premature death	0.00	0.46	0.55	0.09
Retire well before 65	0.12	0.37	0.27	0.07
Buy a better house	0.21	0.20	0.24	0.00
Buy my first house	0.23	0.13	0.00	0.00
Save for my automobile	0.13	0.22	0.00	0.00
Care for my aging parents	0.04	0.18	0.00	0.00
Banking services	u = 1 (27%)	u = 2 (40%)	u = 3 (17%)	u = 4 (16%)
Savings account	0.81	0.85	0.86	0.88
Individual retirement account	0.21	0.76	0.91	0.58
Money market account at a bank union	0.36	0.58	0.86	0.59
Certificates deposit	0.28	0.40	0.55	0.83
Safe deposit box	0.25	0.46	0.69	0.55
Mutual funds	0.09	0.44	0.99	0.23
Individual stocks or bonds	0.04	0.44	0.92	0.27
Treasury bills	0.00	0.31	0.77	0.48
Brokerage account	0.00	0.17	0.70	0.07
Keough or 401 (k) accounts	0.00	0.30	0.45	0.00
Tax-exempt securities	0.00	0.13	0.70	0.19
Money market fund obtained through broker	0.00	0.16	0.60	0.00
Investment counseling or planning services	0.00	0.13	0.32	0.06

Table 11.3 reports the joint probabilities π_{su}, which reveal considerable dependence between the two segment types. In particular, the segment with starter goals overlaps the segments of infrequent and simple investors. Family-oriented and self-oriented

segments both overlap the segments of simple and advanced investors, and the segment of empty nesters overlaps infrequent and conservative investors. These results appear to be intuitively plausible.

Financial institutions might use this joint segmentation scheme to target some subsets of the joint segments. The empty nesters seem to be an attractive segment, which may have been overlooked in the past. Targeting is facilitated by profiling the segments with demographic variables (not shown here).

The joint segmentation mixture model is a flexible approach to identifying the interdependence between different segmentation bases. Even when the purpose of a study is to segment the market on a single basis, the joint segmentation model may provide a more accurate segmentation by capitalizing on the information about the other set of segmentation bases.

Table 11.3: Estimated Joint Segment Probabilities for Banking Services Application (adapted from Ramaswamy et al. 1996)

Goal\ investor segments	$u = 1$ Infrequent	$u = 2$ Simple	$u = 3$ Advanced	$u = 4$ Conservative
$s = 1$ (Starters)	0.15	0.12	0.00	0.00
$s = 2$ (Self-oriented)	0.00	0.11	0.08	0.00
$s = 3$ (Family-oriented)	0.00	0.11	0.04	0.00
$s = 4$ (Empty nest)	0.15	0.06	0.04	0.14

Conclusion

Although the latent Markov model was originally developed for dynamic latent-class analysis, it has another valuable application in marketing in defining market segments based on multiple segmentation bases. Application of this multiple-base segmentation approach will produce market segments that afford the advantages of each segmentation base, as discussed in chapter 2, and are more likely to satisfy the criteria for effective segmentation listed in chapter 1. A problem that still needs to be resolved in joint segmentation, as well as other applications of latent Markov models, is that the number of parameters to be estimated explodes with the number of bases utilized, and the decision on the proper number of segments along each dimension becomes more cumbersome. Identifiability of the joint segmentation was questioned by Walker and Damien (1999). However, in response Ramawamy, Chatterjee and Cohen (1999) showed that while the argument is correct and well known in the mixture literature, it does not apply to their specific model.

12
MARKET SEGMENTATION WITH TAILORED INTERVIEWING

In this chapter we introduce the concept of tailored interviewing as a method to reduce respondent burden and interviewing cost. We then present a tailored adaptive interviewing procedure designed for market segmentation that is based on a mixture model approach. An application to life-style segmentation is provided, in which we demonstrate that the respondent burden is reduced by almost 80%.

Tailored Interviewing

The concept of tailored interviewing is relatively new to marketing researchers. Tailored interviewing procedures for the measurement of attitudes were introduced by Balasubramanian and Kamakura (1989) and extended by Singh, Howell and Rhoads (1990). Tailored interviewing is an approach for measuring one or more underlying traits of consumers, in a way that decreases interviewing costs by using only a fraction of the number of items in a multi-item measurement scale, while controlling for measurement accuracy. It attempts to resolve the common dilemma facing survey researchers: increasing the quantity of information gathered in an interview can increase costs and decrease the quality of that information. Tailored interviewing procedures are based on latent trait models for unidimensional scales, rooted in latent trait theory or item response theory (IRT). Item response theory uses a probabilistic approach to modeling the relationship between observable variables and an underlying construct. It assumes that items and respondents can be ordered along the same unidimensionsal trait continuum, and that the probability of responding positively to the kth binary item for respondent n is a function of the respondent's trait level and the item characteristics:

$$P_{nk} = \frac{1}{1+\{\exp[-a_k(\theta_n-b_k)]\}} \tag{12.1}$$

where a_i is the discrimination parameter indicating the sensitivity of item k, and b_{ik} is the threshold parameter representing the trait level required for a 50% probability of agreeing with the kth item. The individual parameter θ_n measures the trait level for subject n.

The tailored interviewing process operates as follows. All items are first calibrated on a sample of respondents by the method of maximum likelihood. The estimated

Part 3 Special topics in market segmentation

parameters are then used in subsequent tailored interviews. The function that serves as the criterion for the selection of items, given a prior estimate of a subject's trait level, is:

$$I_k(\theta_n) = \frac{P_{k'}(\theta_n)^2}{P_k(\theta_n)(1 - P_k(\theta_n))}. \tag{12.2}$$

This information function is inversely related to the asymptotic variance of the trait estimate. The subsequent interviews are tailored to each respondent, in that s/he has to answer only the items that are most informative of his/her trait. In other words, at each stage of the interviewing process, the most informative item is determined on the basis of equation 12.2. Thus, savings in interviewing time and cost are obtained. The implementation of adaptive interviewing designs requires the following decisions:

1. The quality and size of the *item bank* are critical in the potential payoffs from adaptive designs. The item bank should allow for reliable measurement of trait values over the relevant trait range. It has been suggested that the bank should contain items with sensitivities of at least 0.80, and a wide and even distribution of the thresholds in the [-2.0; 2.0] range.

2. The *entry level* for the items with which the interview is started. Efficient tailored interviewing designs usually employ some information about the sample and purpose of the study to decide which and how many items to use for starting the interview procedure. For dichotomous items, items with medium threshold values are recommended to start the interviews. This rule cannot be applied to graded items, but here separate threshold parameters are estimated for each of the scale points. One of them, for example, in the middle category, may be chosen and the above procedure applied.

3. The *item selection rule* employed. The maximum information approach outlined above for binary items involves two steps. First, the respondent's trait level θ_n is estimated from his or her responses to all preceding questions. Second, the next item is selected that provides the maximum amount of information at the estimated value of θ_n, according to equation 12.2. This procedure ensures that different respondents receive different items in different sequences.

4. The *trait estimation* procedure. At each stage of the interviewing process, an estimate of the respondent's θ_n is to be obtained. In principle this can be accomplished by either maximum likelihood or Bayesian procedures. In previous research, the maximum likelihood method has been predominantly applied.

5. The *termination rule*. Termination criteria must be formulated to decide when

to stop a particular interview. They include measures of required precision for the trait estimate or a minimum level of required information for the next item.

In an application to the measurement of consumer alienation and discontent, Balasubramanian and Kamakura (1989) demonstrated that tailored interviews were about 70% shorter than traditional interviews and the increase in measurement error was minor. Figure 12.1 depicts the results from one tailored interview session conducted by Balasubramanian and Kamakura. It shows the estimates of the respondent's discontent at each stage of the interview, as well as their 95% confidence intervals. The respondent's discontent measure becomes fairly stable after 20 items, and the measurement error is not reduced by asking more questions. In other words, after 20 items, little is gained in pursuing the interview further.

Figure 12.1: Illustration of the Tailored Interview (adapted from Balasubramanian and Kamakura 1989)

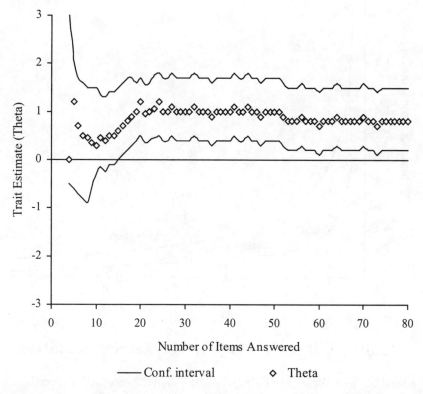

The pattern shown in Figure 12.1 is representative of those for other interviews measuring consumer discontent and alienation; substantial gains in accuracy are attained at the beginning of the tailored interview but, after a certain point, the

Part 3 Special topics in market segmentation

reduction in estimation error becomes small in relation to the additional interviewing costs. Figure 12.2 provides a summary of the cost-accuracy tradeoffs observed in all 137 interviews. Singh et al. (1990) extended the procedure by using Samejima's (1969) model for multichotomous graded-response items. On the basis of a simulation of the tailored interview process to assess consumer discontent, they also demonstrated the potential savings in interviewing time.

Figure 12.2: Cost versus Accuracy Tradeoff: Discontent and Alienation Scales (Balasubramanian and Kamakura 1989)

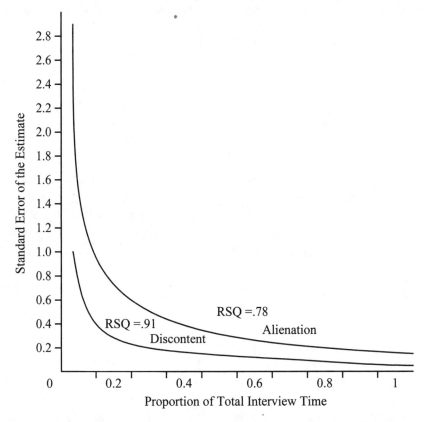

Tailored Interviewing for Market Segmentation

Kamakura and Wedel (1995) proposed a tailored interviewing procedure for segmentation purposes. The procedure shows communalities with the procedures of MacReady and Dayton (1992) and Poulsen (1993). MacReady and Dayton proposed an adaptive testing procedure for measuring constructs composed of multiple nominally scaled dimensions or domains used in psychology. They apply it in an adaptive test of mathematical ability along several domains (arithmetic, algebra, calculus, etc), each with two levels (mastery of the domain, or non-mastery). MacReady and Dayton's algorithm selects a domain (e.g., basic arithmetic) for assessment, and then selects items that are tailored to best classify the subject as a master/non-master on that particular domain. Once that domain is assessed, the algorithm selects the next domain and the process is repeated. Poulsen (1993) also uses a Bayesian classification scheme to develop an adaptive data-collection procedure for binary items. Both of those procedures are based on mixture models. Poulsen has developed computer software for tailored interviewing (for binary items), designated the ABC (for Adaptive Bayesian Classification) system, which is commercially available.

Here we describe the Kamakura and Wedel (1995) procedure. The algorithm is tailored to the classification of respondents into segments and handles items of several scale types. The main goal is the direct classification of subjects into segments. The algorithm takes into consideration the errors incurred in estimating the item parameters during the calibration stage. Previous approaches have treated those parameter estimates as if they were known, error-free values. However, the uncertainty of the parameter estimates in the calibration sample will affect the classification error in the adaptive interview, which is based on these parameters.

Like other tailored interviewing approaches, the procedure requires a bank of pre-calibrated items, or segmentation bases. Those items are pre-calibrated on a sample of consumers to identify a number of segments and to provide the item-parameters within each segment. A mixture model (see chapter 6) is used to obtain these item parameters and to identify the segments. After this calibration stage, the tailored interview procedure can be applied. First, the respondent is assigned to the segments according to *a-priori* probabilities either from the calibration stage or based on other available information. At each stage of the interview, the algorithm selects the one item that contains the largest amount of information on the segment membership of the respondent, given the estimated item parameters and prior probabilities. That item is presented to the respondent, and the answer is used to produce posterior probabilities of segment membership by a Bayesian update of his/her prior-probabilities. With the posterior probabilities as the new priors, a new item is selected, and the process is repeated until an acceptable level of certainty about the subject's classification is reached. The details of each stage in this iterative process are presented below. To explain this model, we introduce some additional notation:

Notation

t	=	$1,...,T_n$ indicate steps in the interview of consumer n,
Q(t)	=	set of all items already used, up to stage t of the interview.

Model Calibration

The first stage in the tailored interviewing procedure for segmentation is the estimation of the mixture model to identify the segments and calibrate the item-

Part 3 Special topics in market segmentation

parameters within each segment. The two most commonly applied types of scale items are accommodated, ordinal and nominal scales. The response by subject n to item k, y_{nk}, can assume a value in the set of categories of that scale $\{l = 1,2,...,L_k\}$, where L_k is the highest response category for scale k. Items are scaled either as graded (i.e., ordinal categories) or nominal variables. The conditional probability (given that s/he belongs to a segment s) that consumer n answers a graded-response item k with response category l is given by:

$$P(y_{nk}=l|\alpha_{ks}) = \binom{L_k-1}{l-1} \frac{(\exp(\alpha_{ks}))^{l-1}}{[1+\exp(\alpha_{ks})]^{l-1}[1+\exp(\alpha_{ks})]^{L_k-l}}. \quad (12.3)$$

where α_{ks} is the item parameter for item k and segment s, reflecting the relative strength of that item in segment s. One of the main advantages of formulation 12.3 is that it allows for a flexible pattern of responses for the item categories, with a very parsimonious, single item-parameter model, while observing the ordinal nature of the response categories. For a nominally scaled item, the response probability is:

$$P(y_{nk}=l|\alpha_{kls}) = \frac{\exp(\alpha_{kls})}{\sum_{l=1}^{L_k}\exp(\alpha_{kls})} \quad (12.4)$$

where α_{kls} reflects the attraction of response category l of item k in segment s, and where for identification $\alpha_{k1s}=0$. The unconditional likelihood is obtained by a weighted sum of the conditional likelihoods, with the prior segment sizes (to be estimated) as weights. Maximum-likelihood estimates of these parameters are obtained on the basis of the observed responses y_n of a sample of N consumers. Details of the estimation procedure are provided in chapter 6.

The asymptotic distribution of these estimates is multivariate normal. The covariance matrix of this multivariate distribution, Σ, is given by the inverse of the Fisher information matrix (see chapter 6). Once the parameters are estimated, consumers in the calibration sample are traditionally assigned in mixture models to the segments on the basis of their posterior probabilities p_{ns} (equation 6.4). A useful indicator of the degree of uncertainty about the respondents' membership in the segments is the entropy of classification, provided in equation 6.11. Once the mixture model has been calibrated, the estimates of the item parameters and class-size parameters as well as the estimate of their asymptotic variance-covariance matrix are the basis for the tailored interviewing process.

Prior Membership Probabilities

To start the actual tailored interviewing process, one must have some prior assessment of the probability of membership in each of the life-style segments. We let the vector of prior probabilities for subject n be denoted by $\tau_n^0 = (\tau_{ns}^0)$. Without any

Chapter 12 Segmentation with tailored interviewing

prior information about the respondent, the prior probability of membership in each segment, computed in the calibration of the mixture model, is the best estimate of this prior probability (i.e., $\tau_n^0 = \pi_n$). If a concomitant variable mixture model is calibrated as described in chapter 9, a respondent's concomitant variables can be used to obtain estimates of the prior probabilities. On the basis of the items presented to the respondent and his/her responses obtained in the successive steps of the tailored interview, the prior probabilities of segment membership are revised as shown next.

Revising the Segment Membership Probabilities

Every answer obtained from the respondent at subsequent stages of the interview presents new information that is used to revise the prior probability of his/her membership in each of the segments. The prior membership probabilities are revised on the basis of the responses obtained up to the current stage of the interview, and the segment size and item parameters. The item parameters are estimated with error however, because of random errors in the calibration data, possible model misspecification, etc. Treating those parameters as values known with certainty may lead to an underestimation of the uncertainty in the classification obtained at each stage of the tailored interview, a too-small number of items being used, and an increased classification error. Therefore, instead of treating the parameter estimates as fixed values, one should use their asymptotic distribution, thus considering their inaccuracy and inter-correlations.

Using the asymptotic (posterior) normal distribution of the estimates, we can show that the posterior membership probability for respondent n, at stage t of the interview is equal to:

$$\tau_{ns}^t = \int \frac{\pi_s \prod_{k \in Q(t)} P(y_{nk}|\alpha_{ks})}{\sum_{s=1}^{S} \pi_s \prod_{k \in Q(t)} P(y_{nk}|\alpha_{ks})} \phi(\Omega, \Sigma) d\Omega, \qquad (12.5)$$

where $\phi(.)$ denotes the multivariate normal density function.

The integration in equation 12.5 does not have a closed form, and cannot be solved analytically. Therefore, the posterior membership probability needs to be evaluated numerically by using a simulation of the multivariate normal distribution. Samples from the distribution $\phi(.)$ are generated, from which equation 12.5 can be calculated to any required degree of accuracy, depending on the size of the random sample (see e.g., Casella and George 1992). These computations must be performed after every answer is given by the respondent, but they are fast enough not to affect the pace of the interview.

Item Selection

At each stage, the tailored interviewing algorithm selects the item to be asked next, given the current knowledge about the respondent's membership in the segments. Here, the most informative item is selected, which is the item that would lead to the clearest allocation of the respondent to one of the segments. This is accomplished by selecting the item k that generates the lowest entropy of classification (equation 6.11) for that respondent. However, item k has not yet been presented to the respondent, and therefore the posterior membership probabilities (which are calculated on the basis of the response to item k) are not known. The solution is to compute the membership probabilities from the probabilities that s/he uses each possible response category of item k. These response probabilities are obtained by using our knowledge of the segment membership of the respondent at that stage of the interview. The probabilities are then used to calculate the *expected* entropy of classification to be obtained with item k. The item selection strategy amounts to finding that item k, among the ones not yet used, that has the lowest value of the expected classification entropy, defined as:

$$E_k = \sum_{l=1}^{L_k} P(y_{nk}=l|\tau_i^{t-1}) \sum_{s=1}^{S} \tau_{is}(y_{nk}=l) \ln \tau_{ns}(y_{nk}=l), \qquad (12.6)$$

where τ_i^{t-1} is the $(S \times 1)$ vector of the membership probabilities obtained in the previous stage of the interviewing process ($\tau_i^0 = \pi$), and $\tau_{is}(y_{nk}=l)$ are the posterior membership probabilities given that respondent n answers the new item k with response category l. The latter quantities are calculated as:

$$\tau_{ns}(y_{nk}=l) = \frac{\tau_{ns}^{tl} P(y_{nk}=l|\alpha_{ks})}{\sum_{s=1}^{S} \tau_{ns}^{tl} P(y_{nk}=l|\alpha_{ks})}, \qquad (12.7)$$

where $P(y_{nk}=l|\tau_n^{t-1})$ is the probability that respondent n would answer item k with category l, given his/her current membership probabilities:

$$P(y_{nk}=l|\tau_n^{tl}) = \sum_{s=1}^{S} \tau_{ns}^{tl} P(y_{nk}=l|\alpha_{ks}). \qquad (12.8)$$

Stopping Rule

A direct and intuitive criterion for stopping the interview is the degree of confidence about the respondent's allocation to one of the segments. The interview is stopped when the respondent can be assigned to one of the segments with a probability equal or greater than 99%.

Application to Life-Style Segmentation

Life-Style Segmentation

Since its introduction by Lazer in 1963, life-style information has become a popular tool in marketing management decision making. Because life-style characteristics provide a rich view of the market and a more life-like portrait of the consumer, they meet the demands of management practice for increasingly sophisticated and actionable marketing information. Life-style segmentation is discussed in more detail in chapter 14. A disadvantage of life-style questionnaires, however, is that because of the broad domain of AIO's that need to be covered, they tend to be very long, mostly involving several hundreds of questions. Often, life-style questions are a part of larger questionnaires with other objectives besides the identification of life-style segments. The problem of questionnaire length is increasingly troubling survey researchers. Here we review the application of tailored interviewing in life style segmentation provided by Kamakura and Wedel (1995).

Data Description

The data used for the application were collected as part of a large study of brand preferences, and life-styles directly related to the consumption of fashion in Brazil. The data were obtained from 800 completed personal interviews among 800 consumers, chosen through a quota sample specified on the basis of socio-economic class and place of residence. Of the 800 interviews, 700 were randomly selected for calibration of the latent-class model and the identification of life-style segments. The other 100 were retained as a holdout sample for the validation of the tailored interview procedure. The following three types of items were used for the application:

a) **Demographic items** - socio-economic class (upper, middle, lower, based on the criteria defined by the Brazilian association of marketing researchers-ABIPEME), gender, age, education, position in the family (head, homemaker, child or grand-child), marital status, and whether the respondent had children.
b) **Interests and opinions** - these were assessed using 40 Likert-type items, rated on 7-point scales.
c) **Activities** - level of participation in 26 activities was assessed on 4-point scales.

Model Calibration

The mixture model described in equations 12.3 through 12.4 was applied to the seven (nominal) demographic items and 66 (ordinal) life-style items, leading to a seven-segment solution. Tables 12.1 and 12.2 report the parameter estimates (α_{ks}) for the 66 life-style items across the seven segments. The results show a clear profile for the segments. To simplify interpretation of the estimates, statistical tests comparing the parameters for each segment against the average across all segments were performed (a conservative test was used: a t-value greater than 3 or smaller than -3 was considered significant). The results of these tests are highlighted in Tables 12.1 and 12.2 in different printing fonts. Parameter estimates shown in **bold** characters indicate that members of the particular segment are more likely to agree with the item than the sample on average. Parameter estimates shown in *italic* indicate that members of the segment are less likely to agree with the item than the population in general.

Profile of the Segments

The results are discussed next for each segment. From Tables 12.1 and 12.2 it appears that members of segments 1 and 2 are more concerned than average about dressing according to their age, less concerned about wearing clothes that appeal to the opposite sex, and have lower interest in social activities than average. Segment 1 is more concerned about *durability* and *easy care*. Members of this segment are the least likely to participate in a large part of the activities included in the survey. That finding is consistent with the demographic profile of this segment: it has a high proportion of members in the lowest social classes, and its members tend to be older, less educated, married with children, and more likely to be heads of the household or homemakers.

Segment 2, in contrast, is more concerned about *comfort* in their clothing. Members of this segment are more likely to read books and newspapers, to attend arts and crafts shows, and to go on family outings. Those activities are consistent with the demographic profile of this segment: it has a high proportion of members in the upper classes and a high proportion of women. Members are more likely to be older, married with children, and homemakers.

Segment 3 is not markedly different from the sample average in terms of its activities and socio-demographic profile, but its members are less likely to be homemakers and more likely to be dependent on parents or grand-parents. This segment expresses very *little interest in fashion*, although they are interested in convenience of care and in comfort.

The members of segment 4 tend to respond negatively to many of the interest and opinion items (Table 12.2). These consumers appear to have *disinterest in clothing* and do not care what others think of their clothes. They do have a favorable attitude toward durability and comfort in wearing. This segment has a higher proportion of men than the other segments.

Members of segment 5 see fashion as an indicator of social status. They also see themselves as opinion leaders. They are less concerned with comfort and convenience than average, and express a strong interest in *extravagant* and *exuberant clothing*. Segment 5 has a higher proportion of members in the lower socio-economic classes, and of women. They are also more likely to be homemakers, to be married, and to have children. Their activities are not markedly different from the average.

Segment 6 has no distinguishing demographic characteristics or activities. Its members show a higher than average probability of agreement with most items, reflecting a *favorable attitude* toward clothing. The answers to the items may be affected to

Chapter 12 Segmentation with tailored interviewing

a certain extent by the response-style of these subjects, as this segment appears to contain "yes-sayers".

Segment 7 expresses little concern about convenience in clothing. These consumers prefer *insinuating clothing* with a *touch of sensuality* that are attractive to the opposite sex. Members of this segment are the most likely to engage in social activities such as going out with friends, going to bars, and practicing sports. That finding agrees with the conclusions drawn from the opinion and interest items and the demographic profile. Segment 7 has a high proportion of members in the upper socio-economic classes and has a higher formal education. They are more likely to be young, single and dependent on parents or grandparents.

Table 12.1: Demographics and Activities by Segment (adapted from Kamakura and Wedel 1995)

Item	VARIABLE	Seg 1	Seg 2	Seg 3	Seg 4	Seg 5	Seg 6	Seg 7	Total
	Segment sizes	13%	15%	10%	14%	15%	14%	19%	100%
	Demographics								
1	Socio-Economic Class - A	3%	21%	9%	13%	13%	14%	23%	14%
	Socio-Economic Class - B	30%	48%	38%	42%	31%	34%	46%	39%
	Socio-Economic Class C&D	67%	31%	53%	45%	57%	52%	31%	47%
2	Gender - male	45%	36%	54%	61%	35%	42%	47%	45%
	Gender - female	55%	64%	46%	39%	65%	58%	53%	55%
3	Age (<19/24/29/39/40+)	76%	78%	55%	63%	68%	52%	31%	59%
70	Last school attended	20%	45%	32%	38%	27%	33%	50%	36%
71	Position in the family - head	34%	29%	26%	33%	25%	21%	3%	23%
	Position in the family -hmkr	48%	51%	19%	29%	51%	32%	9%	34%
	Position in the family -child	18%	20%	56%	39%	25%	47%	88%	43%
72	Marital status - single	17%	16%	49%	38%	18%	46%	85%	40%
	Marital status -marr/div.	83%	84%	51%	62%	82%	54%	15%	60%
73	Children - yes	78%	80%	47%	58%	77%	49%	12%	56%
	Children - no	22%	20%	53%	42%	23%	51%	88%	44%

Part 3 Special topics in market segmentation

Table 12.1 - continued

Item	VARIABLE	Seg 1	Seg 2	Seg 3	Seg 4	Seg 5	Seg 6	Seg 7
	Segment sizes	13%	15%	10%	14%	15%	14%	19%
	Activities							
50	Read books	-0.8	**0.4**	-0.2	-0.1	-0.5	-0.3	-0.1
53	Read newspapers	-0.3	**0.4**	-0.1	0.1	-0.4	-0.3	0.1
66	Go to arts & crafts shows	*-1.7*	**-0.3**	-1.1	-1.1	-0.9	-1.1	-0.9
68	Out with the family	-0.1	**0.7**	-0.1	0.1	0.2	0.1	-0.2
46	Parties or social functions	*-1.0*	-0.5	-0.3	-0.6	-0.6	-0.1	**0.1**
51	Out with friends	*-0.9*	-0.3	0.2	-0.3	-0.6	0.2	**0.9**
54	Travel on vacation	*-0.7*	0.2	-0.1	0.1	-0.3	0.2	**0.4**
58	Go to the movies	*-1.8*	-0.9	-0.6	-0.8	-1.0	-0.6	**-0.1**
59	Go to live music shows	*-2.6*	-1.5	-1.2	-1.5	-1.3	-0.7	**-0.3**
63	Practice sports	*-1.5*	-1.0	-0.6	-0.7	-0.9	-0.4	**0.1**
64	Go to bars	*-1.9*	-1.3	-0.4	-1.1	-1.3	-0.5	**0.3**
65	Listen to music	-0.4	0.4	0.4	0.2	0.2	0.7	**1.2**
44	Theater	*-3.3*	-1.3	-1.6	-1.9	-1.5	-1.2	-1.2
47	Read magazines	-0.6	0.3	-0.2	0.0	-0.2	-0.2	0.3
48	Art exhibits	*-3.3*	-1.6	-2.0	-1.8	-2.1	-1.5	-1.7
60	Dinning out	*-1.3*	-0.5	-0.6	-0.7	-0.8	-0.6	-0.3
61	Gardening	-0.7	-0.3	-1.0	-0.8	-0.2	-0.8	*-1.3*
45	Beach or countryside	-0.7	-0.4	-0.4	-0.3	-0.4	-0.3	0.0
49	Watch soccer games	-2.1	-2.1	-1.6	-1.6	-2.0	-1.4	-1.5
52	Camping	-2.4	-2.1	-1.1	-1.6	-2.0	-1.3	-1.2
55	Visit relatives	-0.4	0.0	-0.2	-0.1	-0.1	-0.2	-0.3
56	Listen to the radio	0.6	1.0	1.2	0.7	1.0	1.5	1.5
57	Fix things at home	-0.5	-0.3	-0.4	-0.1	-0.3	-0.4	-0.8
62	Watch television	1.0	0.8	0.7	0.6	0.9	1.1	0.8
67	Work on artcrafts	-1.0	-0.4	-1.1	-0.7	-0.9	-0.6	-1.0
69	Church	-0.5	-0.3	-1.1	-0.7	-0.5	-0.5	-0.8

Table 12.2: Activities, Interests and Opinions Toward Fashion by Segment (adapted from Kamakura and Wedel 1995)

		SEGMENTS						
ITEM	ATTITUDE TOWARDS FASHION	1	2	3	4	5	6	7
40	Rarely stay home at night	-1.2	-1.3	0.2	-0.7	-0.5	0.6	0.5
42	I like to have people around	0.5	0.5	0.8	-0.1	0.6	2.0	1.4
43	I like parties with music and chatting	0.2	0.5	1.4	-0.1	0.7	2.3	2.2
4	I know if someone is important from the clothes	-0.1	-0.8	-1.8	-0.6	0.3	0.6	-0.7
9	Others treat me better when I dress-up	0.2	0.5	-0.6	0.2	0.5	1.3	0.3
12	Feel more comfortable with others when I dress-up	0.6	1.2	-0.1	-0.0	0.9	2.7	1.0
21	Only those who can't afford repeat the same clothes over and over	1.3	-1.0	-0.6	-0.4	0.4	0.7	-0.7
28	Know financial status of a person from the way of dressing	-0.7	-0.7	-1.2	-1.0	0.5	0.7	-0.8
5	Cannot wear anything not adequate to my a	0.4	1.0	-2.3	-0.4	-0.0	1.1	0.2
19	Be e buying clothing, I verify whether it matches my age	1.2	1.9	-1.0	-0.1	0.1	3.2	0.6
24	Choose clothes according to my age	1.1	1.1	-1.1	-0.4	-0.1	2.1	0.4
6	Extravagant clothes look good on me	-2.0	-3.2	-2.0	-1.8	-0.3	-0.5	-1.3
11	Most important is to dress exuberantly	-2.5	-4.7	-2.6	-2.6	-0.3	-0.6	-1.6
22	I dress to be appreciated by the opposite sex	0.1	0.7	0.5	-0.1	0.6	2.0	1.7
33	Prefer insinuating clothes	-1.9	-2.6	-1.5	-1.7	0.1	-0.2	-0.0
37	Like clothes with a touch of sensuality	-0.8	-1.2	-0.6	-1.2	0.6	0.5	0.6
39	Prefer clothes that are attractive to opposite sex	0.0	-0.1	0.3	-0.7	0.6	1.3	1.3
7	Friends buy clothes as soon as they see me wearing	-2.1	-1.7	-3.7	-1.6	-0.5	0.1	-1.1
10	Others follow me in their way to dress	-2.1	-1.9	-2.9	-1.8	-0.1	0.1	-1.2
13	Among my friends I know most about fashion	-2.0	-1.6	-2.9	-2.0	-0.2	-0.1	-1.1

Table 12.2 - continued

ITEM	ATTITUDE TOWARDS FASHION	SEGMENTS						
		1	2	3	4	5	6	7
20	Friends wait until I wear before adopting new fashion	-2.0	-3.0	-4.1	-1.9	**-0.4**	**0.0**	-1.6
38	Avoid wearing what everyone is using	*-1.3*	-0.6	-2.1	-0.7	**-0.2**	**0.5**	-0.7
41	I am always trying new products	*-1.3*	*-1.0*	-1.1	-1.2	-0.2	**0.9**	**0.2**
8	Most important is how resistant the fabric is	**2.2**	1.1	1.7	0.6	0.7	**1.9**	0.7
17	Cannot have clothes that require special care	0.3	*-0.6*	**0.5**	-0.2	-0.1	**0.5**	-0.4
18	Only buy clothes that will last long	**2.6**	0.6	0.5	*0.1*	-0.2	**1.4**	*0.1*
26	Only buy clothes that do not require special care	**0.6**	-0.3	**0.5**	-0.6	-0.1	**0.6**	-0.9
27	Choose clothes that will last long	**2.4**	0.7	0.8	*-0.1*	0.2	**1.4**	0.3
14	I won't wear anything uncomfortable	1.2	**3.8**	2.1	*1.0*	0.5	*1.0*	1.1
25	Will not wear anything too tight	*0.5*	**2.7**	**1.6**	0.9	0.3	1.2	0.8
34	Will not wear uncomfortable clothes even if lookgood	1.0	**1.8**	1.1	-0.2	0.2	**1.5**	0.4
16	In dressing, I do not care what others think	0.4	1.5	**2.9**	0.6	0.3	1.3	0.7
23	Only go to places where I can dress as I wish	0.4	*-0.9*	1.3	*-0.8*	0.3	**1.2**	-0.2
36	No matter where I go I dress the way I want	**1.3**	*-1.0*	1.6	*-0.3*	0.4	0.6	0.1
30	Dress to cheer up	1.1	**1.9**	0.9	*0.1*	1.1	**2.1**	1.6
32	Clothes help me feel better with myself	1.4	1.3	0.9	0.2	0.9	**2.2**	1.3
35	Dress to improve my self image	0.8	**0.9**	0.0	-0.8	0.5	**1.4**	0.6
29	Like the more classic fashion	0.0	**0.8**	*-1.0*	-0.2	0.1	**0.4**	-0.4
31	Do not feel comfortable with formal wear	**0.7**	*-0.3*	0.3	*-0.3*	0.1	**0.5**	-0.2
15	Think I spend more than I should on fashion	*-1.8*	*-1.9*	*-1.6*	*-1.5*	**-0.2**	**0.4**	0.1

The Tailored Interviewing Procedure

The tailored interviewing procedure was applied to the holdout sample of 100 respondents to test and validate the procedure. The main purpose of a tailored interview in market segmentation is the classification of consumers into segments that have been defined previously (at the calibration stage) by using a set of items that have been previously calibrated. The procedure is tested on a holdout sample because it allows for a comparison with alternative (e.g., random) selection rules, while the "true" classification of respondents (based on all questions) is known. The item pool consists of the parameters for the 73 socio-demographic and life-style items.

Characteristics of the Tailored Interview

In testing the item-selection strategy, the random selection of items was used as a benchmark. Application of the stopping rule led to use of 1600 answers from the 100 holdout respondents instead of the 7300 answers that would have been obtained if all questions for all respondents had been used, an overall reduction of 78% in the total length of the life-style questionnaire. On average, 16 items were used for each respondent, whereas an average of 36 items were used when the item selection was random. Hence, the decrease in the number of items selected was more than two-fold. Figure 12.3 displays the number of items selected in the tailored interview for each of the 100 respondents. The number of items selected is rather skewed. On one extreme, 65 respondents were presented 10 or fewer items; on the other extreme, nine respondents had to go through more than 70 items. This contrasts sharply with the number of items required when they were chosen at random. Figure 12.4 shows that with the random item selection procedure, more than 70 items were used for 18 of the interviews, whereas 10 or fewer items were used for only five of the interviews.

Part 3 Special topics in market segmentation

Figure 12.3: Number of Items Selected in the Tailored Interview until p>0.99 (Kamakura and Wedel 1995)

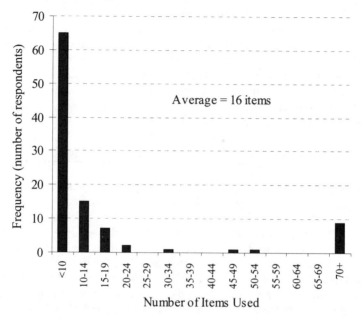

Figure 12.4: Number of Items Selected until p>0.99 with Random Item Selection (Kamakura and Wedel 1995)

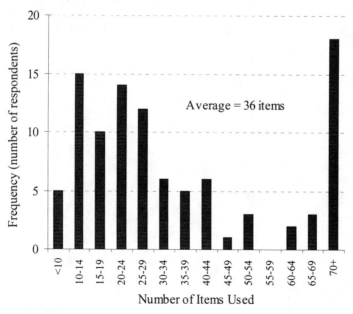

Quality of the Classification

An important question is whether tailored interviewing leads to an efficient classification of respondents into life-style segments. We measure the "quality" of classification as the entropy of the resulting posterior membership probabilities. Here, all items are used for each respondent. The average entropy of classification (computed across all 100 respondents) obtained for each number of selected items is displayed in Figure 12.4 along with its 95% confidence interval. For comparison, the Entropy for a random selection of the items is also depicted.

As one would expect, Figure 12.5 shows that the degree of uncertainty tends to decrease as the interview progresses, because information about the respondent is being gathered with each question. Fluctuations can be attributed to response errors and to numerical inaccuracies. Figure 12.5 also shows that the selection of the most informative item is much more efficient than a random selection of items. The tailored interview procedure approaches the lower-limit of classification uncertainty very early in the interview (after about 10 to 15 questions). The uncertainty about the respondents' classification in the life-style segments is not reduced substantially after that number of items. In comparison, random item selection reaches a lower limit of uncertainty after about 35 to 40 items have been selected in the interview, which is a three- to fourfold increase relative to the proposed tailored interview procedure.

When the stopping rule is employed, despite the considerable reduction of the number of items used in the tailored interviews, respondents are assigned to the segments with a very low degree of uncertainty. The average entropy of classification obtained at the end of the tailored interviews across all 100 respondents was equal to 0.01 (representing 1.0% of the maximum entropy), on the basis of 16 items on average, which is close to the minimum attainable. The average entropy of classification attained with random item selection was 0.03 (3.0% of maximum entropy) on the basis of 36 items on average.

The comparison of the entropy of classification at each stage of the interview does not provide any information about the accuracy of the classification of each respondent to the life-style segments. Figure 12.6 shows the number of respondents at each stage of the tailored interview that were classified to the same segment as they were on the basis of all items.

Figure 12.6 clearly shows that the selection of the most informative item leads to a good agreement with the final classification after the 16th item. After 16 items, more than 70% of the respondents are assigned to the same segment as they were on the basis of all items. When the 99% stopping rule was applied in the tailored interview, the tailored classification agreed with the one obtained with all items for 73 of the 100 respondents. With random item selection these percentages are a little over 50%, and only reach the 70% agreement level after use of more than 55 items. One can see, however that the marginal improvement obtained with the selection of the most informative item decreases as the number of items selected becomes larger.

Figure 12.5: Entropy of Classification of Each Stage of the Interview (Kamakura and Wedel 1995)

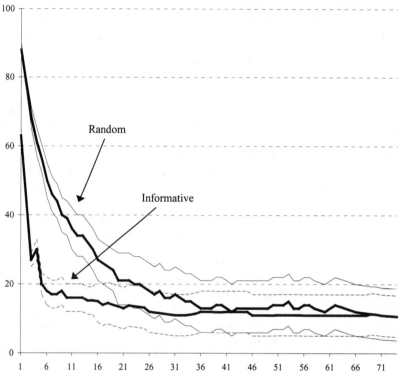

Table 12.3 reports the percentage of respondents correctly classified (relative to the full interview) for the tailored interview, the interview with the 24 items chosen by discriminant analysis to best discriminate across the seven segments, and the random selections of 16 and 24 items. The results show that, overall, the tailored interviews produced results closer to those of full interviews than the larger set of 24 items selected by discriminant analysis or the random selections of 16 and 24 items.

Figure 12.6: Number of Respondents Correctly Classified at Each Stage of the Interview (Kamakura and Wedel 1995)

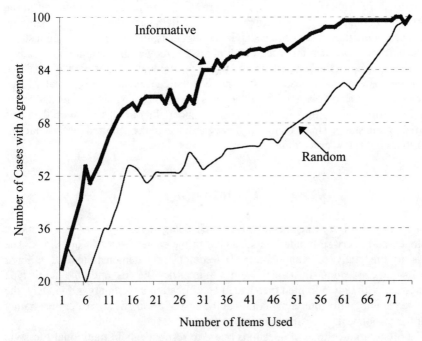

Table 12.3: Percentage of Cases Correctly Classified

MODEL	SEG 1	SEG 2	SEG 3	SEG 4	SEG 5	SEG 6	SEG 7	Total
Tailored Interview	67%	87%	83%	80%	71%	100%	50%	73%
Discriminant Analysis	56%	73%	75%	40%	65%	100%	59%	62%
Random (16 items)	33%	67%	50%	45%	59%	60%	50%	56%
Random (24 items)	33%	60%	42%	55%	65%	60%	73%	59%

One must note that the evaluation of the tailored interviewing process was done in comparison with the final classification based on all items. That final classification is also an estimate, subject to the estimation errors of item parameters incurred in the model calibration. Also, if items are present in the item bank that do not contain any information about the actual membership in the segments, those items may diminish the accuracy of the classification based on all items. Moreover, the validation procedure employed is based on the complete interviews of the holdout subjects. Because tailored interviewing is expected to lead to less respondent fatigue and boredom, actual tailored interviews, in which only a fraction of the items are used, are expected to lead to lower response errors. This again will lead to a lower uncertainty of classification than was reported in this study. Those factors will have attenuated the reported performance of the tailored interview procedure.

Conclusion

In traditional life-style segmentation research, a large number of life-style items are developed and every respondent is subjected to the same battery of questions. The results of the study of Kamakura and Wedel (1995) demonstrate that tailored interviewing can lead to substantial savings in interviewing cost and time in this type of research, with only a limited increase in classification uncertainty. Obviously, the decreased length of the interviews comes at the cost of a somewhat greater uncertainty about the respondents' classification into segments.

The often large volume of questions posed to respondents in traditional life-style segmentation research necessitates the use of mail or personal interviews, and the personal interviews are often considered preferable to monitor and control respondent boredom and fatigue. Apart from the major problems of boredom and fatigue that may seriously affect the quality of the data obtained, lengthy life-style questionnaires may have problems of non-response and partial non-response, as shown in the pilot study reported above. Non-response affects the representativeness of the sample with respect to particular variables of interest. Partial non-response reduces the number of available items that might be informative in the allocation of the respondent to the underlying segments. An important benefit of the tailored interview procedure is therefore not only the reduction of direct costs, but also the potential improvement in the quality of the data obtained.

Because of the reduction of questionnaire length, the tailored interviewing approach lends itself to computer-assisted telephone interviews CATI), which provide a further substantial reduction of the costs incurred with data collection compared to traditional survey methods, as long as a previously calibrated set of items is available. With the growing popularity of CATI systems, the availability of infrastructure no longer hampers the administration of tailored interviews.

Finally, the tailored segmentation approach discussed in this chapter is applicable to other forms of segmentation, as long as the segmentation bases have been previously calibrated. That requirement might restrict the application of the approach, but makes it well suited for marketing research firms that identify a certain typology of segments (e.g., values and life-styles) based on a major study, and then use the calibrated typology in subsequent market segmentation studies for a variety of clients. An interesting and valuable extension of the approach would involve simultaneous calibration of the model and tailored interviewing. Currently that is not feasible because of estimation problems; estimates of the mixture model would be inconsistent at earlier stages of data collection, when data would be available only from a limited sample.

13
MODEL-BASED SEGMENTATION USING STRUCTURAL EQUATION MODELS

We describe a very general approach to response-based segmentation. Market segments are formed in the context of model structures in which multiple dependent variables are involved, and both dependent and independent variables are potentially measured with error: structural equation models. The method subsumes several specialized models, such as mixtures of simultaneous equations and confirmatory factor analysis. We start with a brief introduction to structural equation models.

Introduction to Structural Equation Models

Many excellent textbooks have been written on structural equation models; see for example Bollen (1989) or, for a concise introduction, Long (1983). Here, we provide only an outline of the models to provide a basis for the treatment of the segmentation approach described later. A chapter on structural equation models is not complete without mentioning the important work by Jöreskog (1973, 1978), who laid the foundations of this extremely general form of model and developed the computer program LISREL, the name of which has become almost synonymous with the models.

Structural equation models are used to test the relationship among latent (unobservable) and observable variables through the analysis of the covariance among observed variables. A common application to be discussed in more detail later is the measurement of customer satisfaction (cf. Jedidi, Jagpal and DeSarbo 1997a). In such studies, a number of latent satisfaction variables are hypothesized to underlie consumer satisfaction, which cannot be observed directly. To assess the latent satisfaction factors, a relatively large number of questions are posed to a sample of customers, who indicate their potential satisfaction with the product or service (a home shopping club's services in the application to be discussed next). Such questions may pertain to satisfaction with delivery charge, overall price, the product range, brands, and so on, assessed on 10-point scales. Table 13.1 provides an overview of the variables used in the application.

The relationships among the measured variables are characterized by the covariances among them, and the covariances form the basis for estimating a structural equation model that describes the relations in question according to some set of hypotheses. Therefore, working with structural equations requires a reorientation relative to well-known statistical methods (such as regression analysis) that focus on individual observations. In the population, the covariances among the variables are contained in the matrix Σ.

Part 3 Special topics in market segmentation

Table 13.1: Latent Satisfaction Variables and Their Indicators (Jedidi et al. 1997a)

Satisfaction factor		Manifest indicator variables, satisfaction with:
Overall	(η_1)	general satisfaction (y_1), closeness to ideal outlet (y_2), conformance (y_3), care (y_4)
Price	(ξ_1)	delivery charge (x_1), overall price (x_2)
Variety	(ξ_2)	product range (x_3), brands (x_4), fashion (x_5), exclusivity (x_6)
Catalog	(ξ_3)	overall catalog (x_7), picture quality (x_8), information (x_9), style (x_{10})
Promotion (ξ_4)		letters (x_{11}), gifts (x_{12})
Delivery	(ξ_5)	availability (x_{13}), shipment (x_{14}), speed (x_{15})
Credit	(ξ_6)	procedure (x_{16}), reminders (x_{17})
Quality	(ξ_7)	general quality (x_{18}), product guarantees (x_{19})

The specific structural equation model that is postulated attempts to reproduce the covariances among the observed variables as accurately as possible from a set of parameters (collected in the vector $\boldsymbol{\theta}$), so that the fundamental hypothesis is: $\boldsymbol{\Sigma} = \boldsymbol{\Sigma}(\boldsymbol{\theta})$. The sample covariances among observed variables provide the sufficient statistics for the model parameters. $\boldsymbol{\Sigma}(\boldsymbol{\theta})$ is called the implied covariance matrix. The hypothesized relations among the variables are depicted in a path diagram. The path diagram for the satisfaction example is shown in Figure 13.1, where we depict only the relations between latent variables in the model.

Two types of observed variables are distinguished: indicators of exogenous and endogenous latent variables. Those indicator variables are denoted by **x** and **y**, respectively. An example of an indicator for an endogenous variable in the application mentioned above is general satisfaction (y_1); an example of an exogenous indicator is satisfaction with letter (x_{11}).

Figure 13.1: Path Diagram for latent Variables in the Satisfaction Study (Jedidi et al. 1997a)

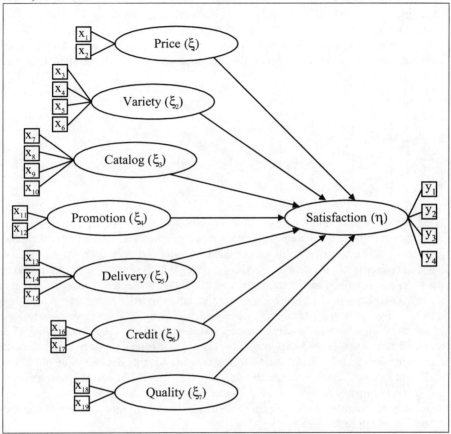

The structural equation model comprises two sub-models. The first is called the *measurement model* and relates the observed indicators (**x** and **y**) to a set of unobserved or latent variables. In the example above, satisfaction with variety is not observed directly, but is measured by the indicators shown in the table. This part of the model is also called a confirmatory factor model when considered in isolation. The measurement models for the endogenous and exogenous indicator variables are formulated in their general form respectively as:

$$y = \Lambda_y \eta + \varepsilon \tag{13.1}$$

and

$$x = \Lambda_x \xi + \delta. \tag{13.2}$$

Here, $\boldsymbol{\eta}$ is a (m × 1) vector containing the latent endogenous variables (i.e., the variables that are explained within the model), $\boldsymbol{\Lambda}_y$ is the (p × n) matrix of factor loadings, showing which indicator variable loads on which factor, and $\boldsymbol{\delta}$ is a vector of error terms with expectation zero and uncorrelated with the vector of latent variables. The equation for **x** is similar, with q manifest and n latent variables. In the customer satisfaction application, for example, the measurement model for overall satisfaction is:

$$\begin{bmatrix} y_1 \\ y_2 \\ y_3 \\ y_4 \end{bmatrix} = \begin{bmatrix} \lambda_{11}^x \\ \lambda_{21}^x \\ \lambda_{31}^x \\ \lambda_{41}^x \end{bmatrix} [\eta_1] + \begin{bmatrix} \varepsilon_1 \\ \varepsilon_2 \\ \varepsilon_3 \\ \varepsilon_4 \end{bmatrix}. \tag{13.3}$$

A similar set of equations determines the measurement model for the X-variates. The error terms in both measurement models may be correlated, with the covariance matrices of the vectors of error terms denoted by $\boldsymbol{\Theta}_\varepsilon$ and $\boldsymbol{\Theta}_\delta$ respectively. $\boldsymbol{\Phi}$ denotes the covariance matrix of the latent exogenous variables $\boldsymbol{\xi}$. (Covariances of the endogenous variables are explained by the structural part of the model described below). To make the measurement model identified, the latent variables must be assigned a scale. One way of doing that is to arbitrarily fix one of the loadings for each latent variable to one, so the latent variable has the same scale as that respective indicator. In equation 13.3 above, for example, λ_1 could be set to one. There are several necessary and sufficient conditions for (global) identification of the measurement model. An example is the sufficient three-indicator rule, which states that there should be one nonzero element per row in the matrix $\boldsymbol{\Lambda}$, three or more indicators per factor, and $\boldsymbol{\Theta}$ should be diagonal. We will not elaborate on identification rules here, but refer to the structural equation modeling literature for details.

The second part of the model, the *structural equation model*, specifies the relationships between exogenous and endogenous (latent) variables:

$$\eta = \mathbf{B}\,\eta + \boldsymbol{\Gamma}\,\xi + \zeta. \tag{13.4}$$

The (m × m) matrix **B** specifies the relationships among the m latent endogenous variables. Its diagonal equals zero (as the endogenous variables cannot affect themselves). If **B** is a lower triangular matrix, there are no reciprocal causal effects (e.g., variable η_1 influences variable η_2 but not vice versa), and the structural model is said to be *recursive*. The (m × n) matrix $\boldsymbol{\Gamma}$ captures the effects of the exogenous variables on the endogenous variables, and ζ is a vector with residuals with expectation zero and uncorrelated with the endogenous and exogenous latent variables. The error terms in ζ may be correlated and have a covariance matrix denoted by $\boldsymbol{\Psi}$. For the customer satisfaction example, the structural equation is:

Chapter 13 Model-based segmentation using structural equation models

$$[\eta_1] = [0][\eta_1] + [\gamma_{11} \quad \gamma_{12} \quad \gamma_{13} \quad \gamma_{14} \quad \gamma_{15} \quad \gamma_{16} \quad \gamma_{17}] \begin{bmatrix} \xi_1 \\ \xi_2 \\ \xi_3 \\ \xi_4 \\ \xi_5 \\ \xi_6 \\ \xi_7 \end{bmatrix} + \begin{bmatrix} \zeta_1 \\ \zeta_2 \\ \zeta_3 \\ \zeta_4 \\ \zeta_5 \\ \zeta_6 \\ \zeta_7 \end{bmatrix} \quad (13.5)$$

There are several necessary and sufficient conditions for the identification of the structural part of the model. Examples of sufficient rules are the null-rule, which states that the matrix **B** should contain only zeros as in the example above, and the recursive rule, which states that **B** can be written as a lower triangular matrix and **Ψ** should be diagonal. An important rule for identification of the entire structural equation model is the two-step rule, which states that if the measurement and structural equation models are identified separately, the whole model is identified. For details we refer to the literature.

The above form is a very general one, and encompasses a wide variety of specific models. For example, the confirmatory factor analysis model arises if all parameters in equation 13.4 are set to zero. Likewise, second-order factor models are a special instance of the general model outlined above. Simultaneous equation models arise as a special case when there are no latent variables, and each manifest variable is set identical to a corresponding latent variable. Other well-known models that arise as special cases are panel data models, errors-in-variables models and seemingly unrelated regression models, among others.

The application of structural equation models in any form requires the use of efficient numerical methods to estimate all parameters in equations 13.1 through 13.5 simultaneously. A major breakthrough has been made by Jöreskog, who developed the flexible LISREL program, that made such estimation routines commonly available. Parameters can be estimated by optimizing a variety of fitting functions, including maximum likelihood (ML), which estimate the parameters in such a way that some "distance" measure between the sample and implied covariance matrices is minimal. The ML approach requires all variables to be normally distributed in order for significance tests of the parameter estimates based on asymptotic standard errors to be valid. Further, the approach makes likelihood ratio tests of nested models possible. A variety of criteria are available to judge the goodness of fit of the estimated models, including the LR-chi-square test for goodness of fit, and goodness-of fit (GFI) and adjusted goodness of fit (AGFI) indices, which measure the relative amount of the covariance predicted by the model. In addition, a host of incremental fit indices can be calculated from the chi-square statistics of the model in question and a baseline model, and their degrees of freedom. For details we refer to the works cited in the beginning of this chapter.

A-Priori Segmentation Approach

Typically, heterogeneity in structural equation models has been addressed by assuming that consumers can be assigned to segments *a priori*, on the basis of, for example, a demographic variable. In the customer satisfaction example above, one may expect that determinants of overall satisfaction differ according to the usage frequency of services: heavy or light user segments may be distinguished. Questions that arise are: Do the indicator variables measure the same latent constructs across the two groups? Are the mean satisfaction levels the same for heavy and light users? And, are the determinants of overall satisfaction the same for these two groups? When the parameter values differ across groups, the aggregate-level model is misspecified and one runs the risk of drawing erroneous conclusions. The procedure used to overcome those problems is called *multi group structural equation modeling* (Jöreskog 1971). Here, the structural equation model is specified for S segments, assumed to be known a priori:

$$y^{(s)} = v_y^{(s)} + \Lambda_y^{(s)} \eta^{(s)} + \varepsilon^{(s)}, \quad s = 1,...,S \tag{13.6}$$

$$x^{(s)} = v_x^{(s)} + \Lambda_x^{(s)} \xi^{(s)} + \delta^{(s)}, \quad s = 1,...,S \tag{13.7}$$

$$\eta^{(s)} = \alpha^{(s)} + B^{(s)} \eta^{(s)} + \Gamma^{(s)} \xi^{(s)} + \zeta^{(s)}, \quad s = 1,...,S. \tag{13.8}$$

Equations 13.6 through 13.8 are the same as equations 13.1 through 13.5, but here they are specified within each of the S groups separately. In addition, intercepts are included; for example, $v_y^{(s)}$ is a ($p \times 1$) vector containing the means of **y**. Similarly, $\boldsymbol{\alpha}^{(s)}$ is a vector of intercepts in the structural part of the model, and $E(\xi^{(s)}) = \boldsymbol{\tau}^{(s)}$ is the expectation of the latent exogenous variables. In many applications the means $v_y^{(s)}$, $v_x^{(s)}$, $\boldsymbol{\alpha}^{(s)}$ and $\boldsymbol{\tau}^{(s)}$ are not estimated; the indicator variables are then assumed to be centered. However, the inclusion of such means may be of interest, especially if segments are to be compared, because segments may differ in the mean values of the variables.

In the example above s = 1,2 could denote heavy and light users. The covariance matrices are also indexed by s in this model: $\Theta_\delta^{(s)}$, $\Theta_\varepsilon^{(s)}$, $\Phi^{(s)}$ and $\Psi^{(s)}$. In principle, models of an entirely different form can be postulated for the S segments, for example, a model with seven latent satisfaction factors for the heavy users and a model with only three latent satisfaction factors for the light users. However, in most applications researchers have assumed that the form of the models is the same across the segments, while allowing (certain sets of) parameter values to differ across the segments. By employing a general specification of the model across the segments and testing whether certain sets of parameter values are equal to zero in either of the segments, models of different forms may actually be obtained. Any of the above sets of parameters can be allowed to differ across the segments. A possible hierarchy of hypotheses may be used to investigate the degree to which the structural equation model differs among segments: $H_v \rightarrow H_{v\Lambda} \rightarrow H_{v\Lambda B\Gamma} \rightarrow H_{v\Lambda B\Gamma\Theta} \rightarrow H_{v\Lambda B\Gamma\Theta\Psi} \rightarrow H_{v\Lambda B\Gamma\Theta\Psi\Phi}$. Here, H_v represents the hypothesis: $v_x^{(1)} = ... = v_x^{(S)} = v_x$, and $v_y^{(1)} = ... = v_y^{(S)} = v_y$ and so on. The model is estimated to the S sample covariance matrices of the segments. Because the hierarchy of hypotheses presents a series of nested models, the successive hypotheses can be tested with LR tests.

Post Hoc Segmentation Approach

As already outlined in chapter 3, a limitation of the a-priori approach is that heterogeneity often is not captured adequately by demographic and psychographic variables. However, in many situations, substantive theory on the variables causing heterogeneity is either not available or incomplete. Then, such groups should be estimated from the data *post hoc*. Some researchers have used a two-step procedure, in which the sample is first partitioned into segments by means of some cluster algorithm (e.g., K-means) applied to all variables, and then a structural equation model is brought to the data within each of the resulting segments. That procedure is sub-optimal in general because, without a statistical model, clustering methods such as K-means are not a priori believable to adequately identify the cluster structure in the population. The problem is compounded when the data contain measurement error, which may seriously affect cluster solutions, as discussed in chapter 5. Model-based clustering procedures in the context of simultaneous equation models and structural equation models have been proposed by Jedidi, et al. (1996), by Jedidi, Jagpal and DeSarbo (1997a,b) and by Dolan and Van der Maas (1998). These approaches extend the a-priori approach for addressing heterogeneity discussed above to a post-hoc approach in which the heterogeneous groups are identified simultaneously with the structural equation model, using the mixture model framework. Before discussing that approach, we provide the consumer satisfaction application to illustrate it.

Application to Customer Satisfaction

The data for the application were collected by telephone survey of 1564 customers for a customer satisfaction study conducted for a home-shopping club in Europe. We summarize the results presented by Jedidi, Jagpal and DeSarbo (1997a). Overall satisfaction (η_1) was hypothesized to be determined by seven latent predictors, as shown in Table 13.1 and Figure 13.1.

The approach used was to simultaneously estimate a number of unobserved segments, and within each of those segments the structural equation model depicted in Figure 13.1. The method allows one to restrict the measurement model to be identical across segments. Models with various numbers of segments and with and without the restriction of identical measurement model parameters were estimated. The CAIC statistic pointed to an $S = 3$-segment solution, in which the measurement part of the model was identical across latent segments. The parameter estimates showed that all manifest indicators loaded significantly on the respective latent constructs. The loadings are not reported here, but Table 13.2 shows the mean levels for the satisfaction constructs across segments (the values for segment 1 are not given, as they are restricted to zero for reasons of model identification). The results show that segment 3 has the highest and segment 2 the lowest mean satisfaction level across all seven constructs and the mean differences are significant across the segments.

Table 13.2: Mean Factor Scores in Three Segments (adapted from Jedidi et al. 1997a)

Factor	Segment 2	Segment 3
Overall satisfaction	-0.6	0.85
Price	-2.53*	2.77*
Variety	-1.27*	1.84*
Catalog	-1.56*	1.87*
Promotion	-0.76*	0.62*
Delivery	-0.86*	1.07*
Credit	-0.83*	0.58*
Quality	-2.97*	3.96*

* $p < 0.01$.

Table 13.3 gives the structural parameter estimates for the three segments. Delivery, credit, and quality satisfaction influence overall satisfaction in segment 1. In segment 2 those factors also influence overall satisfaction, but catalog and promotion satisfaction have significant effects. Finally, in segment 3 only delivery and credit satisfaction have an effect on overall satisfaction. The results thus show that although the confirmatory measurement model for overall satisfaction and the satisfaction components was homogeneous across segments, the structural model was not, and different segments of customers were present in which overall satisfaction is differentially influenced by satisfaction components. Estimating an aggregate structural equation model across all three segments yields incorrect results, as that model indicates significant effects of all satisfaction components, with exception of price (the results are not shown here). The implications of the findings are that managers should target the segments differentially to optimize overall satisfaction for all customers. Targeting the segments would be facilitated by profiling the segments with consumer background variables. However, in the application such variables were not available.

Chapter 13 Model-based segmentation using structural equation models

Table 13.3: Estimates of Structural Parameters in Three Segments (adapted from Jedidi et al. 1997a)

Factor	Segment 1 (43.7%)	Segment 2 (39.0%)	Segment 3 (17.3%)
α	0	0.830	-0.171
Price	0.093	-0.006	0.011
Variety	0.031	0.051	0.061
Catalog	0.058	0.275*	0.115
Promotion	0.075	0.079*	0.000
Delivery	0.183*	0.188*	0.163*
Credit	0.167*	0.119*	0.163*
Quality	0.331*	0.213*	0.099

* $p < 0.01$.

The Mixture of Structural Equations Model

The post-hoc mixture segmentation approach of Jedidi, Jagpal and DeSarbo (1997a,b) starts from the existence of S unobserved segments with prior probabilities $\pi_1, ..., \pi_S$, as used throughout this book. Conditional upon segment membership, the model structure as described by equations 13.6 to 13.8 is assumed; that is for each unobserved segment a structural equation model holds, where the different models have the same basic form but possibly different parameter values. Consider the general model defined by the three equations. Then, the mean vector of the observed [(p+q) × 1] vector of indicator variables (x',y')' conditional upon segment s is:

$$\mu_s = \begin{bmatrix} v_y^{(s)} + \Lambda_y^{(s)} B^{(s)-1}(\alpha^{(s)} + \Gamma^{(s)} \tau_\xi^{(s)}) \\ v_x^{(s)} + \Lambda_x^{(s)} \tau_\xi^{(s)} \end{bmatrix}. \quad (13.9)$$

Its [(p + q) × (p + q)] conditional covariance matrix is given by:

$$\Sigma_s = \begin{bmatrix} \Lambda_y^{(s)} B^{(s)-1}(\Gamma^{(s)} \Phi^{(s)} \Gamma^{(s)'} + \Psi^{(s)}) B^{(s)-1'} \Lambda_y^{(s)'} + \Theta_\varepsilon^{(s)} & \Lambda_y^{(s)} B^{(s)-1} \Gamma^{(s)} \Phi^{(s)} \Lambda_x^{(s)} \\ \Lambda_x^{(s)} \Phi^{(s)} \Gamma^{(s)'} B^{(s)-1'} \Lambda_y^{(s)'} & \Lambda_x^{(s)} \Phi^{(s)} \Lambda_x^{(s)'} + \Theta_\delta^{(s)} \end{bmatrix}. \quad (13.10)$$

The model can be estimated by maximum likelihood. It is assumed that conditional upon segment s, the vector of observed indicators has a multivariate normal distribution, with mean vector and covariance matrix as shown in equations 13.9 and 13.10. Then, the likelihood is:

$$L(y, x | \mu, \Sigma) = \prod_{n=1}^{N} \sum_{s=1}^{S} \pi_s f_s(y, x | \mu_s, \Sigma_s). \qquad (13.11)$$

The likelihood needs to be maximized with respect to the free parameters given the data and a pre-specified number of segments S. The maximum likelihood estimates of μ_s and Σ_s are functions of the postulated measurement and structural model components, as shown in equations 13.9 and 13.10. Note that in estimating the model, the determinant of Σ_s needs to be positive for all s, for which a minimum sample size of $(p + q)(p + q + 1)/2$ in each segment is needed. The likelihood is maximized by using the EM algorithm, as described in chapters 6 and 7. Here, each M-step comprises a complex numerical maximization process. For a description of the M-step we refer to Jedidi, Jagpal and DeSarbo (1997a).

In estimating the mixture of structural equations model, one assumes that it is identified. The authors show that it is necessary to assume a specific distribution for the data for identification. Suppose the a priori multi group structural equation model is identified, then the finite mixture structural equation model is identified provided that the data within the unknown segments follow a multivariate normal distribution. Information criteria such as BIC and CAIC can be used to select the appropriate number of segments. Segment separation is investigated by using the entropy statistic (see chapter 6).

Special Cases of the Model

The mixture of structural equations described above is very general and subsumes a variety of models as special cases. For example, if the structural model parameters in equation 13.8 and the measurement model parameters in equation 13.6 are zero, the model reduces to a mixture of confirmatory factor models. If the model consists of equations 13.6 and 13.7, a higher order confirmatory factor model results. The latter model was applied to consumer satisfaction by Jedidi, Jagpal, and DeSarbo (1997a). The results of the application were discussed previously in this chapter. The authors also show that although the three consumer segments vary considerably in terms of importances they attach to the various dimensions of satisfaction, an aggregate-level analysis reveals that all satisfaction dimensions are important to all consumers.

If all measures are error-free and the corresponding parameters in equations 13.5 and 13.6 are set to zero, the system reduces to a mixture of simultaneous equations or multivariate regressions. Jedidi et al. (1996) applied the latter model to examine price-quality relationships, using a sample of 2404 business in the PIMS (Profit Impact of Marketing Strategy) database that contains four-year averages of data on business-market characteristics and competitive strategies. The endogenous variables in the model were relative perceived quality and relative price, and the model included several exogenous variables related to marketing strategy, market structure and internal structure. In the application an S = 3- segment solution appeared to be optimal. The results showed that among all three segments, higher quality brands tended to be higher priced, which implies that price conveys important information about quality. However, the interdependence between price and quality was much weaker in one of the three segments (15.7% of the firms). From an analysis of the posterior probabilities it appeared that businesses in the early stage of the PLC were most likely to belong to this segment. The segment also tended to consist of smaller numbers of competitors, and businesses that have relatively wide product and consumer scopes. In this application, the mixture approach was effective in revealing heterogeneity in the sample of businesses, which invalidates aggregate-level conclusions.

Analysis of Synthetic Data

To investigate the performance of the mixture of structural equations algorithm in recovering a known structure, Jedidi and co-authors performed two Monté Carlo studies. In the first study, synthetic data were generated for 1500 subjects according to a confirmatory one-factor model. Data were generated for S = 2 and S = 4 segments and p = 3 and p = 6 indicator variables. The loadings were generated randomly from U(0.5, 1.5); for each segment, one loading was set to unity for identification. The means of the exogenous variables were generated randomly from U(-s, s) for segment s to provide separation between the segments. Values chosen for the error variances were consistent with findings in behavioral research. Each dataset was analyzed five times with different sets of random starting values (each replication was generated with different sets of random parameter values). In addition, the models were estimated with S = 1,...,5, to investigate the recovery of the true number of segments.

The model performance criteria included GFI as a measure of goodness of fit, and correlations of the actual and estimated segment memberships (CORP) and measurement model parameters (CORM). In addition, Jedidi et al. (1997a) report the proportion of the cases in which the information criteria BIC and CAIC correctly identified the true number of segments. Table 13.4 replicates some of the results.

The results of the model selection criteria indicate that BIC is slightly superior to CAIC in the case of exactly identified models (the model with p = 3 indicators is exactly identified). For over identified models (p = 6), both criteria perform very well. The goodness of fit index (GFI) indicates that the fit of the mixture structural equation model is excellent. Measurement model recovery, as evidenced by the correlations between true and estimated parameters, is also very good. Segment membership recovery is quite good, and improves when the model is over identified. The results were compared with those of a standard multi group structural equation model in which segment memberships were known, and were found to be only slightly poorer than the results of that model (not shown here).

Table 13.4: Performance Measures in First Monté Carlo Study (Jedidi et al. 1997a)

Measures	S = 2, p = 3	S = 2, p = 6	S = 4, p = 3	S = 4, p = 6
GFI	0.988	0.995	0.998	0.990
CORM	0.996	0.998	0.995	0.996
CORP	0.890	0.931	0.884	0.903
CAIC	0.90	1.00	0.80	0.90
BIC	0.90	1.00	0.60	1.00

A second experiment was conducted that examined a more complicated nonrecursive model. Here the number of segments (S) was assumed to be known a priori and the study focused on the robustness of the procedure under violations of the assumption of multivariate normality. The correlations between exogenous constructs were drawn from U(-0.8, 0.8), the parameters in **B** and **Γ** were drawn from U(-1,1), and the values of variance parameters in the model were generated to conform to findings in behavioral research. Again, five replications with different parameter values were generated per treatment. The treatments consisted of different levels of skewness (0.0, 0.75 and 1.0) and kurtosis (0.0 and 2.75) and of the number of segments (2 and 4). Additional performance measures calculated in the study were the correlation of true and estimated structural model parameters (CORS) and the proportion of local optima (LOC). We summarize the results in Table 13.5.

The GFI seems to be quite stable, independent of violations of distributional assumptions and numbers of segments. For S = 2, the recovery of segment membership (CORP) seems to be rather insensitive to violations of distributional assumptions, but S = 4 models seem to be somewhat more sensitive. The measurement model seems to be quite robust for both S = 2 and S = 4: the correlation between true and estimated parameters is almost perfect. The structural part of the model for S = 2 also seems to be quite robust, but the structural parameters for the S = 4 model seem to be less accurate and more sensitive to skewness and kurtosis of the distribution of the data. Problems with local optima seem to be minor for S = 2, but more severe for S = 4. One can conclude that parameter recovery is in general satisfactory and that the methodology is reasonably robust, except when skewness and kurtosis are severe, but problems increase as the number of segments increases.

Table 13.5: Performance Measures in Second Monté Carlo Study (Jedidi et al. 1997a)

	S = 2		S = 4	
Skewness\Kurtosis	0/ 0	1/ 2.75	0/ 0	1/ 2.75
GFI	0.994	0.994	0.988	0.988
CORM	0.998	0.998	0.997	0.997
CORS	0.976	0.966	0.945	0.883
CORP	0.985	0.988	0.950	0.931
LOC	0.120	0.080	0.360	0.280

Conclusion

The mixtures of structural equations methodology is very general and subsumes a large number of other models, including mixtures of confirmatory factor models, higher order confirmatory factor models, simultaneous equations and seemingly unrelated regressions. We have shown that estimation of this class of models is possible by using the EM algorithm to maximize the likelihood. The results have strong implications for users of LISREL and related programs: aggregate analyses performed with these programs may be seriously misleading when there are significant differences in model structure across unobserved segments. In addition, two-stage procedures using, for example, K-means in the first step and a multi group structural equation model in the second step are not recommended to detect market segments. A potential caveat is that very large samples are needed to estimate these models. The methodology awaits further application in marketing research.

14
SEGMENTATION BASED ON PRODUCT DISSIMILARITY JUDGEMENTS

Dissimilarity judgements have been represented in the marketing and psychometric literatures as either spaces or trees. The dominant tree representations have been the ultrametric and the additive trees. We describe an extension of those approaches to accommodate differential points of view of consumer segments, by allowing for different segments that have different perceptions of the stimuli, represented as mixtures of spaces, or mixtures of trees, or mixtures of both.

Spatial Models

Perceptions are frequently studied using graphical representations of dissimilarity judgments of stimuli that either take the form of trees or spaces. The assumption underlying the analysis of dissimilarity judgements is that subjects compare the stimuli on the basis of a number of attributes. These attributes can be either discrete features or continuous dimensions (Tversky 1977), recovered through the analysis of dissimilarity judgments with models that represent them as a tree (Sattath and Tversky 1977) or as a space, respectively (Carroll and Green 1997). The choice between trees and spaces is based on a) prior theory on the attribute-types discerned by subjects for that particular type of stimuli, b) the basis of the relative fit of the two models, and/or c) diagnostic measures such as the skewness of the dissimilarity judgments (cf. Ghose 1998).

We start with explaining the spatial representation of paired comparison data. Such a spatial representation relates to the mixture unfolding models described in Chapter 8. Let i, j, k = 1,...,I denote stimuli. We assume a judgement process based continuous dimensions represented by a spatial MDS model. The data, d_{ijn}, is the dissimilarity judgement of stimuli i and j by subject n. Although many nonmetric representations have been proposed (e.g. Cox and Cox 1994) we deal with the stochastic nature of the respondents decision process by making distributional assumptions, which enables us to adopt the approach of maximum likelihood estimation. In particular we assume the $P = I(I-1)/2$ dissimilarity judgments for subject n to follow a normal distribution. Here δ_{ij} is the expected value of d_{ijn}, and σ^2 its variance.

It is assumed that the dissimilarities are produced by a T-dimensional spatial model where the location of stimulus i on dimension t is represented by x_{it}:

$$\delta_{ij} = \mu + \sum_{t=1}^{T} (x_{it} - x_{jt})^2 \;, \tag{14.1}$$

with μ an intercept term. Note that this representation is homogeneous: it is assumed that the parameters apply to the dissimilarity judgements of all subjects n. Because of the distribution assumptions on the observed dissimilarities, the model parameters can be estimated by maximum likelihood. The MDS model has TI associated parameters, but is invariant to centering, scaling and rotation, which subtracts T(T+1)/2 parameters, while there is one variance parameter, so that TI-T(T+1)/2+2 effective parameters are to be estimated.

We provide an example by analyzing the data published by Schiffman, Reynolds, and Young (1981, pp. 33-34). We derive the MDS space using maximum likelihood estimation. In a sensory experiment, 10 subjects (non-smokers, aged 18-21 years) tasted ten different brands of cola: 1. Diet Pepsi (DP), 2. RC Cola (RCC), 3. Yukon (Y), 4. Dr. Pepper (DRP), 5. Shasta (S), 6. Coca-Cola (CC), 7. Diet Dr. Pepper (DDP), 8. Tab (T), 9. Pepsi-Cola (PC), 10. Diet Rite (DR). Each subject provided 45 dissimilarity judgments by means of paired comparisons on a graphical anchored line-scale. The judgments were transcribed on a scale from 0-100 representing same (near 0), and different (near 100). For the sake of illustration, we restrict ourselves to a two-dimensional representation. Figure 14.1 presents the T=2 spatial representation of the ten cola brands. The horizontal dimension seems to separate the diet versus the regular brands: On the left hand side of the horizontal axis one observes Diet Dr. Pepper, Diet Pepsi, Tab, and Diet Rite, on the right hand side the regular brands Yukon, RC Cola, Pepsi, Shasta and Coca-Cola. Dr. Pepper seems to take a somewhat intermediate position. The vertical axis separates the Dr. Pepper brands at the top from the other brands, with Yukon taking an intermediate position. This is apparently due to the specific cherry-taste of this brand. Among the regular cola brands on the right hand side of the plot there seems to be fairly little distinction, among the diet colas Diet Dr. Pepper stands out. The two dimensions in the plot can thus be interpreted as a diet versus regular and a cherry taste dimension, respectively. These attributes are however discrete features, rather than continuous dimensions, reason why we turn our attention to a tree-representation of these data.

Tree models

Tree models that have been proposed include traditional hierarchical clustering, ultrametric trees, and additive trees, as briefly discussed in Chapter 5. Tree models of dissimilarity judgements represent a set of brands as nodes in a connected, undirected graph. The proximity between the brands is represented in the tree by the height distance, or path-length, between the nodes. The two dominant types of tree structures heeded by psychometricians are ultrametric and additive trees. Those are characterized respectively by the ultrametric inequality -defined for all triples of stimuli and the additive inequality -defined for all quadruples of stimuli, as explained in more detail below. One may interpret ultrametric trees in terms of a common features model of dissimilarity, in which the features have a hierarchical structure, additive trees may be interpreted in terms of a distinctive-features model (Corter 1996). Extensive reviews have been provided by DeSarbo, Manrai, and Manrai (1993) and Corter (1996).

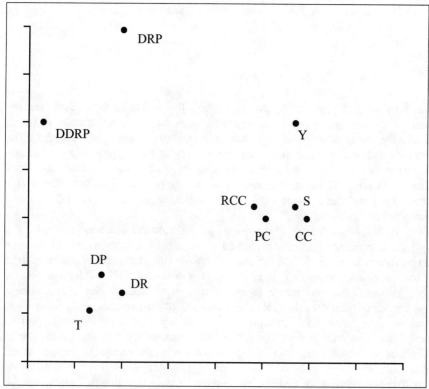

Figure 14.1: T=2 dimensional Space for the Schiffman et al. (1981) Cola Data

Assume that the dissimilarity judgements for al subjects are derived from a single ultrametric tree, so that each triple of expected dissimilarities satisfies the ultrametric inequality:

$$\delta_{ij} \leq \max(\delta_{ik}, \delta_{jk}), \ \forall\, (i, j, k) \tag{14.2}$$

The ultrametric inequality ensures that for any three objects, labeled i, j, and k, for which (14.2) holds, i and j are less distant from each other than each of them is from k. It can be shown (Corter 1996) that the inequality in (14.2) implies that the largest two of any three distances are equal. The constraints in (14.2) impose the ultrametric inequality on the expected distance judgements. Note that again this is a homogeneous representation: a single ultrametric tree is assumed to underlie the dissimilarity judgements of all subjects; deviations are assumed to be measurement error. The so called node-height convention states that for I stimuli there are I-1 parameters corresponding to the heights of the I-1 nodes in an ultrametric tree (Corter 1996, p. 16). In addition there is one variance parameter, so that the effective number of parameters estimated is I.

If it is assumed that the dissimilarity judgements are derived from an additive tree, each quadruple of expected dissimilarities satisfies the additive or four-point inequality:

$$\delta_{ij} + \delta_{kl} \leq \min(\delta_{ik} + \delta_{jl}, \delta_{il} + \delta_{jk}) \ \forall \{i,j,k,l\}.$$ (14.3)

It can be shown (Corter 1996) that this inequality is identical to restricting the largest two of the three sums of distances between the four objects to be equal, or equivalently, to imposing the ultrametric inequality on pairs of distances. The constraints in (14.3) impose the additive inequality at the aggregate level: a single additive tree is assumed to underlie the judgements of all subjects and deviations are measurement error. The effective number of parameters for an additive tree is 2I-3 (Corter 1996) and there is one variance parameter, which results in 2I-2 effective parameters for this model.

In order to illustrate tree models, we estimate the ultrametric tree for the cola data. The resulting tree is presented in Figure 14.2. One branch of the tree contains three diet colas, which have relatively low distances (the distance to their least common ancestor node): Diet Pepsi, Diet Rite and Tab. Under the feature matching model, the path-length from the root of the tree to the least common ancestor node of these tree diet cola's is a measure of the importance of the features shared by these stimuli. The interpretation of the common ancestor node of these three colas as diet/regular feature is hampered somewhat by the fact that the fourth diet cola in the stimulus set, Diet Dr Pepper is joined to the regular Dr Pepper, albeit with a relatively large distance. The common ancestor node of these two brands seems to be a brand taste feature: Dr Pepper versus other brands, which can be interpreted as the presence or absence of the characteristic cherry flavor. The sub-tree in the middle of the ultrametric tree in Figure 14.2 shows a set of nodes that can be interpreted as representing brand-specific features, distinguishing the five remaining non-diet brands.

Figure 14.2: Ultrametric Tree for the Schiffman et al. (1981) Cola Data

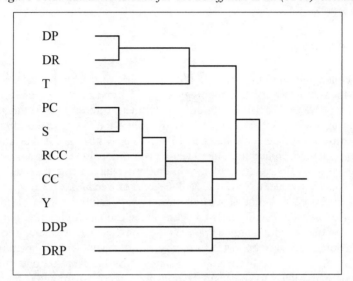

Mixtures of Spaces and Mixtures of Trees

The above models do not account for heterogeneity. However, using the mixture model framework, it is relatively straightforward to do so. Assume that the sample of N subjects is comprised of S unobserved segments in proportions π_s, as usual. The observed dissimilarity judgements in different segments are assumed to be characterized by different spatial or tree representations. Given segment s:

$$d_{nij} = \delta_{ij|s} + \varepsilon_{nijs}, \tag{14.4}$$

where $\delta_{ij|s}$ is the expectation of d_{nij} given segment s, i.e., the expected distance between stimuli i and j in segment s. Assuming the expected distances in each segment to conform to (14.1), a mixture of spaces arises:

$$\delta_{ij|s} = \mu_s + \sum_{t=1}^{T}(x_{its} - x_{jts})^2 \ . \tag{14.5}$$

Note that the locations of the brands on the dimensions, x_{its}, differ by segment, which corresponds to a different set of brand perceptions. Each segment thus has its own perceptual orientation (cf., De Sarbo, Manrai and Manrai 1994). The number of parameters is S(TI-T(T+1)/2)+3S-1.

Alternatively, one may assume that the unobserved segments have different ultrametric trees underlying their brand dissimilarity judgements. For

Part 3 Special topics in market segmentation

segment s, each triple of expected distances then satisfies the ultrametric inequality as shown in (14.2):

$$\delta_{ij|s} \leq \max(\delta_{ik|s}, \delta_{jk|s}), \ \forall\ (i,j,k) \tag{14.6}$$

The set of constraints in (14.6) imposes the ultrametric inequality for each segment separately, thus allowing the tree topology to differ by segment. Recently, Wedel and DeSarbo (1998) present this mixture of ultrametric trees. They estimate their model using an EM algorithm (see Chapter 6), where they impose the ultrametric constraints on the estimated distances in each M-step with a sequential quadratic programming method. In addition, they deal with additional external constraints on the tree topology and correlation of the dissimilarity judgements that will not be discussed here. This model has S(I-1) +2S-2 degrees of freedom.

To illustrate this model, we reiterate its application to the Cola data by Wedel and DeSarbo (1998). Figure 14.3 presents the estimated ultrametric trees for the S=2 model, that was found to be optimal based on the CAIC statistic. Recall that the S=1 solution in Figure 14.2 presents a somewhat mixed picture: the interpretation of a diet/non-diet feature is hampered by the fact that one diet cola, Diet Dr Pepper is joined to the regular Dr Pepper, these two brands sharing the cherry taste brand feature. In the S = 2 solution in Figure 14.3, segment 1 perceives taste differences between diet and non-diet Cola's. Diet Pepsi and Diet Rite are close, indicating very similar tastes. These two brands form one sub-tree, together with Diet Dr. Pepper and the diet version of Coca-Cola: Tab. The node separating these four brands from the others is therefore best interpreted as a diet/non diet feature. Diet Dr. Pepper shares the least number of features with the other three brands: that node can be interpreted as cherry/non-cherry flavor. In the non-diet sub-tree various colas are joined at different path-lengths, but here too Dr Pepper has a large distance from the other brands. In the second segment Dr. Pepper and Diet Dr Pepper, respectively Coca-Cola and Tab, are joined at relatively small distances: indicating that this segment primarily tastes differences among brands. Note that particularly the Dr. Pepper brands stand out, indicating that the cherry/non cherry flavour feature is an important distinguishing characteristic. However, the exception to this grouping of brands is that Diet Pepsi and Regular Pepsi are joined in a sub-tree with several other brands. Wedel and DeSarbo conclude that the specific brand tastes are the dominant features determining dissimilarity judgements in this segment, but that Pepsi has not been able to provide its diet version with its specific brand taste. Comparing the S = 2 results with the S=1 in Figure 14.2, it is seen that the heterogeneity causes the mixed structure in the S = 1 solution. The two segments are reported to be very well separated and all posteriors are virtually equal to zero or one. Interestingly, all of the subjects in segment one have the ability to taste a compound called *phenylthiocarbamide* (PTC), which tastes bitter. Such subjects are called *tasters*. All subjects in segment two do not have the ability to taste PTC (Schiffman, Reynolds and Young, 1981, p. 227). Subjects' ability to taste PTC apparently determines the extent to which they use diet/ non-diet, or brand specific tastes as the dominant feature to determine similarities between cola's.

Wedel and DeSarbo (1998) report the ability to taste PTC to be determined by one particular gene, which if present causes the ability to taste this bitter compound. This finding illustrates the potential of the mixture approach to identify a discrete latent trait in the population: on the basis of sensory judgements: the mixture model has identified two segments of subjects which differ in one particular inheritable characteristic.

Figure 14.3. Mixture of Ultrametric Trees for the Schiffman et al. (1981) Cola Data (adapted from Wedel and DeSarbo 1998).

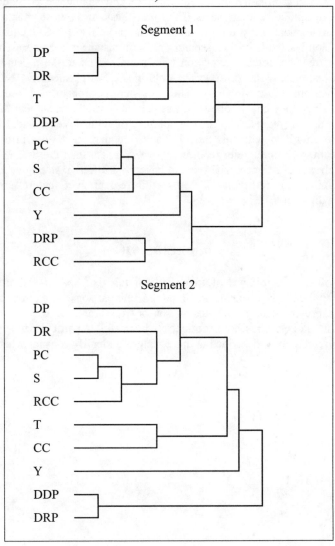

Mixtures of Spaces and Trees

It has been argued that trees and spaces are complements rather than substitutes (Carroll 1976; Ghose 1998). Therefore, several hybrid models for dissimilarity judgements have been proposed. Carroll and Pruzansky (Carroll 1976; Carroll and Pruzansky 1980) developed a hybrid model that combines multiple tree structures and a single spatial configuration. Carroll and Chaturvedi (1995) proposed a hybrid model that accommodates individual differences. More recently, Wedel and Bijmolt (1998) proposed a mixture of ultrametric tree and spatial MDS models for the analysis of dissimilarity judgements. Their model assumes two segments, where the distances in segment one are represented by a space, i.,e. for S=1 equation (14.5) with T=2 applies, and for the second segment distances are represented by an ultrametric tree, i.e. for S=2 equation (14.6) applies. Once the parameters of the model are estimated, the posterior probabilities, π_{ns}, are calculated by means of Bayes' Theorem (equation 6.4). These posterior probabilities are important quantities, since they enable one to assess post hoc whether a subject has used the tree or the spatial representation, or a mixture of both. Wedel and Bijmolt also analyse the cola data with this model. They find that the S=2 hybrid model fits better that the S=2 tree model reported above, but also than the S=2 MDS model. Different classes of subjects thus seem to have used different strategies to judge the taste dissimilarity of the colas, based on discrete product features, respectively continuous dimensions.

Conclusion

Mixtures of spaces, mixture of trees and mixtures of spaces and trees can be gainfully applied for the analysis of paired comparisons data. Such data seem currently more frequently collected in sensory research, but we argue that their collection in marketing research continues to be useful, in particular when combined with mixture models as described in this chapter, to identify consumer heterogeneity in judgement.

PART 4
APPLIED
MARKET SEGMENTATION

In this part of the book we review four major application fields of market segmentation that have recently received increasing attention in applied research. The application areas are in the classes of general observable bases (geo-demographics, discussed in chapter 15), general unobservable bases (values and lifestyles, discussed in chapter 16), product-specific observable bases (response-based segmentation, chapter 17) and product-specific unobservable bases (conjoint analysis, in chapter 18).

15
GENERAL OBSERVABLE BASES: GEO-DEMOGRAPHICS

In this chapter, we describe an increasingly popular form of market segmentation that combines multiple segmentation bases: demographics, geography, lifestyle and consumption behavior. Another distinct characteristic of this type of market segmentation is that its focus is on neighborhoods rather than individual consumers.

"*Birds of a feather flock together.*" That old adage perfectly reflects the essence of geo-demographic segmentation, which is based on the assumption that people are similar to their closest neighbors in their sociodemographic characteristics, lifestyles and consumption behavior (Mitchell 1995). Given that assumption, geo-demographers identify segments by clustering neighborhoods rather than individual consumers. Therefore, one can view geo-demographic segmentation as a post hoc segmentation approach similar to those discussed in chapters 2 and 3, differing mostly in its focus on small neighborhoods rather than individual consumers. Because geo-demographic segments are tied directly to geographic location in the form of census block groups, carrier routes, postal code areas (ZIP, ZIP+4), and so on, the *identifiability* and *accessibility* of segments (see chapter 1) can be considerably enhanced by classifying consumers into geo-demographic segments, or by cross-linking the segments obtained from other bases with those defined geo-demographically.

Geo-demographic segmentation had its origin in the early 1970s, when a marketing researcher (Jonathan Robbin) had the idea of combining survey data from various sources with data from the 1970 U.S. Census at the ZIP code level. Rather than defining market segments by clustering individual consumers or households, he form ed clusters of small neighborhoods, adopting the principle that individuals of similar sociodemographic characteristics and lifestyles tend to aggregate in the same neighborhoods.

In each decennial census, the U.S. Census Bureau collects data from households for about 150 variables including individual demographics (age, education, income, occupation, employment, etc.), household information (number and age of children, marital status, etc), and dwelling characteristics (owner/renter, single/multiple units, etc.), which are reported only at a certain level of aggregation (i.e., census block group, as defined later) to protect the privacy of the reporting households. Combined with additional data on consumer behavior from private sources (e.g., consumer surveys), these data represent a wealth of useful information to marketers. To identify homogeneous clusters of neighborhoods in terms of their socioeconomic and lifestyle profiles, Robbin first subjected the data to principal component analysis and then used the uncovered factors as the bases for segmenting neighborhoods. Dozens of cluster solutions were considered, some with up to 100 clusters, but a 40-cluster solution was chosen as sufficiently large to represent the distinguishing characteristics of the neighborhoods and small enough to be meaningful and useful for marketing purposes. The result from that exercise was PRIZM™ (Potential Rating Index for ZIP Markets). ZIP codes provided the missing link relating demographic and lifestyle information to geography.

As soon as it was launched as a marketing service by Claritas in 1978, PRIZM™ was helping magazines such as *Time* and *Newsweek* target differentiated versions of their weekly issues, featuring ads tailored to specific PRIZM™ clusters, and helping advertisers such as GM and American Express pitch their messages to the various lifestyles in the forty clusters. Geo-demographic segmentation was further popularized with the publication of an interesting paperback (Weiss 1988) presenting vivid portraits of the 40 clusters across the United States. On the basis of Weiss' (1988, pg. xv) own results, the geo-demographic concept might have received more exposure in the Money and Brains cluster (where the purchase of books is 54% greater than national average) than in the Hispanic Mix cluster (with only 56% of the national average in book purchases). The evidence presented in Weiss' (1988) book, although annedoctal, is highly consistent across a variety of lifestyle characteristics and consumption behaviors, strongly supporting the face and external validity of this form of market segmentation.

Applications of Geo-demographic Segmentation

Geo-demographic segmentation has many uses in various areas of marketing, some of which are fairly obvious; marketing problems that require answers to "who are my customers" (identifiability) or "where are they" (accessibility) are potential applications of this form of segmentation. Geo-demographic systems will show the type of people who own a piano and/or a gun, where the dog owners live, and so on. A marketer of dog foods knows that among affluent Americans, the highest dog ownership is among PRIZM™ clusters Kids & Cull-de-Sacs and God's Country, whereas in the middle class the best bet is the Shotguns & Pickups cluster (*Chicago Tribune* November 5, 1995).

The popular presentations of geo-demographic data through detailed maps may give the misleading impression that geo-demography must involve sophisticated cartography. For most applications it does not. Direct marketers, for example, need only ZIP or ZIP +4 codes as the link for fusing their mailing lists to a geo-demographic system for list-segmentation purposes. Because ZIP codes are partially nested within Census Block Groups (CBGs), the fusion of the two databases can be either done directly at the ZIP code level, or indirectly through the CBGs (i.e., mailing list ZIP, CBG, geo-demographic clusters). Once the two databases are fused, the direct marketer can identify the geo-demographic clusters with the highest concentrations of highly

Chapter 15 General observable bases: geo-demographics

responsive members of the mailing list, and then select new ZIP codes within those clusters to be targeted. A direct marketer using a rented list may run a test mailing to a random sample from the list. Once sales (or enquiry) results are obtained from the test run, the rented list would be fused to the geo-demographic system to identify clusters of high response and then select the ZIP codes to be targeted in the final run.

The usefulness of geo-demographic segmentation for marketing purposes is well demonstrated by the following example, based on actual PRIZM™ case studies. Home Pro, a 20- year-old 800-store home improvement chain known for "no-frills good value," was facing growing competition from home improvement superstores and other retailers catering to "do-it-yourself" customers. Home Pro used a PRIZM™ report on consumers who buy and use home improvement products and services to identify five target segments. For example, one of the segments, which they named "This Old Housers," contained four PRIZM™ cluster groups, made up of college-educated white-collar professionals with incomes over $75,000, married with children, living in single-family homes, who like to work on home projects as a pastime. However, the four clusters differed on the following home-improvement-related lifestyles:

Starter Families: Own dogs, play softball, join a health club, read "Metropolitan Home".
Boomers and Babies: Go jogging, have mortgage, watch MTV, read "Sunset" and "Personal Computing"
Upward Bound: Visit Seaworld, play tennis, read *"Working Mother"* watch *"Seinfeld."*
Suburban Sprawl: Enjoy home remodeling, own home gym, read *"Family Handyman"* and *"Home Mechanics."*

Home Pro conducted focus groups with selected participants from these PRIZM™ clusters to gather more detailed information about their members. On the basis of all the qualitative and quantitative information, it decided to expand the store assortment, add in-store consultation services, install interactive computerized help, and offer remodeling seminars in the evenings. It also focused its promotional message with (1) direct mailing of a free instructional video to prospects from PRIZM™ coded lists in the trade areas for its stores, (2) print ads in the magazines favored by the target segments, and (3) newspaper ads in major metropolitan and suburban newspapers selected through a PRIZM™ link to Scarborough and US Suburban Press databases.

The usefulness of geo-demographic systems is further enhanced by their linkage to other secondary data sources. For example, by linking geo-demographic information to media and audience sources (e.g., Arbitron, A.C. Nielsen, CableTrack, Scarborough, etc.) marketers can target their message more effectively. Using PRIZM™'s clusters, Buick decided where to rent billboard space and Isuzu chose lifestyle magazines over news weeklies (Mitchell 1995). Cable Networks, Inc. uses cluster system maps to help advertisers find where their best prospects are located within their cable systems (Mitchell 1995). Cluster systems can also help the media to understand their audience better. The publishers of *Apartment Life*, for example, repositioned the magazine to improve revenues by redirecting subscriptions to an urban, but more affluent readership. Once the format and content were changed, the magazine was renamed *Metropolitan Home* (to reflect the more upscale urban focus) and PRIZM™ was used to classify the subscribers of *Apartment Life* and the new subscribers of *Metropolitan Home*. The analyses showed that the penetration of *Metropolitan Home* among the affluent Blue Blood Estates and Money & Brains clusters was more than three to four times higher than in the national average, bot only two times higher for *Apartment Life*. The reverse was observed in less affluent segments such as Emergent Minorities and Gray Power, where the penetration of *Apartment Life* was more than twice as high as that for *Metropolitan Home* (Curry 1993).

Commercial Geo-demographic Systems

Because of its wide range of marketing applications, geo-demography has evolved into a significant service within the marketing research industry. Since the development of the first system (PRIZM™) in the early 1970s, many others have arisen. A detailed description of all those services is beyond the scope of this book (we refer to Curry 1993 for a comprehensive discussion of geo-demography and database marketing). Here, we briefly describe two of the major geo-demographic systems in the United States. Similar systems can be found in other countries, such as the Dutch system described later in this chapter.

PRIZM™
(Potential Rating Index for ZIP Markets)

This system, developed and marketed by the U.S. firm Claritas, combines census data with a wide variety of private information. From a factor analysis of census data, Claritas identified six main geo-demographic factors: *social rank* (income, education, occupation), *household composition* (age, marital status, dependents), *mobility* (length of residence, ownership), *ethnicity* (race, language, ancestry), *urbanization* (population density, urban classification) and *housing* (owner/renter, home value). Those factors were combined with data from automobile registrations, magazine subscriptions, credit applications, consumer product-usage surveys and other sources to identify homogeneous clusters of census block groups, with the use of traditional clustering techniques. The system is fine-tuned with annually updated census block and block group data, in combination with more than a half-billion consumer purchase records from private sources.

The pioneer of the geo-demography industry, PRIZM™ is now in its third generation; the original typology was first revised after the 1980 census and more recently with the 1990 census data. The last major revision produced a more complex typology, increasing from 40 previous clusters to 62 new ones describing a more ethnically and racially diverse population. Of the 62 new clusters, five are predominantly Hispanic. Despite the surge of immigration from Asia in the last decade,

Chapter 15 General observable bases: geo-demographics

none of the new clusters are predominantly Asian, but that group shows an above-average concentration in the most upscale clusters. Another distinction between the new typology and the older ones is the identification of an extra level of urbanization, distinguishing between neighborhoods in large cities and small cities. The latter, named "second cities," also include new cities that emerged on the fringes of large metropolitan areas.

In response to clients' concerns about the *substantiality* of some of the clusters in the previous systems, this third generation of the PRIZM™ system contains clusters of comparable sizes, each representing no more than 2% of the population of U.S. households. The 62 PRIZM™ clusters are further classified into 15 broader social groups defined in terms of their degree of urbanization and affluence, as shown in Table 15.1. Each of the broad social groups contains between two to five clusters (indicated in parentheses in Table 15.1), also ordered in terms of affluence. Among all 15 groups, Elite Suburbs (S1) contains the most affluent consumers, whereas Urban Cores (U3) is the group with the lowest income and highest poverty ratios.

Each of the 62 clusters is richly described in Claritas' literature in terms of its demographic profile, geographic concentration, lifestyle and other characteristics of interest to marketers. Below are brief descriptions of a few clusters in some of the extreme groups of Table 15.1 (i.e., U1, U3, T1 and T3):

- U1 *Urban Uptown:* Cluster 06 – "Urban Gold Coast" is the most densely populated among all 62 clusters, and has the highest per-capita income. This cluster has the greatest concentration of singles renting in multi-unit high- rise buildings, the lowest incidence of auto ownership, and the fewest children among all clusters. The neighborhoods are predominantly white, with a higher than average concentration of Asians. Half of the cluster lives in New York City.

- U3 *Urban Cores*: Cluster 47 – "Inner Cities" is concentrated in large Eastern cities and among America's poorest neighborhoods. It has twice the nation's level of unemployment, with many residents receiving public assistance. The majority of the households (80%) in this cluster are African-American and 70% of those with children have single parents.

- T1 *Landed Gentry*: Cluster 15 – "God's Country" consists of neighborhoods in the exurbs of major metros, the outskirts of second cities and scenic towns, populated by educated, upscale married executives and professionals with multiple incomes and large families. Residents are owners of single-unit homes, and their life centers around the family and outdoor activities.

- T3 *Working Towns*: Cluster 55 - Mines & Mills is populated by an older, predominantly white, poor, uneducated and mostly single population living in old mine or steel towns.

Part 4 Applied market segmentation

Table 15.1: PRIZM™ Classification by Broad Social Groups

Affluence	Urbanization				
	Rural	Town	2nd City	Suburban	Urban
High $	T1(4) Country Families		C1 (3) 2nd City Society	S1 (5) Elite Suburbs	U1 (5) Urban Uptown
		T2 (4) Exurban Blues		S2 (5) Affluentials	
	R1(4) Country Families		C2 (5) 2nd City Centers	S3 (4) Inner Suburbs	U2 (5) Urban Midscale
	R2 (2) Heatlanders	T3 (4) Working Towns			
	R3 (5) Rustic Living		C3 (4) 2nd City Blues		U3 (3) Urban Cores
Low $					

Source: Claritas Promotional material. © 1997 Claritas, Inc. PRIZM and Claritas are registered trademarks of Claritas, Inc. The 62 PRIZM cluster nicknames ("Blue Blood Estimates", "Starter Families", etc) are trademarks of Claritas, Inc.

Chapter 15 General observable bases: geo-demographics

These narrative profiles give managers rich metaphors to use in conceptualizing and communicating among themselves about their target markets. More concrete, quantitative details are also available for each cluster in terms of their lifestyle activities and interests, financial activities, consumption behavior, media exposure and so on.

In addition to the original PRIZM™, defined at the ZIP code level, Claritas offers PRIZM+4™, which works with neighborhoods of fewer than 10 households on average, defined by ZIP +4 postal codes. It also offers a system (TelePRIZM™) defined at the (area code)-prefix combination, which is of special interest to telemarketers.

ACORN™ (A Classification of Residential Neighborhoods)

This geo-demographic system, offered by CACI Marketing Systems in the United States, classifies 226,000 census block groups into 40 residential clusters (plus three nonresidential neighborhood segments) based on 61 characteristics from general (income, age, household type) to very specific (e.g., type of cars owned, home value, preferred radio formats, magazine readership). Like other GD systems, ACORN™ goes beyond simple demography and geography, showing the human beings behind the numbers: who they are, how they spend their money and where they can be found.

The two-step method of analysis used to develop ACORN™ is similar to the one described before. First, the most pertinent consumer characteristics were identified by principal component analysis and graphical methods. Then, in a second independent stage, market segments were created by cluster analytic techniques. The stability of the segments was verified and confirmed by replicating the clusters with independent variables. The validity of ACORN™ is constantly verified by relating the 40 clusters to external variables not used to generate the clusters, such as recent consumer survey data, certifying that the 40 clusters are able to distinguish differences in consumer behavior as expected. This validation process has already been repeated with more than 140 indicators of consumer choice. Each of the 40 ACORN™ segments is richly described in its User's Guide in both narrative and table formats, along five main dimensions: *lifestyle* (activities and interests), *demographic* (e.g., age, marital status, ethnicity), *socioeconomic* (e.g., disposable income, employment, education), *residential* (e.g., renter/owner, single/multiple family units) and *location* (e.g., degree of urbanization, geographic location).

Cluster profiles also include a Purchase Potential Index (PPI) for a wide range of products and activities. PPI measures the potential demand for a product or service in geographical area in relation to the national average. The index is calculated by linking ACORN™ segments to survey data such as those of Mediamark Research Inc. or Simmons Market Research Bureau, leading to a consumption rate for each product and ACORN segment. On the basis of the ACORN™ composition within a given geographic area, the demand for the product, in relative to the national average is computed. With those indices, a marketer can find useful consumption-related information such as: (1) young, single households and working parents are more likely than average to buy fast foods, (2) suburban and rural families are more likely to own pets, and (3) dual earners and affluent households are more likely to own multiple cars,

Part 4 Applied market segmentation

families own minivans, and urbanites are less likely to own cars. Even between the two most affluent ACORN™ clusters one finds clear distinctions in consumption behavior. Consumers in the Top One Percent cluster, for example, are more likely to join country clubs, whereas those in Wealthy Seaboard Suburbs prefer health clubs.

Table 15.2 lists all ACORN™ segments, organized into eight major residential and one nonresidential group. The table also includes a few examples of the many indicators available for each cluster. The socioeconomic ranking is based on a variety of indicators such as income, education and housing value. Table 15.2 gives the ranking of the clusters on two of the many PPI indices available for a wide range of products and services. A comparison of the two indices illustrates ACORN™'s discriminant validity. Wealthiest Seniors and Successful Suburbanites show the highest demand for books and are ranked low on demand for lottery tickets; in contrast, the less affluent Social Security Dependents and Rustbelt Neighborhoods show the highest demand for lottery tickets.

The Geo-demographic System of Geo-Marktprofiel

Geo-demographic segmentation is being applied in Europe as well as in the United States. For example, the Dutch firm Geo-Marktprofiel has developed a geo-demographic segmentation system for the Netherlands. The system is based on postal codes. Each ZIP code in their system refers to 16 households on average, and the Netherlands is divided into 414,000 ZIP-code areas. For each ZIP code, the databases contain information from five sources.

1. Street and number information for all 441,000 ZIP code areas in the Netherlands, containing 16 households per ZIP code area on average.

2. A large number of inhabitant and housing characteristics for each ZIP code area. This information is obtained and updated through telephone interviews with two people in each ZIP code area, who are questioned about the type of housing in that area. Responses of the two subjects are checked for consistency, and if inconsistent additional interviews are conducted in the same ZIP code area.

3. Information obtained from the RAI Data Center, the automobile license -plate register with information on all privately owned cars in the Netherlands. Characteristics include number of cars, year of make, and current value of the cars.

4. Information on individual consumers obtained from the Response Plus survey. The survey provides information on more than 150 life-style variables of more than one million households. Included are characteristics such as purchases from mail order companies, private or public medical insurance, participation in lotteries, and magazine readership.

Chapter 15 General observable bases: geo-demographics

5. At certain ZIP-code levels, the data are further supplemented with product, media and interest information obtained from surveys of the market research agency Interview, among 40.000 respondents.

To make the 414,000 ZIP codes manageable, they are divided into 336 classes, called Geo-Market Profiles (GMP). A specific GMP is a combination of three main dimensions: socio-economic status (six classes), family lifecycle (seven classes) and degree of urbanization (eight classes). Table 15.3 describes these three main dimensions of the system.

The 42 main segments that result from cross-classifying socioeconomic and lifecycle dimensions, show meaningful differences in car ownership, car price, type and cost of housing, mobility, interest in sports and culture, subscriptions to journals and magazines, religion and possession of durables. For example, segment 1, labeled Starters with High Socioeconomic Status, live in houses that are renovated, have central heating and a bath tub, but less often have a garden. Prices of the houses are above Dfl 200.000. Mobility is high, and these individuals tend to live in the center of large cities. Their education is high, often at the university level. They are interested in finance, fitness, running, tennis, skiing, photography, wines, museums, restaurants theater and concerts, and short holidays. Often, these individuals are either reformed or not religious. Possession of durables is limited, with exception of surfboards and Walkmans. Insurance is predominantly private and few donations are given to charity.

Geo-demographic cluster 34 (Retired with Low Socioeconomic Status) contains older households in the lowest socioeconomic class, many with retired individuals. Very few have a car, and if they own a car it is likely to be more than seven years old. These individuals live in poorly maintained flats with two or three rooms, commonly built between 1860 and 1980. Most houses are rented for less than Dfl 500 a month. This geo-demographic cluster is located in the eastern, northern and southern parts of the Netherlands, in the center of cities with more than 5,000 inhabitants. Mobility is very low. Education is often only primary school. Interest focuses on artcrafts such as needlework. Daily newspapers are seldom read. Members of this cluster are unlikely to own pets and tend to vote for left-wing parties. They pay cash for their groceries, are publicly insured, and donate above average to charity

Part 4 Applied market segmentation

Table 15.2: ACORN™ Cluster Classification

Cluster Name	Socio-econo-mic Rank	Median Hhold. Income $1000	Geogra-phic Area	Percent of Hholds.	Ranking by PPI Read Books	Ranking by PPI Lottery Tickets
1. Affluent Families						
1A Top one percent	1	95.4	Suburb	1.1	3	23
1B Wealthy Seaboard Suburbs	3	60.6	Suburb	2.5	8	14
1C Upper Income Empty Nesters	5	54.2	Suburb	2.0	7	18
1D Successful Suburbanites	2	68.6	Suburb	2.0	2	22
1E Prosperous Baby Boomers	6	46.6	Suburb	3.3	15	17
1F Semirural Lifestyle	7	48.0	Suburb	4.4	5	9
2. Upscale Households						
2A Urban Professional Couples	8	40.4	Urban. area	4.2	4	12
2B Baby Boomers with Children	10	37.1	Suburb	3.7	12	10
2C Thriving Immigrants	11	40.3	Central city	1.7	34	33
2D Upscale Urban Asians	12	37.2	Central city	0.7	22	38
2E Older Settled Married Couples	13	38.5	Suburb	4.3	11	6
3. Up & Coming Singles						
3A High-Rise Renters	4	38.2	Central city	2.4	16	26
3B Enterprising Young Singles	9	31.4	Urban. area	3.7	18	24
4. Retirement Styles						
4A Retirement Communities	14	32.5	Urban area	1.2	21	30
4B Active Senior Singles	18	28.6	Urban area	3.2	20	5
4C Prosperous Older Couples	15	35.8	Urban area	3.4	13	7

Chapter 15 General observable bases: geo-demographics

Table 15.2 - continued

Cluster Name	Socio-economic Rank	Median Hhold. Income $1000	Geographic Area	Percent of Hholds.	Ranking by PPI	
					Read Books	Lottery Tickets
4D Wealthiest Seniors	16	35.2	Town	0.9	1	39
4E Rural Resort Dwellers	19	24.7	Rural	1.0	24	15
4F Senior Sun Seekers	31	22.2	Town	1.8	9	8
5. Young Male Adults						
5A Twenty Somethings	26	19.1	Central city	2.0	10	37
5B College Campuses	28	15.6	Urban. Area	0.9	6	36
5C Military Proximity	22	23.6	Urban. area	2.1	29	32
6. City Dwellers						
6A East Coast Immigrants	23	26.8	Central city	2.1	27	16
6B Middle-Class Black Families	30	29.6	Central city	1.1	38	4
6C Newly Formed Households	21	26.6	Urban. area	5.3	14	3
6D Settled Southwestern Hispanics	35	19.7	Urban. area	2.4	35	35
6E West Coast Immigrants	34	22.4	Central city	1.2	39	28
6F Low income: Young & Old	37	16.1	Urban. area	2.6	33	25
7. Factory & Farm Communities						
7A Middle America	20	30.4	Rural	7.6	25	19
7B Young Frequent Movers	24	26.6	Rural	2.8	26	20
7C Rural Industrial Workers	33	19.7	Rural	5.3	31	34
7D Prairie Farmers	25	23.2	Farm	0.9	30	40
7E Small Town Working Families	27	25.4	Town	1.7	28	27

Part 4 Applied market segmentation

Table 15.2 – continued

Cluster Name	Socio-economic Rank	Median Hhold. Income $1000	Geographic Area	Percent of Hholds.	Ranking by PPI	
					Read Books	Lottery Tickets
7F Rustbelt Neighborhoods	29	25.1	Town	3.7	19	2
7G Heartland Communities	32	19.4	Town	3.7	23	29
8. Downtown Residents						
8A Urban Hispanics	38	18.6	Central city	1.2	37	21
8B Social Security Dependents	17	10.3	Urban	1.0	17	1
8C Distressed Neighborhoods	40	8.6	Central city	1.2	32	11
8D Low income Southern Blacks	39	11.7	Central city	1.7	40	31
8E Urban Working Families	36	21.5	Central city	1.8	36	13
9. Non-residential Neighborhoods						
9A Business Districts	NA	NA	NA	0.0	NA	NA
9B Institutional Populations	NA	NA	NA	0.0	NA	NA
9C Unpopulated Areas	NA	NA	NA	0.0	NA	NA

Source: ACORN™ User's Guide. 8 1997 CACI Marketing Systems.

Chapter 15 General observable bases: geo-demographics

Table 15.3: Description of the Dimensions of the GMP System

Class	Description	Number hh.	Percent hh.
Socioeconomic Dimension			
High	Income over two times average, expensive housing	299,570	4.83
Above average	Income average to two times average, above average prices of housing	1,307,920	21.09
Average	Average income	2,611,750	42.11
Low	Income below average, cheaper rental houses	1,329,953	21.44
Minimum	Income low or no income, high unemployment rate	411,348	6.63
Various	Both high and low income, varying prices of housing	213,735	3.45
Lifecycle Dimension			
Starters	Young singles up to 35 years	130,027	2.09
Young families	Young couples up to 35 years without children	524,365	8.45
Families with young children	Families with children below 10 years	1,658,303	26.74
Families with older children	Families with children above 10 years	2,080,972	33.55
Older families	Couples older than 35 years	1,298,826	20.94
Older singles	Older singles, mostly retired	388,297	6.26
Bipolar	Both young and older families without children	93,864	1.51
Urbanization Dimension			
Large cities 1	Over 500.000 inhabitants: Amsterdam, Rotterdam	581,908	9.38
Large cities 2	250.000 to 500.000 inhabitants: The Hague	215,908	3.48
Middle sized cities	100.000 to 250.000 inhabitants	915,944	14.77
Smaller cities	50.000 to 100.000 inhabitants	1,018,873	16.43
Larger commuter cities	20.000 to 50.000 inhabitants	1,190,215	19.19
Smaller commuter cities	10.000 to 20.000 inhabitants	727,678	11.73
Villages	5.000 to 10.000 inhabitants	640,005	10.32
Small villages	Less than 5.000 inhabitants	912,011	14.70

Methodology

The starting point for most geo-demographic systems in the United States is the data supplied by the U.S. Census Bureau from its decennial censuses and periodic revisions. Those data are combined with personal information from credit-card applications, marriage licenses, birth registries, auto registrations, survey data and other sources to add lifestyle information to demographic data. The merging of the public and private databases is done at a certain level of geographic aggregation, rather than at the individual or household level.

The U.S. Census Bureau defines geographic units on two bases: (1) political - representing the U.S. political system, and (2) statistical, the basis for statistical counts. For marketing purposes, the following statistical areas are of most direct relevance.

1. *Census blocks*- each of these 8.5 million geographic units is roughly equivalent to a city block. They are usually compact areas bounded by streets and other prominent physical features (e.g., lakes, rivers and railroads), as well as certain legal/political boundaries. Those in urban areas are grouped into *census block groups*.

2. *Census block groups* (190,000 units in the U.S.)- These units are the smallest ones for which complete statistics are available, to ensure privacy. They are the basic unit of analysis in most geo-demographic systems. Typically, a CBG comprises about four to six city blocks with about 350 households. Most geo-demographic systems also include about 60,000 more statistical units in non-urban areas, where CBGs are not defined.

3. *Census tracts* (48,200 units)- These county sub-units are locally defined statistical areas containing collections of census block groups, with generally stable boundaries. They were originally defined to have relatively homogeneous demographic profiles and to contain an average of 4,000 residents.

4. *Block numbering areas* (11,200 units)- Similar to census tracts, these are defined for the purpose of grouping blocks in counties without census tracts.

5. *Counties* (3,141 units)- sub-units of the 50 states and the district of Columbia.

Census block groups (containing about 350 households) are the preferred bases for geo-demographic segmentation, because they closely relate to actual neighborhoods delineated by natural boundaries such as major streets. Further, they are the smallest geographic units for which census data are available, thus being a logical common bases for combining census data with consumer information from private sources. Another important geographic unit for combining those data is the postal ZIP code area, which defines consumers' geographic location in most commercial databases. As the original ZIP codes contain roughly 2,500 households each on average, the linkage is not perfect, aside from the fact that CBGs are not necessarily nested within ZIP areas. With the implementation of ZIP +4 postal codes, the situation improved, but the relationship between CBGs and postal codes reversed; ZIP +4 areas (containing fewer than 10 households on average) are contained (but not necessarily nested) within CBGs. Although constructed at the CBG or ZIP level, the major geo-demographic systems also relate their segments to other geographic units of direct interest to marketers, such as Nielsen's Designated Market Areas (DMA) and Arbitron's Area of Dominant Influence (ADI).

Another important factor contributing to the development of geo-demographic systems and their wide acceptance in marketing was the availability of the TIGER (topologically integrated geographic encoding and referencing) system for commercial use. It was the foundation for the automated geographic operations at the U.S.Census Bureau. This machine- readable mapping database contains information on the boundaries of counties, census tracts, block numbering areas, and county subdivisions, down to the CBG level. Even though TIGER contains only geographic information (i.e., individual streets and geographic features digitally coded by latitude and longitude), it can be easily merged with statistical data to represent geo-demographic information in graphic form.

The starting point for most of the commercial systems currently available is a large database at the neighborhood level (CBGs or ZIPs) combining public (e.g., census) and private data from various sources. In principle, this information could be used to link with a particular marketer's customer database or another secondary source, as long as those sources have their entries defined at the same level of geographic aggregation. However, the databases are likely to be less reliable at the micro-level; whenever consumers move, change jobs, graduate from college, etc., their entries in the databases become less accurate. More aggregate clusters of homogeneous geographic units have relatively more stable profiles. Furthermore, a smaller number of segments (40 to 60) with distinctive profiles and buying behaviors is more meaningful to marketers than hundreds of thousands of neighborhoods.

The typical methodology used to uncover geo-demographic clusters is the popular *tandem approach* combining principal components analysis with some clustering algorithm. First, a vast number (hundreds) of variables is reduced to a more manageable and meaningful set of latent factors. Then, the CBGs or ZIP areas are grouped into a smaller number of homogeneous segments through traditional clustering procedures. However, this two-stage process has been the subject of recent criticisms (Arabie and Hubert 1994) for not making full use of all available information. More effective approaches include the formation of segments by mixture or clustering techniques applied to all variables, followed by a factor analysis to enhance interpretation of the clusters or latent classes. Alternatively, one could form the geo-demographic segments while simultaneously identifying the underlying dimension used to determine their homogeneity. This can be accomplished with STUNMIX models. As described in chapter 8, these mixture unfolding models can be used to identify latent segments along underlying dimensions that are determined at the same time as the segments.

Adaptations of this general family of models could be used for full-information maximum likelihood estimation of latent segments in a geo-demographic context. Currently, a major problem is posed by the computational effort involved in applying these methods to real-life geo-demographic databases.

Most geo-demographic segmentation systems form their clusters along multiple bases, including urbanization, demographics, consumption behavior and other characteristics. In fact, one of the major commercial systems (PRIZM^{TM}) uses two main dimensions - degree of urbanization and degree of affluence - to summarize its 62 clusters. With this purpose in mind, one might want to apply the joint segmentation approach discussed in chapter 11, which would lead to geo-demographic segments directly associated to each of the segmentation bases. The approach would produce joint segments defined by combinations of basic segments formed along each of the main underlying bases such as ACORN^{TM}'s lifestyle, demographic, socioeconomic, residential and location, or PRIZM^{TM}'s urbanization and affluence. Again, more computationally efficient algorithms are needed to make such applications feasible at a large scale.

An additional problem posed by geo-demographic data is its inherent spatial correlation structure. Typically, neighboring cluster block groups or ZIP+4 areas tend to be more similar to each other than to areas farther away, thus leading to spatial correlation. From research in time-series analysis, it is well known that ignoring such correlational patterns across observations can lead to an overestimation of heterogeneity. Further developments are necessary to incorporate the effects of unobserved spatial correlation in clustering and mixture models.

Linkages and Datafusion

The first step in using any of the commercially available geo-demographic systems is to merge the cluster classification database with the firm's own database or those from secondary sources such as other syndicated services or consumer surveys. Once the databases are merged, one can identify which clusters perform below/above average in terms of sales, contribution margin or other measures of performance. Geography plays an important role in this merging process; at the most micro level, the dabatases may be merged on the basis of ZIP+4 codes, comprising no more than 10 households on average. Most commonly, customer databases contain ZIP code information. Direct marketers can use geo-demographic systems and implement their program at the ZIP code, carrier route or ZIP+4 level.

The practice of linking geo-demographic systems to other databases is so widespread that most systems have already been cross-referenced to other syndicated services and many partnerships have been formed, offering joint services. Once their customer database is merged with a GD system, users can not only relate their sales to the geo-demographic segments, but also have available all the other information provided by the linked syndicated services. A sample of syndicated services linked to one or more geo-demographic systems is listed in Table 15.4.

Recently, Kamakura and Wedel (1997) proposed a mixture model based methodology for datafusion. In practice datasets are often fused with so-called hot-deck procedures. These procedures duplicate data based on heuristic rules: when a value is missing from sample A (the recipient sample), a reported value is taken from sample B (the donor sample, so that each recipient subject is linked to one (or more) donor subjects on the basis of common variables, present in both files (often, those are demographic variables). A major limitation of hot-deck procedures are their heuristical nature, due to which statistical properties of the fused dataset are unknown. Kamakura and Wedel (1997) address the situation where a researcher wants to cross-tabulate discrete variables collected in the two independent samples. The datafusion method they propose is based on a mixture model and allows for statistical tests of association based on multiple imputations of the missing information in the files. Here, the data files are combined multiple times on the basis of the estimated mixture model, which enables one to assess the variability of the fusion procedure. A comparison revealed that this method outperforms the traditional hot-deck procedures for data fusion. An elaborate description of the approach is beyond the purposes of the present chapter.

Conclusion

Geo-demographic segmentation has been a highly useful and popular tool for marketers. It is one of the most effective approaches to segmentation based on general observable bases. In addition, by combining demographic, lifestyle and consumption information with geographic location, geo-demographic systems address at least two important criteria for effective market segmentation: *identifiability* and *accessibility*. However, the popularization of geo-demography may overly trivialize GD segmentation. Fancy names and colorful descriptions can mislead the careless user to believe that differences across segments and within-segment homogeneity are much greater than they actually are. The names and narrative descriptions can be useful metaphors with which managers can communicate with each other about their target segments, but users are well advised to apply them only after understanding the real, quantifiable differences between their target segments and the national averages.

One should also be well aware of the fact that most geo-demographic segments are clusters of neighborhoods, rather than households or consumers. The unit of analysis is usually the census block group or the ZIP+4 postal region. Therefore, aside from the usual assumption of homogeneity within clusters, one must also take the aggregate characteristics of the CBG or ZIP+4 area as representative of each household in that area. Most commercial geo-demographic systems use a wealth of data from individual consumers, to supplement the demographic data from the Census Bureau with detailed life-style information. However, systems that are built from geographic units such as the census block group must match individual-level data with the more aggregated units, resulting in clusters of neighborhoods rather than individual households.

Finally, we believe that geo-demography is a fruitful area for the application of mixture models. The techniques currently used in the development of geo-demographic systems do not make optimal use of the information contained in the available data. The

Part 4 Applied market segmentation

effectiveness of geo-demography can be further enhanced by the application of techniques such as STUNMIX for the simultaneous identification of underlying dimensions and latent classes, and the development of new mixture models to account for unobserved spatial correlation within and between census cluster groups. However, for such large scale applications to become feasible, more efficient estimation algorithms must be developed. A step in that direction is the Generalized EM algorithm (GEM).

Table 15.4: Media and Marketing Databases Linked to Geo-demographic Systems

Vendor	Services
Consumer Research	
Information Resources Inc (IRI)	Scanner data
MRCA Information Services	Diary panels
A.C. Nielsen	Scanner and panel data
Gallup	Omnibus panel
NPD Group	Diary panels
NFO Research	Mail panel
SRI International	Values and lifestyles
Survey Sampling	Sampling systems
Media Measurement and Planning	
Arbitron	Radio broadcasting ratings
Interactive Market Systems (IMS)	Media research and planning
Mediamark Research Inc (MRI)	Media and consumption measurement
Nielsen Media Research	TV cable and broadcasting ratings
Scarborough Research	Multi-media in local markets
Simmons Market Research	Media and market research
Telmar	Media research and planning
U.S. Suburban Press Inc. (USSPI)	Suburban newspapers
Direct Marketing	
ADVO	Direct marketing services
Database America	Direct marketing services
Donneley Marketing	Direct marketing services
Metromail	Direct marketing services
Polk Direct	Direct marketing services
TRW Target Marketing	Direct marketing services
Industry Specific Services	
DATAMAN	Mortgages
EPRI	Electric power
Footwear Market Insights	Footwear
Inforum	Health Care
Hooper-Holmes	Credit analyses
J.D. Power & Associates	Automotive
PNR & Associates	Telecommunications
Polk Automotive	Vehicle registration database

16
GENERAL UNOBSERVABLE BASES: VALUES AND LIFESTYLES

Lifestyle segmentation has been one of the most popular forms of market segmentation in the literature. We devote this chapter to a review of psychographics as it is applied to values and lifestyle segmentation.

Since its inception in the 1960s, psychographics and lifestyles have received wide attention among consumer researchers. The terms "lifestyles" and "psychographics" are used interchangeably in the marketing literature, although a few authors make a distinction between the construct (lifestyles) and its operationalization (psychographics). The main purpose of psychographics is to obtain a better understanding of the consumer as a person by measuring him/her on multiple psychological dimensions as well as on the way s/he lives, things in which s/he is interested, and his/her opinion on a diverse range of current topics. Its popularity as a basis for market segmentation can be attributed to several factors (see also chapter 2). First, the quality of the motivational insights obtained is useful in understanding the underlying reasons for observed consumer behavior. Second, the generality of the consumer profiles obtained through psychographic segmentation makes it applicable to a wide range of products and services. Third, the conceptual framework is flexible enough that measurement instruments can be tailored to specific domains of application (van Raaij and Verhallen 1994). Fourth, the consumer profiles obtained are implementable, in that they provide guidance for new product development, and the execution of advertising messages.

Psychographic research has its origin in the confluence of two other schools of psychological research in marketing: motivational and personality research (Wells 1974). Motivational research had its roots in psychoanalysis, making use of projective techniques and in-depth qualitative analysis of observations obtained from very small samples (Dichter 1958). Motivational research produced colorful portraits of consumers' needs and desires, but was criticized for its subjectivity and lack of generalizability. Personality research in marketing had its origins in clinical psychology, making use of psychometrics, a quantitative approach to the measurement of individual traits, including personality (Koponen 1960). Personality research produced objective, quantifiable measures. However, the measurements seldom correlated with consumer behavior at levels that would be useful for marketing decision making. Psychographics then arose as an approach to produce a rich and detailed portrayal of consumers' motives and reflect various facets of consumers' lives with measurements that could be subjected to statistical analysis.

Part 4 Applied market segmentation

Activities, Interests and Opinions

Early lifestyle segmentation studies (starting in the early 1970s) operationalized the life-style construct through a large battery of Likert-type statements covering the following categories (Plummer 1974):

Activities: Reported behavior related to club membership, community, entertainment hobbies, shopping, social events, sports, vacation and work.

Interests: Degree of excitement about and attention to achievement, community, family, fashion, food, home, job, media and recreation.

Opinions: Beliefs about business, culture, economy, education, future, politics, products, self and social issues.

Aside from the Likert-type items about activities, interests and opinions (AIO's), most AIO studies include demographic variables. The broad range of areas listed above is commonly used in AIO studies that can be applied to more than one product market. Such studies may include between 200 to 300 AIO statements. Commonly, a data reduction technique such as factor analysis is first used to translate the large battery of items into a small, more meaningful and interpretable number of underlying psychographic dimensions (Alpert and Gatty 1969; Darden and Reynolds 1971; Reynolds and Darden 1972; Moschis 1976). For example, in an early study, Wells and Tigert (1971) used 300 Likert-type items that were then factor analyzed and reduced to 22 lifestyle dimensions.

The underlying dimensions are then used in two main types of application. They may be used directly as the basis for post-hoc lifestyle segmentation, as inputs in cluster or mixture models. The objective here is to use psychographics as a segmentation basis and create a typology of consumers in terms of their lifestyles (Gutman and Mills 1982; Kinnear and Taylor 1976; Peterson 1972; Ziff 1971). Another common application for the underlying psychographic dimensions is in describing market segments obtained by a priori segmentation on a different basis (e.g., buying behavior, demographics, etc.). In this second type of application, the objective is to use lifestyles to discriminate among segments obtained from other bases, thus producing a richer profile of their members (Bearden, Teel and Durand 1978; McConkey and Warren 1987; Teel, Bearden and Durand 1979).

Rather than using the broad range of AIO's listed above, some researchers have used more specific items that tap the activities, interests and opinions associated with aspects of life more directly related to the product or brand under study (Ziff 1971). Other researchers combine general AIO items with product-specific items. In a study about health care services, for example, Blackwell and Talarzik (1977) included items that measured respondents' opinions about their doctor and about the medical profession in general, among other items. Those authors found significant correlations between the product-specific items and some of the general AIOs.

The combination of general and product-specific AIO items is likely to produce a lifestyle typology closely related to consumer behavior in the product category under study, thus providing more useful insights to the marketing manager. However, it will also lead to even larger sets of AIO items, with negative consequences for data collection, especially in view of consumers' growing resistance to long surveys. The length of AIO scales is a major constraint for their inclusion in most surveys. A common approach is to first develop a typology based on a large battery of items (often more than 200) and then trim the set to items that best discriminate across the identified segments. That approach, however, is not necessarily efficient, because it will still apply a standard questionnaire to every consumer, even if some of the items might provide little information about a specific person (Kamakura and Wedel 1995). In chapter 12 we presented a model that allows the researcher to tailor the interview so that each respondent will be asked only a subset of the general AIO scale that provides the maximum amount of information about his/her lifestyles. For one application, on average 16 items suffice for accurate classification in lifestyle segments.

Values and Lifestyles

The measurement of activities, interests and opinions is only one of several attempts to produce general lifestyle typologies that are applicable to a wide spectrum of consumption behaviors. Some researchers argue that constructs such as activities and attitudes are immediately affected by the environment, and are therefore neither stable nor generalizable. If the purpose is to create a typology that transcends products and brands, one must look at the innermost drivers of behavior (Valette-Florence 1986). The work by Vinson, Scott and Lamont (1977) suggested that values, AIO variables and product attributes (benefits) are ordered hierarchically from most centrally held and farthest removed from specific behavior to most closely related to actual behaviors. Similar suggestions have been made by Rokeach (1973) as well as in means-end chain theory (Reynolds and Gutman 1988), where consequences intervene between values and product attributes. Researchers in this area argue that values are a useful basis for psychographic segmentation because values are less numerous, more central and more closely related to behavior than are personality traits, and more immediately related to motivations than are attitudes (Valette-Florence 1986). Recently, means and chain theory was applied to international segmentation by ter Hofstede, Steenkamp and Wedel (1999).

Rokeach's Value Survey

Rokeach defines a *value* as "an enduring belief that a specific mode of conduct or end-state of existence is personally or socially preferable to an opposite or converse mode of conduct or state of existence" (Rokeach 1973, p. 5). A *value* is a single central belief that transcends any particular object, in contrast to an attitude, which is a belief about a specific object or situation. A *value system* is an enduring organization of values along a continuum of relative importance to the individual. Value systems are used by individuals as general plans for resolving inner-conflicts and for decision making. Therefore, the identification of value systems and the classification of consumers according to them is likely to result in homogeneous segments in terms of the main motives driving their general behavior. Rokeach was one of the pioneers in the quantitative measurement of values and value systems. He identified two main types of values: *terminal values,* "...beliefs or conceptions about ultimate goals or desirable end-states of existence...", and *instrumental values,* "... beliefs or conceptions about desirable modes of behavior that are instrumental to the attainment of desirable end-states" (Rokeach 1973, p. 48). Rokeach's Value Scale (RVS) contains the 18 terminal and 18 instrumental values listed in Table 16.1.

Given the hierarchical nature of value systems, measures are usually obtained by asking consumers to rank the 18 values in the terminal and instrumental sets in order of importance. Several researchers have modified the RVS scale to yield interval measures of value importance (see Alwin and Krosnick 1985 for a comprehensive review). Monadic ratings, however, are susceptible to response styles (Alwin and Krosnick 1985) and social desirability effects (Rokeach 1973), and reduce respondents' willingness to make judgments about the *relative* importance of the 18 values in each set.

The structural relationships among the 36 terminal and instrumental values in Rokeach's scale were investigated theoretically and empirically by Schwartz and Bilsky (1987, 1990) in several cross-cultural comparisons of value structures. Those authors found that although different cultural groups might show distinct value priorities, they can be mapped into a single universal space composed of only seven motivational domains, organized according to whether they serve individualistic, collectivist/societal or mixed interests, as listed in Table 16.2.

Rokeach's Value Scale has been commonly used to describe the aggregate value structure of populations. For example, Rokeach (1973) describes the value system in the United States in 1971. Rokeach and Ball-Rokeach (1989) compare the aggregate value system in the U.S. in several years from 1968 to 1981, showing the relative stability of values on the long run. Bond (1988), Penner and Ahn (1977) and Kamakura and Mazzon (1991) compare the aggregate value system in the US with those in China, Vietnam and Brazil, respectively. Although different cultures have been found in these studies to differ in their value systems, the universality of human values (i.e., the underlying value dimensions) has been clearly documented (see also Schwartz and Bilsky 1987, 1990).

Table 16.1: Rokeach's Terminal and Instrumental Values

Terminal Values	Instrumental Values
Family Security (taking care of loved ones)	Honest
Happiness (contentment)	Loving
A World at Peace (free from war and conflict)	Responsible
Self-respect (self-esteem)	Helpful
True Friendship (close companionship)	Cheerful
Freedom (independence, free choice)	Polite
Inner Harmony (freedom from inner conflict)	Clean
Equality (brotherhood, equal opportunity for all)	Forgiving
A World of Beauty (beauty of nature and the arts)	Broad-minded
A Comfortable Life (a prosperous life)	Intellectual
Wisdom (a mature understanding of life)	Self-controlled
Pleasure (an enjoyable, leisurely life)	Independent
A Sense of Accomplishment (lasting contribution)	Courageous
Mature Love (sexual and spiritual intimacy)	Capable
Social Recognition (respect, admiration)	Obedient
National Security (protection from attack)	Ambitious
An Exciting Life (a stimulating, active life)	Imaginative
Salvation (being saved, eternal life)	Logical

Part 4 Applied market segmentation

Table 16.2: Motivational Domains of Rokeach's Values Scale

Nature of Interests	Motivational Domain	Description of the Motivational Domain
Individualistic	*Enjoyment*	Values reflecting physiological gratification translated into socially acceptable terms
	Self-direction	Reliance and gratification from one's independent capabilities
	Achievement	Social recognition and admiration
Collectivistic or Societal	*Prosocial*	Values expressing a concern for the welfare of others
	Restrictive Conformity	Values (mostly instrumental) emphasizing conformity to social norms
Mixed (individual & societal)	*Maturity*	experiencing and coming to terms with life (Schwartz and Bilsky 1987, p.553)
	Security	basic needs for safety at the family, national and global levels.

Rokeach's Value Scale was also used in the past to explain differences in value systems among market segments defined a priori on another basis such as media usage, brand preferences, and drug and alcohol addiction (Becker and Connor 1981, Pitts and Woodside 1983, Toler 1975, respectively), among others. In those applications, the RVS scale was used to describe segments, rather than as a basis for market segmentation. A model for the explicit purpose of market segmentation using the RVS scale was proposed more recently by Kamakura and Mazzon (1991). The purpose of their mixture rank-order logit model (discussed in chapter 13) is to identify groups of consumers that are homogeneous in terms of their terminal value priorities, or terminal value systems. Application of the model to data obtained from the RVS scale leads to the identification of multiple value systems within a population. This choice-based model is also in tune with Rokeach and Ball-Rokeach's (1989, p. 776) call for measuring values "in a manner that will faithfully reflect the phenomenological reality of people engaging in value choice behavior". The model was later extended for the simultaneous identification of value systems defined by both terminal and instrumental values (Kamakura et al. 1994). There the mixture of rank logits is adapted to consider two sets of value rankings (terminal and instrumental). The two sets of rankings are locally independent (because each respondent ranks the values within each set in separate tasks), but are clearly correlated across the sample because theory predicts a logical connection between the desired end states in life (terminal values) and modes of conduct that will lead to those goals (instrumental values). Kamakura et al. (1994)

also identify the value segments based on an incomplete ranking of terminal and instrumental values. Because of the large number of values in the two sets (a total of 2 H 18 items to be sorted in order of importance), collecting the complete rankings represents a substantial burden to the respondent, especially because values are only one of many constructs under study in a typical survey. A desirable feature of the model is that it is able to measure the relative importances of all terminal and instrumental values in each value system, even though each respondent has supplied only a partial ranking within each set. That is possible because the value system is identified on the basis of the information (albeit incomplete) provided by all members of the segment about both terminal and instrumental values. The researchers used a *combined approach* (see chapter 11), in which each segment is identified on the basis of both sets of values.

The List of Values (LOV) Scale

Kahle's (1983) LOV is an abbreviated scale that contains only terminal values. Moreover, the scale is limited to motivational domains that relate to individualistic or mixed (individual and societal) interests. The scale is composed of the following nine values: *self-respect, self-fulfillment, accomplishment, being well respected, fun and enjoyment, excitement, warm relationships with others, a sense of belonging,* and *security.* Two of those items are identical to Rokeach's terminal values. The other seven either combine several of Rokeach's items or generalize a specific RVS item. The nine items cover the three individualistic and the two mixed motivational domains listed in Table 16.2.

The main distinction between the RVS and LOV scales, from a theoretical point of view, is that the latter does not include any value related to societal interests. Beatty et al. (1985) argue that person-oriented (i.e., individualistic) values are of greater relevance than societal values in a consumer-behavior context. However, their argument implies that consumption decisions are not influenced by a consumer's concern for the welfare of others or conformity to social norms. An empirical comparison between the RVS and LOV scales on a convenience sample of 356 residents of a college town (Beatty et al. 1985) showed limited evidence of convergent validity between the two scales.

An obvious advantage of the LOV scale is the simplicity of data collection. A full ranking of the scale items does not demand as much processing effort from the respondent as the ranking of the 18 values in the terminal and instrumental sets of Rokeach's scale. Moreover, in many applications of the LOV, subjects are placed into segments on the basis of only the single value chosen as the most important (Beatty et al 1985; Kahle 1983; Kahle et al. 1986, among others), thus requiring just a single question for classification. However, the classification of consumers into segments on the basis of a single observation is highly vulnerable to response errors, and is not consistent with Rokeach's and Maslow's hierarchies of values and motivations; important information about the respondent's value system is left out. A recent comparison between this simple classification and one based on complete value systems (Kamakura and Novak 1992) employing the mixture of multinomial logits showed a substantial advantage of the latter in predicting a wide range of reported behaviors. Furthermore, the clusterwise rank-logit model can be easily adapted for value-system segmentation with incomplete and tied rankings, thus requiring only a partial ranking of Rokeach's terminal and instrumental values (Kamakura et al. 1994).

The Values and Lifestyles (VALS™) Survey

Certainly the most widely applied psychographic segmentation instrument is SRI's Values and Lifestyles (VALS™). The VALS™ lifestyle typology draws from the same theoretical concepts used by Rokeach in his value scale. More specifically, the original VALS typology starts from Maslow's needs hierarchy, defining four major groups: *need-driven*, *outer-directed*, *inner-directed* and *integrated*. Need-driven consumers are at the bottom of the hierarchy and have consumption patterns driven by basic need rather than preference. At the top of the hierarchy are the integrate's, "self-actualizing" consumers whose value systems combine outer-directed social concerns with inner-directed personal concerns. VALS™'s complete typology includes a subdivision of the need-driven and inner-directed groups into three hierarchical segments each, as well as a breakdown of outer-directeds into two hierarchical segments. More details about this typology are not relevant at this point, because it has been replaced by a new one.

When it was introduced in 1978, the original VALS™ system was criticized by academic consumer researchers, not only for its questionable validity or reliability, but also for the secrecy of its conceptual development and methodology, which prevented researchers from testing it. Nevertheless, the VALS™ system was well received in the popular press for its colorful descriptions of consumers in terms of their cultural and societal values. The critics granted that the VALS™ typology, given its basis in social values, was helpful in developing and executing advertising messages tailored to the intrinsic motivations of the target consumer. However, they complained that its dependence on general social values led to a typology that did not necessarily apply to consumers' behavior toward a specific brand or product category. VALS™ was criticized for being based solely on criteria (i.e., values) far remote from actual product-related behavior. When relating actual buying behavior to the VALS™ segments, some users found that their primary customers were classified into segments with diametrically opposite VALS™ profiles. Another common complaint was that the nine segments in the VALS™ typology differed markedly in sizes, with some segments (e.g., experientials) being less than 5% of the population.

Using a small convenience sample of students in Oregon, Kahle, Beatty and Homer (1986) compared the VALS™ classification with the one obtained with the LOV scale, coming to the conclusion that the LOV classification predicted consumer behavior trends more often than the VALS™ system. However, when their study was later replicated (Novak and MacEvoy 1990) with a large national probability sample, demographics alone had more predictive power than the LOV scale. Furthermore when demographics were not included, the VALS™ scoring system produced better predictions of consumer behavior.

While the criticisms were circulating, a new VALS™ typology (VALS2™) was under development to address many of the concerns. VALS2™ departs from the original system in two ways. First, in contrast to the original typology based on the hierarchy of social values, VALS2™ starts from multiple personality constructs. Second, the typology was developed with explicit consideration of buying behavior. The starting point for VALS2™ was a large survey (400 items) of more than 2,300 consumers, measuring 65 psychological constructs, including a wide range of personality traits and interests. In a second survey, the psychological constructs were combined with respondents' media behavior and consumption patterns over 170 product categories; items that did not correlate with behavior were eliminated, leading to a final scale with four demographic (age, education, income and sex) and 42 psychographic items measuring intellectual, status and action orientation (Piirto 1991).

The new VALS2™ defines lifestyle segments along two constructs: *self-orientation* and *resources*. Even though the new typology started from the measurement of personality, demographics and interests, its main dimension, self-orientation, refers to consumers' three basic motivation: (1) principles or beliefs, (2) status and (3) action. Principle-oriented consumers try to behave in accordance to their beliefs. Status-oriented consumers make consumption decisions to either maintain or improve their self-perceived social status. Action-oriented consumers seek to affect the social and physical environment in which they live.

The resources dimension reflects consumers' ability to act on their motivations. This second dimension acknowledges financial and psychological constraints that prevent consumers from acting on their values and motivations. VALS2™ defines eight segments of fairly similar sizes (each representing 8 to 16 % of the population), in contrast to the original typology of nine highly skewed segments. The two most extreme segments on the resource dimension are not distinguished in terms of self-orientation. *Actualizers*, who have the most abundant resources, are able to indulge in any of the three self-orientations. *Strugglers*, on the opposite extreme, are poor and uneducated, and have their lives focused on meeting basic daily needs, thus showing no clear self-orientation. The remaining six segments are clearly defined on self-orientation and resources, as shown below:

	Principle Oriented	Status Oriented	Action Oriented
High Resource	Fulfilleds	Achievers	Experiencers
Low Resource	Believers	Strivers	Makers

As with AIOs and the RVS, the VALS™ typology is often used not to define the market segments for a brand or product, but to better describe segments formed a priori on another basis. The purpose generally is to obtain a better understanding of the segments formed on another basis. For example, segments may be formed first on the basis of brand usage, and then described according to the VALS™ typology (Piirto 1991). Such classification of heavy/light users by VALS2 typology is reported in

Simmons' Mediamart for a large number of product categories, and for the leading brands in those categories.

Aside from the typology of eight lifestyle segments, VALS2™ provides information about the images and words that most appeal to members of each segment. Therefore, once the target market for a brand is identified in terms of the VALS2™ typology, its manager can choose the images and words that best reflect the motivations of the members of that group in an attempt to produce more effective promotional messages. Critics of the VALS2™ system still argue that only a small portion of a brand's target market will fit into a single VALS2™ group (Piirto 1991), leading to multiple (possibly conflicting) messages for the same target market.

Applications of Lifestyle Segmentation

Here we briefly review a few of the numerous studies applying AIOs for lifestyle segmentation. For more details, we refer the reader to the comprehensive discussion of psychographics by Gunter and Furnham (1992). Many of the applications of AIOs in market segmentation use activities, interests and opinions not as a basis for segmentation, but as descriptors of segments defined in terms of product use, media exposure, shopping behavior, etc. For example, Wells and Beard (1974) use lifestyle dimensions to produce a psychographic profile of heavy users of various products. McConkey and Warren (1987) use general AIOs to compare the psychographic profiles of heavy, light and non-buyers of state lottery tickets, finding significant differences, especially between heavy and non-buyers.

Several similar descriptive studies using AIO comparisons can be found in the retailing and advertising literatures. Peterson (1972) compares the psychographic profiles of the audiences for newspapers, magazines, TV and radio, whereas Teel et al. (1979) use AIOs to describe the lifestyle of radio and television audiences defined a priori by time slots. Potter et al. (1988) identify post hoc segments of VCR owners based on their reported usage behavior, and then produce a psychographic profile for each of the behavioral segments based on a small set of 19 AIO items. A more elaborate approach is used by Bass et al. (1969), who first identify post hoc magazine readership segments by using cluster analysis, and then discriminate among those readership segments by using psychographic and demographic variables. In retailing, AIOs have been used to compare customers and noncustomers of fast food restaurants and department, discount and convenience stores, as well as to compare shoppers of different types of stores (Bearden et al. 1978). They have also been used to contrast local shoppers (those who tend to shop in their trade area) with out-shoppers (those who make longer shopping trips), defined for various product categories.

The descriptive studies reviewed above highlight an important limitation of general AIOs (as well as other general bases for lifestyle segmentation): the typologies obtained from the general bases are useful for marketing decision making only if they are related to other segments obtained on the basis of product usage, shopping behavior, media exposure, etc., as shown above. In other words, general lifestyle typologies are rarely used as market segments per se; they are often used as descriptors of segments obtained from other segmentation bases. In these cases, AIOs might be more useful as multidimensional measures of the various aspects of lifestyle, such as the 22 general dimensions of lifestyle identified by Wells and Tigert (1971), than as bases for segmentation.

On the other hand, general AIOs may produce actionable lifestyle segments when combined with items that are specific to the product/service under study. For example, Ziff (1971) reports a study in which general lifestyle items were combined with items specific to various product categories. For prescription drugs, Ziff found four segments with clear health-related lifestyles: *(1) realists*, who emphasize convenience, do not worry excessively about their health, and do not see a strong need for prescription drugs; *(2) authority seekers,* who are doctor and prescription oriented; *(3) skeptics*, who have very low health concerns, and are skeptical about over-the-counter remedies; *(4) hypochondriacs*, who are extremely concerned about their health and heavy users of over-the-counter drugs. In a similar vein Blackwell and Talarzyk (1977) also combine general AIOs with items more specifically related to the service under study (health care services). Lesser and Hughes (1986) combine general AIOs with items more directly related to consumer shopping behavior, to produce a typology of shopping lifestyles.

Applications of Rokeach's Value Survey for market segmentation have been relatively less common. In most applications, the RVS scale is used to describe segments formed on another basis. For example, Pitts and Woodside (1983), found that values were not strongly related to product usage or brand preferences for cars, deodorants and vacations, but that values were related to the criteria used in making choice decisions. Becker and Conner (1981) found differences in RVS values across heavy users of various advertising media. In contrast to those descriptive applications, Kamakura and Mazzon (1991) use the RVS instrument as the basis for segmentation, and identify six value segments in a sample of 800 adults in Brazil.

In another application of the RVS as a segmentation basis, Kamakura et al. (1994) constructed a cross-national typology of value systems from a sample of 1573 housewives in three European countries (UK, Italy and West Germany). Their values segmentation model was applied to the sample irrespective of nationality to identify value systems that were not necessarily constrained by geopolitical boundaries. That approach was used to test whether groups of consumers of distinct nationality shared the same value system. On the basis of the CAIC criterion the researchers identified five segments, described in Table 16.3 in terms of their latent values, measured as deviations from the column and row means. The value segments derived with mixture models often show a dominant ordering of values across all segments, with only subsets of value systems differing among segments. The authors solved this problem by double-centering the values, which tends to show the differences among the segments more clearly. The double-centered values indicate whether a particular terminal or instrumental value (marked by bold numbers) is more/less important to a segment than to the sample as a whole. The distinctions among those segments can be visualized in Figures 16.1 and 16.2, which show a singular value decomposition of the results from Table 16.3.

Contrary to their expectations, Kamakura et al. found a strong country effect in the value segments. Segments D and E were clearly Italian, containing the majority (87%) of them. Segments A and B contained 77% of the Germans, and more than half (54%) of the British respondents were members of segment A. Nevertheless, nearly half of the British, 23% of the Germans and 13% of the Italians belonged to "cross-national" segments. The geographic orientation of those segments can be seen in Figures 16.3 and 16.4, which show the centroid position for each of the regions in Italy, Germany and the UK, based on the relative size of each segment within those regions among the 1573 respondents. Once again, one can clearly see that the value segments are closely related to specific nationalities; the regions of the UK mostly contain members of segment C, all regions in Germany have a even combination of segments A and B, and the Italian regions contain different combinations of segments C and D.

Several of the published studies on the LOV scale have compared it with other psychographic instruments (Beatty et al. 1985; Kahle et al.1986; Novak and McEvoy 1990). Early applications of the LOV scale defined the lifestyle typology solely on the basis of the highest ranked value (Kahle 1983; Kahle et al. 1986). The empirical results of Kamakura and Mazzon (1991) illustrate the potential errors in classifying consumers into value segments solely on that basis; two pairs of segments show the same value as the most important, but differ substantially on the relative importance of the other 17 terminal values. Predictive tests also show correlations with general activities and fashion-related beliefs that are of the same order of magnitude as in previous validation studies. Moreover, arbitrary decisions were previously made to collapse the nine segments into a smaller number of segments. For example, Kahle (1983) collapsed *fun and enjoyment* with *life with excitement*. Kahle et al. (1986) collapsed those same segments with *being well respected*, thus mixing segments that emphasize quite distinct value domains - enjoyment and achievement (Schwartz and Bilsky 1987). More recently, LOV users developed a preference for monadic ratings. For example, Madrigal and Kahle (1994), measured the LOV with a rating scale in a study of vacation activity preferences. They factor-analyzed the rating scales, reducing the nine values to four underlying components, and then subjected the factor scores to a cluster analysis, to produce value segments. They found that one of the four segments had consistently high scores on all four factors, whereas another segment had consistently low scores. Those results indicate that the first segment finds all aspects of life very important and the second finds them unimportant, apparently in contradiction to Rokeach's notion of value-systems, i.e., the ordered nature of human values (Rokeach 1973, Rokeach and Ball-Rokeach 1989). Those results may have been brought about partly by flaws in the segmentation methodology, that is, the use of rating scales to measure value systems and the application of factor analysis prior to cluster analysis, which may eliminate part of the variance in the data that could have been useful in identifying segments (see chapter 5).

Table 16.3: Double-Centered Values (adapted from Kamakura et al. 1994)

Rokeach's Values	Segments				
	A	B	C	D	E
Terminal Values					
Family security	-0.4	**1.0**	**1.4**	**-0.7**	0.1
Happiness	0.4	**0.9**	**1.4**	-0.5	**-0.8**
World peace	**0.6**	**1.1**	**-0.5**	**-0.6**	0.3
Self-respect	0.1	-0.3	0.2	0.2	0.1
True friendship	0.4	0.2	0.4	-0.4	-0.2
Freedom	0.1	**-0.9**	-0.1	**0.6**	0.3
Inner harmony	**0.6**	0.4	**-0.9**	0.2	0.0
Equality	0.4	-0.2	**-0.7**	0.0	0.5
A world of beauty	**0.8**	**0.9**	**-0.9**	-0.2	-0.4
A comfortable life	**-1.0**	-0.3	**0.5**	**0.5**	-0.1
Wisdom	0.1	**-1.0**	-0.1	0.1	**0.6**
Pleasure	**-0.8**	-0.4	**0.8**	0.3	-0.4
Accomplishment	0.3	0.4	0.0	-0.1	-0.4
Mature love	**0.7**	0.5	-0.1	-0.2	-0.5
Social recognition	-0.2	-0.3	-0.4	0.5	0.2
National security	**-0.6**	0.3	-0.1	-0.2	0.4
Exciting life	**-1.1**	**-1.8**	0.5	**1.0**	**-0.8**
Salvation	-0.4	-0.5	**-1.3**	-0.5	**1.2**
Instrumental Values					
Honest	0.1	**1.4**	0.3	**-0.9**	0.5
Loving	0.5	0.3	**0.6**	-0.5	-0.1
Responsible	**0.7**	**0.8**	0.2	-0.5	-0.4
Helpful	**0.8**	**1.2**	-0.1	**-0.8**	-0.2
Cheerful	0.5	0.3	0.4	-0.2	-0.3
Polite	-0.5	0.1	0.4	-0.3	0.4
Clean	**-1.2**	0.1	0.5	-0.2	0.4
Forgiving	**0.7**	0.4	0.1	**-1.0**	0.1
Broadminded	**1.2**	0.2	-0.2	0.1	-0.5
Intellectual	**0.7**	0.3	**-1.7**	**0.6**	-0.2
Self-controlled	0.0	0.3	-0.1	0.0	0.1
Independent	-0.2	**-0.7**	-0.1	**0.7**	0.2
Courageous	-0.4	**-1.1**	-0.3	**0.6**	**0.6**
Capable	**-0.7**	**-2.3**	**0.6**	**0.6**	0.0
Obedient	**-2.4**	0.3	-0.3	-0.2	**0.7**
Ambitious	**-1.0**	-0.2	0.3	**0.7**	-0.3
Imaginative	**1.0**	**-0.7**	-0.3	**0.6**	**-0.9**
Logical	0.1	**-0.9**	-0.3	**0.8**	0.0

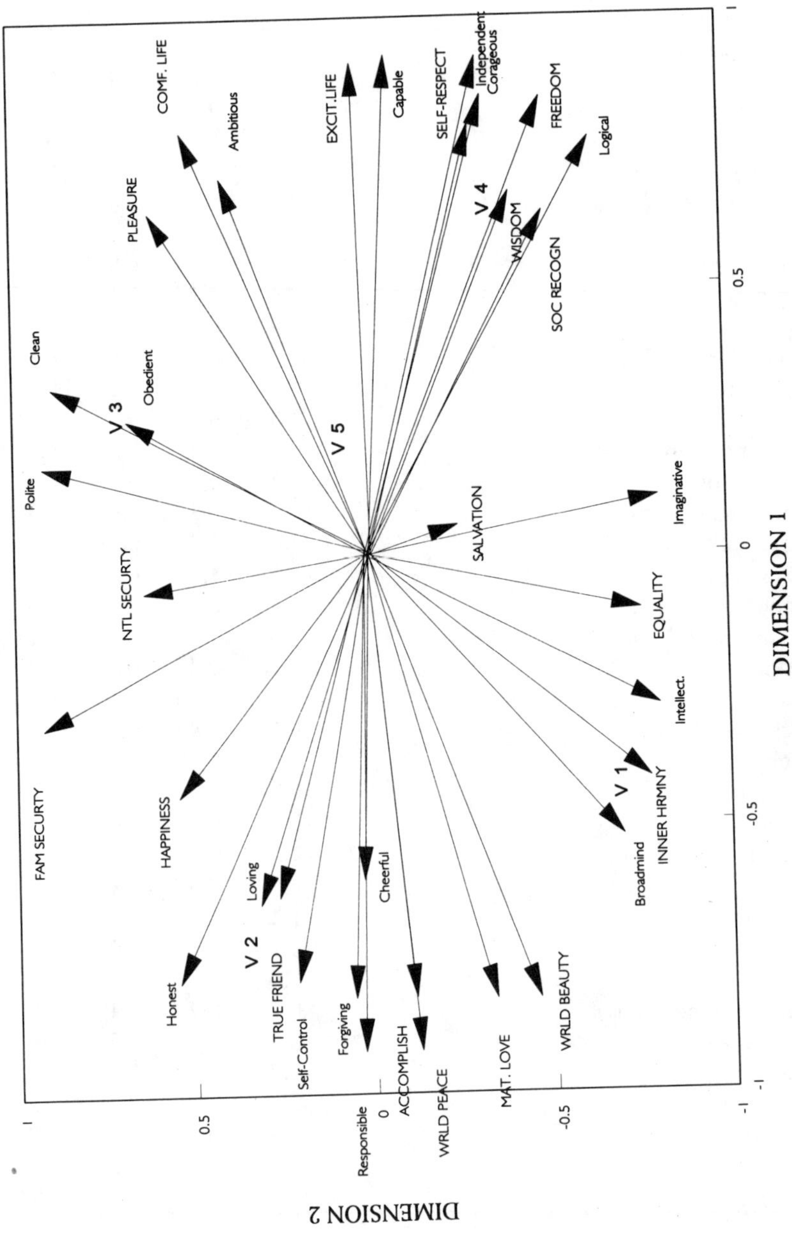

Part 4 Applied market segmentation

Figure 16.1: Values Map: Dimension 2 vs. Dimension 1 (Kamakura et al 1994)

Chapter 16 General unobservable bases: values and lifestyles

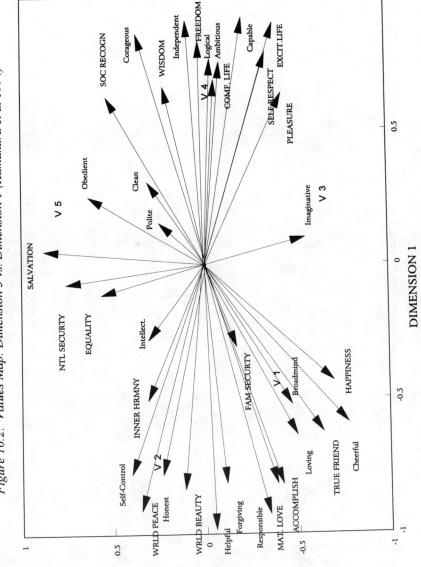

Figure 16.2: Values Map: Dimension 3 vs. Dimension 1 (Kamakura et al 1994)

273

Part 4 Applied market segmentation

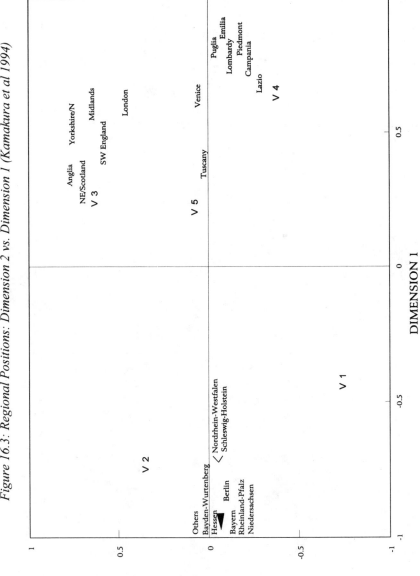

Figure 16.3: Regional Positions: Dimension 2 vs. Dimension 1 (Kamakura et al 1994)

Chapter 16 General unobservable bases: values and lifestyles

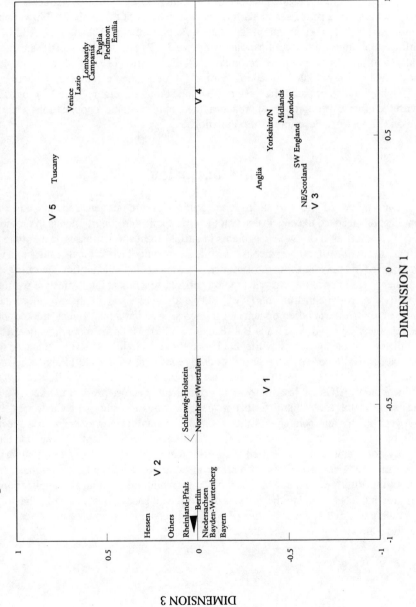

Figure 16.4: Regional Positions: Dimension 3 vs. Dimension 1 (Kamakura et al 1994)

Full descriptions and discussions of VALS™ applications are not common in the marketing literature because of its proprietary nature and commercial purposes. A few attempts have been made to compare the original VALS™ with the LOV, as discussed previously in this chapter. Subscribers of syndicated services such as Simmons and Mediamark can find cross-tabulations of the eight VALS2™ lifestyle segments against a wide range of a priori segments formed on the basis of other observable variables such as product and brand usage, demographics, media habits, etc. Additional information about other applications of the VALS2™ typology is mostly anecdotal, limited to brief descriptions of commercial studies. Piirto (1991) gives an example of how that typology would be used in a particular product market (analgesic) where users of leading brands of pain relievers are described along the eight VALS2™ groups. As with the other psychographic approaches, the emphasis in most applications seems to be on the post-hoc description of populations.

Conclusion

Our review of the various psychographic measurement instruments and their applications leads to the conclusion that their most valuable contribution is in providing a richer description of market segments identified from another bases, rather than the direct formation of homogeneous lifestyle segments. Two critical issues must be considered in applying those measurement instruments to the identification of lifestyle segments: stability of the lifestyle typology and its generalizability. Lifestyle segments based on value-type instruments (RVS, LOV and VALS™) are likely to be more stable over time than those based on activities, interests and opinions. Values are known to be relatively stable (Rokeach and Ball-Rokeach 1989). AIO's are more directly affected by the constraints of everyday life and hence expected to be less stable. Because they are closer to the core of consumers' cognitive system, values are likely to be more generalizable across situations. Because they are more directly affected by the environment, AIOs are less likely to be generalizable across product markets. Future research in this area should further investigate ways to establish a theoretical link between values and behavior. Although values are stable, consumers with a specific purchase or consumption pattern may fit in different value segments. Means-end chain theory, for example, might be used to assess and explain such findings, by showing that the same behavior may have different consequences to different individuals, thus satisfying different value systems. On the methodological side, the application of mixture models tailored to the analysis of two sets of bases (lifestyles and behavior) is necessary to simultaneously derive lifestyle and behavioral segments and to reveal the relationship among them (see chapter 11).

17
PRODUCT-SPECIFIC OBSERVABLE BASES: RESPONSE-BASED SEGMENTATION

The applications of market segmentation discussed in the preceding chapters are based on the assumption that values, lifestyles and geo-demographic profiles are associated with consumer behavior. That assumption is crucial if the segments identified on the basis of those variables are to differ in their response to the marketing mix, thus enabling managers to target specific marketing strategies to chosen segments. In this chapter we review applications in which segments are formed directly on the basis of consumer preferences and their response to price and sales promotions.

The Information Revolution and Marketing Research

Diffusion of Information Technology

In the 1970s laser and computer technology first enabled retailers in the United States to electronically record purchases made in stores. Since then, the adoption of scanning has increased rapidly. Although scanning was initially adopted as a means of saving labor and costs, the accumulation of point- of- sale information in databases has provided accurate and detailed consumer behavior information to marketers and marketing researchers. Several years after those developments had taken place in the United States, scanning was introduced in Europe. In 1990, 18,530 stores were equipped with scanning devices in the US, and 62% of the commodity volume turnover was scanned; in Europe 16,230 stores had scanners, and 27% of the commodity volume was scanned. There are large differences between countries in Europe. For example, in 1994, only 10% of the volume was registered through scanning in Switzerland, versus 85% in Sweden (compared to 71% for the US and 50% for Canada in the same period). In most Western countries the penetration of scanning is now sufficiently high to provide reliable information for decision making in the groceries market. Scanning affords the advantages of greater accuracy and standardization of measurement, relatively low costs of data collection, short and exact data intervals, higher speed of reporting and a more detailed information. Scanning data have been used in many fields of managerial decision making, including brand and category management, the design and timing of promotional efforts, distribution and shelf space allocation, and efficient consumer response. Foekens (1995) provides a detailed overview of the development of scanning in the US and Europe.

Part 4 Applied market segmentation

With the rapid implementation of information technology in the form of checkout scanners, large customer databases, and geo-demographic databases, there is a growing need to coordinate and integrate the accumulating information. The need for integration is intensified by the growing complexity of the marketplace, where the increasing number of market segments and the proliferation of products and services are making it more difficult for managers to decide what to offer, how best to place the offerings and how to communicate with narrower, more specialized market niches. Advertisers must now reach a much more diverse consumer population through a broad range of specialized media (cable TV, catalogs, specialized magazines, geographically targeted print media, Web pages, E-zines, etc.) to be heard and noticed in a more cluttered environment. Brand managers must target promotions more effectively in a market with multiple targeting options, competing against a larger number of brands for consumer attention and trade acceptance; for example, the average supermarket in the US carries more than 20,000 items and a large store carries in excess of 60,000 items.

Traditional sources of marketing data such as diary panels, store audits, advertising recall surveys and TV audience panels have provided useful information for marketing decision makers on important variables such as advertising exposure, message recall, store environment (shelf facings and position, prices, promotions, displays, etc.) and purchase behavior. However, the information was collected in different samples that were often not directly related to each other because of differences in samples, volumetric definitions, time aggregation, etc. Diary panels provided information about *who* purchased *what*, but not about *why*. Store audits provided information on aggregate demand and on the store environment, but that information could not be related directly to the behavior of consumers at the individual or segment level.

Early Approaches to Heterogeneity

Given the constraints on traditional sources of consumer and retail data, the early behavior-based segmentation models did not consider consumers' response to marketing activity. Early consumer choice researchers (Herniter and Magee 1961, Massy 1966) used Markov processes to explain consumers' brand-switching over time, without considering causal factors, such as price and sales promotions, that could be inducing the switching behavior. In the early studies of consumer choice behavior, researchers were concerned with the spurious effects of unobserved heterogeneity. By analyzing the aggregate and individual choice behavior of consumers from a diary panel, Massy (1966) demonstrated that the aggregation of zero-order brand switching matrices across consumers with diverse brand preferences would lead to an aggregate switching matrix of a higher order.

One of the earliest models accounting for unobserved heterogeneity was the beta-Bernoulli model proposed by Frank (1962), in which the probability of buying or not buying a brand was modeled as a Bernoulli process at the consumer level, with a beta distribution on the purchase probabilities across consumers. Other authors (Massy, Montgomery and Morrison 1970, pp. 94-112) proposed heterogeneous Markov models that allow for individual differences in brand-switching probabilities. For example, Morrison (1966) proposed a brand-switching model with two segments, a "stayer"

Chapter 17 Product-specific observable bases: response based segmentation

segment that never switches brands and a "mover" segment that switches brands according to a first-order Markov process. More recently, Grover and Srinivasan (1987; cf. Poulsen 1993) proposed a model with multiple "stayer" segments and multiple zero-order switching segments; they applied Clogg's (1977) MLLSA for a latent-class analysis (Goodman 1974a) of a brand-switching matrix, identifying the multiple "stayer" and "mover" segments. Grover and Srinivasan (1989) later extended their approach to consider the impact of marketing activity on market shares. They first aggregated household-level choices to produce aggregate brand-switching matrices, and then applied the latent class analysis of brand switching matrices to identify market segments. In a third stage, they classified a panel of consumers into the identified market segments and then applied an MCI logit model (Nakanishi and Cooper 1982) to relate the segment-level market shares to price and sales promotions. That latent-class approach for brand switching analysis has been extended by Poulsen (1990), who proposes a finite-mixture of Markov chains that allows for differences in switching probabilities across segments, thus allowing for first-order Markov brand-switching behavior within each segment. These models are described in more detail in chapters 6 and 10 of this book.

Household-Level Single-Source Data

The integration of response data (purchase incidence, brand choice) with causal factors at the store level (store display, feature, discounts) and at the household level (advertising exposure, coupon use) into single-source databases led to a revolution in the marketing research industry and in academic research in the field of marketing. An important step in the development of technology has been the collection of household-level scanner data. Here, individual households identify themselves at the checkout counter through an identification card, so that purchases made can be traced back to individual households over time. Such household level scanner data have been available in the US for some time, but in Europe they are available only on a limited scale. The combination of store environment data from a store tracking panel with household-level data on purchase history, demographic and psychographic profile, and advertising exposure results in powerful single-source databases. The availability of single-source data, at both store and household levels, created the opportunity for more accurate and more direct measurement of consumer response to the marketing mix, especially for the components of the marketing mix with immediate impact on consumer behavior, such as price and sales promotions. Those integrated sources offer many benefits over the traditional independent sources, such as:

(1) Measurement of causal factors at the point of their effect (e.g., displays and in-store promotions at the store level; coupon usage and observed exposure to advertising at the household level).

(2) Non-obtrusive measurements of actual purchase behavior at either the household or store level and greater reliability of measurement.

(3) Timely (typically on a weekly basis) measurements of market response at a highly disaggregate level (either the household or store level) and high speed of reporting.

(4) Estimation of consumer and market response to marketing mix by relating observed behavior with causal factors collected from the same sampling units, or from directly matched samples.

(5) Convergent volumetric projections at the regional (50 to 66 markets in the US) and market levels based on both household and store tracking panels.

Despite the richness of information obtained from household scanner panels, it is generally believed that more accurate information about sales volume can be obtained from store-tracking panels, because of their broader market coverage and higher level of aggregation over consumers. For example, Leeflang and Olivier (1985) have shown that sample survey data from consumer panels and retail store audit data are subject to different biases, leading to substantial differences in projected sales, while also differing from internal factory shipment data. Those authors have shown that substantial reductions in the biases can be obtained by using coverage factors as weights in the analyses. Providers of household panel data attempt to overcome such problems through elaborate projection systems that account for the over/under representation of various types of consumers in their samples. The weights necessary to project results from the household sample to the population have been largely ignored in academic research, possibly because these projections systems are proprietary. The theoretical equivalence between micro-level models of household purchase behavior and macro-level models of brand shares at the store have been demonstrated by Russell and Kamakura (1994) for a heterogeneous model of consumer purchase behavior and by Gupta et al. (1996) for a homogeneous model of brand choice. Gupta at al. (1996) also compare own-elasticities obtained from simple IIA models (homogeneous multinomial logit) estimated from household and store-tracking panels for two product categories, finding statistically significant differences in the choice behavior, but only small differences in the estimated own-elasticities. Foekens (1995) presents an overview of problems due to various forms of aggregation on scanner data. Further research is needed in this area, to isolate the aggregation bias implicit in the IIA restriction from the sample biases.

Single-source databases also minimize problems in reconciling definitions of products, units of measurement, time periods, and market boundaries across data sources, and provide richer information for projecting volumes from the household level or store level to regions and total market. The highly disaggregated level of the data (at the SKU, household, store and weekly levels) from integrated sources enables the analyst to estimate market response to marketing mix variables at the micro-level, and to form consumer segments on the basis of preferences for brands, sizes, flavors, etc. and consumer responses to in-store and out-of-store causal factors. The following information is usually available from a single-source database from household and store panels in the United States:

Chapter 17 Product-specific observable bases: response based segmentation

Households:
- Household composition (demographic profile of each household member)
- Dwelling (characteristics of the residence)
- Geographic (ZIP 4+, ADI and other classifications)
- TV viewership (exposure to programming and advertising, by household member)
- Shopping trip (timing, location and nature of the shopping trip)
- Purchases (UPCs chosen, units and volume purchased, prices paid, coupon usage)
- Projection weights

Stores:
- Store characteristics (store type, ACV by category, market geo-demographic)
- Shelf stock (UPCs carried each week, shelf facings and location)
- Prices (regular and discounted, by SKU)
- In-store promotions (displays, special packages)
- Weekly store and feature advertising

Currently, the two main providers of single-source databases for packaged goods in the US and Europe are A.C. Nielsen and Information Resources Inc. (IRI). Nielsen originally collected sales data through store audits. Nielsen's store auditing service evolved into Scantrack, a store tracking panel containing more than 4,800 stores in 50 major metropolitan areas by mid-1990s, reporting weekly sales data directly collected at the checkout scanners. Nielsen's Scantrack service started with an in-store ID card panel of 27,000 households, but moved to an in-home hand-held scanner panel with 40,000 demographically balanced US households that covers purchases in all channels (food and non-food stores) over 1,147 counties. The household sample is enhanced for local special studies (e.g., test markets) in 16 major markets. Nielsen ties part of the household panel to its Monitor Plan TV tracking panel, covering 23 major TV markets.

IRI (founded in 1978) started as a market testing service by equipping all grocery stores in a market with checkout scanners, building a panel of households that made purchases at those stores using a scannable ID card, and controlling the exposure of those households to TV advertising through their cable service. IRI moved from a test marketing service to a national tracking service (Infoscan) in 1987. By the mid-1990s IRI provided store-level data from more than 3,000 grocery stores, 550 drug stores and 288 mass merchandisers, combined with household-level behavior data collected from a sample of 60,000 households through an ID card scanned at the store, and from a sample of 20,000 household using hand-held scanners at home.

In Europe, for example GFK (previously AGB) has implemented a single-source data collection system called Dating-2000 based on home scanning. Here, households scan purchases when they return from a shopping trip by using a hand-held scanning device, which is downloaded to the computers by telephone. The data are integrated with demographic, psychographic and media exposure data obtained through questionnaires.

Consumer Heterogeneity in Response to Marketing Stimuli

Store-level tracking data have been widely used to assess the competitive effects of price and sales promotions at the aggregate (overall market, region or store) level (Blattberg and Wisnewski 1989; Cooper and Nakanishi 1988). Because the sampling unit is often a store, chain or sales region, these aggregate analyses reflect the actual sales and shares in the markets under study, leading to more valid results for the specific markets. However, the aggregation of consumer behavior at the store or market level may produce results of limited usefulness to managers, because of aggregation biases. For example, Russell, Bucklin and Srinivasan (1993) report results from aggregate analyses with a large number of counter-intuitive cross-price elasticities. Moreover, results obtained from store-level data can be related to consumer characteristics only indirectly, through the profile of the typical shopper at each store. Store-tracking data provide limited, if any, insights into brand preferences and response to price and sales promotions within consumer segments. A possible exception is the unfolding logit approach by Zenor and Srivastava (1993).

In contrast, the integration of household scanner panel data and store environment data from tracking panels leads to a better understanding of the relationship between purchasing decisions (timing, incidence, choices), marketing mix variables (e.g., price and sales promotions) and consumer characteristics. The availability of actual behavioral data from individual households triggered a new wave of academic research modeling consumers' shopping and choice behavior in the packaged-goods industry. However, because of data limitations (less than a dozen purchase cycles on average from each consumer), models have often been estimated by pooling the data across consumers. The fact that consumers differ in both brand preferences and response to the marketing mix has critical implications for the results of these aggregate models, to both marketing researchers and managers. As in the case of the stochastic brand-switching models mentioned previously in this chapter (Frank 1962, Herniter and MaGee 1961, etc.), the estimation of aggregate models on the basis of individual data produces results that misrepresent consumer purchase behavior and the pattern of brand competition in the marketplace. That critical issue was raised by Hausman and Wise (1978), who demonstrated that the behavior of individuals following a simple choice process with Independence from Irrelevant Alternatives, when aggregated over individuals, will appear as a more complex process in which similar alternatives would compete more closely for choice shares than dissimilar ones. Empirical evidence of this biasing effect was found by Fischer and Nagin (1981), who compared individual-level estimates with those obtained from pooled data across individuals.

Because of the biasing effects of consumer heterogeneity in brand preferences and response to causal factors, several models have been proposed to account for the unobserved differences across consumers. Those efforts can be grouped into three major categories: (1) *models with exogenous indicators of preferences,* (2) *fixed effects models*, (3) *random-effects models*.

Models with Exogenous Indicators of Preferences

The biasing effects of unobserved heterogeneity in consumer preferences were a major concern since the very first attempts to model consumer choice behavior with single-source databases. In their seminal work on the estimation of multinomial logit models of brand choice with supermarket scanner data, Guadagni and Little (1983) used two predictors to account for individual differences in preferences for brands and package sizes. Those "brand loyalty" and "size loyalty" variables were defined as the exponential smoothing of past choices of brands and package sizes at each choice occasion, and were used as explanatory variables in the deterministic component of utility in a multinomial logit model. A similar measure of brand loyalty was used by Krishnamurthi and Raj (1988), for example, but on the basis of the recent choice shares for each brand; those authors also added income as a predictor to capture variations in purchase behavior across consumers.

In both cases, the individual-level measures of brand and size preferences change from one choice occasion to the next, depending on the recent purchase history. Therefore, while accounting for differences in preferences across consumers, those measures of brand and size loyalty impose a serial dependence among consecutive choice events. On one hand, the measures minimize the problem of *spurious state dependence* due to unobserved heterogeneity, but on the other, they imply a choice process with *structural state dependence* (Heckman 1981), for which the traditional maximum-likelihood estimator for the multinomial logit model no longer applies because the likelihood function for the observed choice histories can no longer be built as the product of choice probabilities for independent events. Nevertheless, the approach of including loyalty measures is widely used to account for spurious state dependence (as shown later in this chapter), because it is believed to reduce the bias and improve the consistency of the estimates for the response coefficients in choice models. However, although accounting for heterogeneity in brand preferences, the approach still assumes a single response coefficient across all consumers.

Fixed-Effects Models

These models consider differences in brand preferences across consumers by attempting to estimate certain parameters at the individual level, or to at least account for the individual-level effects in the estimation. A traditional approach in econometrics has been to include an intercept for each household or individual in the logit model (cf. Amemiya 1985 p. 353). Traditionally, heterogeneity has been considered in econometrics as a nuisance that should be addressed, but that is not the focus of research. A potential problem with the fixed effects estimator is that the estimates of the response parameters are not consistent as the sample size N goes to infinity. The reason is that the regularity conditions of the likelihood equations are violated. However, if both N and T (the number of time points) go to infinity, the estimators are consistent. Estimation procedures for fixed effect models include conditional and concentrated likelihood methods (cf. Amemiya 1985). Discussion of those methods is beyond the purposes of this chapter. Jones and Landewehr (1988) apply the fixed effects approach, using Chamberlain's (1980) maximum likelihood estimator to estimate the aggregate choice response to the store environment conditional on sufficient statistics for the household-specific fixed effects. The price parameter is assumed to be the same for all households, but is estimated conditional on the potential differences in brand preferences across consumers. Predictions of consumer response to the causal factors will be conditional on the unobservable individual effects (i.e., brand preferences). A more flexible fixed-effects model is proposed by Rossi and Allenby (1993). Their two-stage Bayesian approach uses the asymptotic distribution of the estimates from an aggregate logit model (after a correction for sample size) as a prior to be updated on the basis of the observed data from each household, producing household-level estimates of the posterior means for both the fixed effects (i.e., brand intercepts) and the response coefficients at the household level.

Random-Intercepts and Random Coefficients Models

Rather than attempting to obtain estimates for each household as in the fixed-effects approach or to directly explain differences with the use of exogenous variables, random-effects models account for unobserved heterogeneity across consumers by assuming that the parameters of the individual-level model vary across consumers according to a probability distribution. The so-called random intercepts models are a special case of this class of models. The random intercepts approach has been commonly applied in econometrics, where the intercepts in the choice model are assumed to follow some (usually a normal) distribution (cf. Heckman 1981). The advantage of assuming such a distribution for the intercepts is that the ML estimators for all parameters are consistent, as shown by Heckman (1981). Chintagunta, Jain and Vilcassim (1991) applied several of such models to household-level scanner data.

The random intercepts model is in fact a special case of random coefficients models, in which all coefficients are allowed to vary across the sample. That approach is more in line with the interest in response heterogeneity in marketing research. One of the earliest random coefficients choice models was the random-coefficients probit model developed by Hausman and Wise (1978), which assumed a probit choice model at the individual level and a multivariate normal distribution of the response coefficients across individuals. They showed that heterogeneity in "tastes" (i.e., multivariate normal distribution of response coefficients) leads to aggregate predictions that depart from the counterintuitive *independence from irrelevant alternatives* (IIA), even in situation where every individual is assumed to follow an IIA choice process. Hausman and Wise's (1978) model was extended to ideal-point preferences by Kamakura and Srivastava (1986), who show the competitive/cannibalization effects in product positioning due to preference heterogeneity. That random-ideal-points model was later extended by Kamakura (1991) in a finite-mixture model that allows for a more flexible distribution of ideal points in a product-positioning map as a finite mixture of either multivariate normals or uniform distributions with predefined ranges. These probit models were previously limited to problems of up to three choice alternatives because of computational problems in evaluating cumulative multivariate normals. However, that limitation has been overcome by the estimation of multinomial probit models with the simulation methods (McFadden 1989). The simulation methods involve an approximation of the integrals by repeatedly drawing from the distributions to be integrated, and summing across the draws.

A random-coefficients formulation similar to Hausman and Wise's (1978) was proposed by Gonul and Srinivasan (1993), but they used a multinomial logit model at the individual level, with a normal mixing distribution for the parameters across consumers. A different approach was taken by Steckel and VanHonacker (1988) in their heterogeneous conditional logit model. They assumed a logit model at the individual level but allowed for differences across consumers in the shape parameter of the extreme-value distribution of random utilities. Therefore, their heterogeneous logit model accounts for heterogeneity in the variance of the random component of utility, that is,, the degree of stochasticity in choice behavior.

Other authors (Heckman and Singer 1984) argue that the imposition of a parametric density function on the random effects might bias the estimates for the aggregate (common) response coefficients, and therefore advocate the use of nonparametric formulations. Heckman and Singer (1984) proposed a non-parametric estimation method that approximates the probability distribution of the intercepts across individuals by a point-mass discrete distribution. That approach was used by Gonul and Srinivasan (1993) to account for preference heterogeneity in a multinomial logit model. Chintagunta, Jain and Vilcassim (1991) compared the semiparametric with several parametric specifications of the random intercepts model and found superior fit for the semi-parametric specification. Similar to some of the models discussed previously, this "point-mass" nonparametric approach has the drawback of allowing for differences only in the intercepts, but not in the response coefficients. In fact, the non-parametric random intercepts approach is a mixture of logits model that fits in the GLIMMIX framework described in chapter 7 of this book, where only the intercept is allowed to vary across the mixture components.

Response-Based Segmentation

A full semiparametric random-effects approach was proposed by Kamakura and Russell (1989). Their finite mixture of multinomial logits assumes that consumers are grouped into segments that are relatively homogeneous in brand preferences and response to causal factors. Their article provides one of the benchmark works on heterogeneity and market segmentation in marketing. Their model, a member of the GLIMMIX family (chapter 7) of models, allows for consumer heterogeneity in both brand preferences and response to price and other causal factors. As explained in chapter 7, estimation of that finite mixture of logits leads to the simultaneous classification of consumers into market segments and estimation of a multinomial logit model for each segment. In contrast to the continuous mixtures of multinomial logits or probits, the semiparametric model affords reasonable flexibility in the distribution of brand preferences and response coefficients across consumers because it does not impose any strict functional form or any particular correlation structure among the segment-level parameters. The model, like many other mixture models in this book, assumes perfect homogeneity within each segment or mixture component. That assumption can be verified through the entropy measure (equation 6.11); if the derived segments are not well separated (i.e., the posterior segment membership probabilities are too "fuzzy"), neither of the segment-level multinomial logit models dominates the others in explaining each consumer's observed behavior. The authors propose a correction to account for within-segment heterogeneity in the computation of price elasticities. Nevertheless, further research is needed to develop response-segmentation models that allow for within-segment heterogeneity.

Example of Response-Based Segmentation with Single-Source Scanner Data

Kamakura and Russel's model was one of the first GLIMMIX models for response-based segmentation. They applied their finite mixture of multinomial logits to choice data obtained from a household scanner panel. The basic assumption of their model is that consumers differ in their brand preferences and response to the marketing mix, but can be grouped into relatively homogeneous market segments. Therefore, conditional on the consumer belonging to a particular (but unobservable) segment, his or her history of brand choices can be modeled by the well-known multinomial logit model. Before we describe the Kamakura and Russell model in more detail, we define the following notation.

Notation		
n	=	1,...,N indicate consumers
j	=	1,...,J indicate brands
t	=	1,...,T indicate choice occasions
y_{jnt}	=	1 if brand j was chosen by consumer n at occasion t or zero otherwise
s	=	1,...,S indicate segments
X_{jnt}	=	vector of predictors for brand j and consumer n at choice occasion t
β_s	=	vector of response coefficients for segment s
π_s	=	relative size of segment s

The likelihood of the observed choice history by consumer n, y_n, conditional on consumer n belonging to segment s is given by,

$$L(y_n|\beta_s) = \prod_{t=1}^{T} \prod_{j=1}^{J} P_j(\beta_s)^{y_{jnt}} \qquad (17.1)$$

where

Chapter 17 Product-specific observable bases: response based segmentation

$$P_j(\beta_s) = \frac{\exp(X_{jnt}\beta_s)}{\sum \exp(X_{j'nt}\beta_s)} \qquad (17.2)$$

The unconditional likelihood is given by

$$L(y_n|\beta;\pi) = \sum_{s=1}^{S} \pi_s L(y_n|\beta_s) \qquad (17.3)$$

The model above can be estimated either by a gradient search or through the EM algorithm described in Appendix A2, in which the M-step consists of the weighted (by the posterior class memberships) maximum likelihood estimation of a multinomial logit model for each segment. Market-level (across segments) cross-elasticities for the marketing mix variable X_j (for brand j) on the market shares for another brand j' can be calculated by

$$\eta_{j'j} = \sum_{s=1}^{S} \eta_{j'j}^s \pi_s S_{js} / S_j \qquad (17.4)$$

where

$\eta_{j'j}^s = -\beta_s S_{j's} X_j$ is the cross-elasticity of variable X_j for brand j on the market shares of brand j' within segment s,

S_{js} is the market share for brand j within segment s, and

S_j is the aggregate (across all segments) market share for brand j.

An interesting feature of this model is that even though each segment shows a simple cross-elasticity structure of a simple-effects market share model (Cooper and Nakanishi 1988), the aggregate (across all segments) cross-elasticities show a more complex competitive structure with differential effects.

The authors apply their model to 78 weeks of choice data for four brands of a frequently purchased product by a sample of 585 households. Using Akaike's information criterion (AIC), they decided on five segments, one of which did not respond to price. They also identified four "perfect loyal" segments that purchased only one brand during the sampling period. The estimated price elasticities for each of the four price-sensitive segments and for the aggregate (across all segments) market are shown in the first four columns of Table 17.1. Each of the columns shows the estimated percentage change in market share for the brand in the row in response to a 1% change in the price for the brand in the column. The impact of a brand's price within each segment is essentially the same across all competitors, thus showing the well-known proportional draw implied by the multinomial logit model. Moreover, the brand with the largest market share within a segment (e.g., brand A in segment 1, or Brand B in segment 2) has the highest impact on competitors in that segment. To measure the overall impact a brand has on competitors, Kamakura and Russell (1989) use a variation of the *clout index*, originally developed by Cooper and Nakanishi (1988),

$$\text{Clout}_{j'} = \sum_{j \neq j'} \eta_{jj'}^2 \qquad (17.5)$$

The extent to which brand j' is affected by competitors is measured by the *vulnerability index*,

$$\text{Vulnerability}_{j'} = \sum_{j \neq j'} \eta_{j'j}^2 \qquad (17.6)$$

The clout index and vulnerability index for each brand are listed in the last columns of Table 17.1. They show the strong position of brand A in segment 1, where it has a high clout with a minimum vulnerability to competitors. That same brand occupies a weak position in segment 4, however, having the lowest clout and highest vulnerability in that segment.

The market-level cross-elasticities show a more complex pattern of brand competition, which depends on how closely the brands compete with each other within particular segments. For example, a price cut by brands B, C or private labels would have less market wide impact on brand A than on the other competitors.

Extensions

Despite (and because of) its shortcomings, the model proposed by Kamakura and Russell (1989) has been the subject of several extensions. For example, Gupta and Chintagunta (1994) combine this model with Dayton and Macready's (1988) concomitant-variables approach into an integrated model that identifies consumer segments on the basis of their similarities in brand preferences and response coefficients, while simultaneously relating segment membership to consumers' demographic profile. This and other concomitant-variables mixture models are discussed in detail in chapter 9.

By combining the Kamakura and Russell model with Katahira's (1990) joint-space market response analysis, Chintagunta (1994) develops a model for both market segmentation and brand positioning. Instead of estimating a set of brand intercepts for each latent segment, Chintagunta uses Katahira's factor structure consisting of a joint space where brands are represented as points and the segment-level brand preferences are modeled as vectors. That approach is an interesting combination of the models discussed in chapters 7 and 8, including the features of a GLIMMIX model for the response to price and sales promotions and of a STUNMIX model for the mapping of brand preferences.

Chapter 17 Product-specific observable bases: response based segmentation

Table 17.1: Estimated Price Elasticities (adapted from Kamakura and Russell 1989)

	Brand A	Brand B	Brand C	Private Label	Market Share	Clout	Vulnerability
Segment 1							
Brand A	-1.5	0.5	0.4	0.3	79%	90.9	0.5
Brand B	5.7	-5.0	0.3	0.1	9%	0.5	32.0
Brand C	5.2	0.4	-4.6	0.2	7%	0.3	26.9
Private Label	5.7	0.3	0.3	-4.5	5%	0.1	32.4
Segment 2							
Brand A	-4.0	2.8	0.3	0.2	22%	3.7	7.8
Brand B	1.1	-1.5	0.4	0.2	65%	21.9	1.5
Brand C	1.0	2.8	-3.6	0.1	9%	0.4	9.1
Private Label	1.2	2.5	0.3	-3.3	4%	0.1	7.7
Segment 3							
Brand A	-8.5	2.1	4.1	0.6	16%	8.4	21.1
Brand B	1.5	-6.7	4.8	0.4	26%	14.4	26.0
Brand C	1.6	2.5	-4.1	0.5	52%	56.8	9.0
Private Label	1.9	1.9	4.1	-6.9	6%	0.7	24.3
Segment 4							
Brand A	-15.1	3.2	3.6	4.7	9%	7.5	46.1
Brand B	1.6	-11.3	4.8	4.4	24%	36.9	44.7
Brand C	1.5	4.0	-10.2	4.9	30%	55.4	41.8
Private Label	1.7	3.2	4.4	-8.0	37%	65.5	32.6
*Total Market**							
Brand A	-2.7	0.6	0.7	0.3	35%	4.3	1.1
Brand B	1.1	-3.9	2.1	0.8	26%	8.6	6.1
Brand C	1.3	2.3	-4.5	1.2	25%	10.7	8.4
Private Label	1.2	1.7	2.4	-4.6	14%	2.3	10.3

*Includes four brand loyal and one price-insensitive segments, in addition to segments 1 through 4.

One of the limitations of the finite-mixture of multinomial logits is that it still assumes *independence from irrelevant alternatives* within each market segment, implying a simple cross-elasticity with proportional draw across brands within each segment. To allow for a more complex competitive structure within a market segment, Kamakura, Kim and Lee (1996) propose a finite-mixture of nested logit models, based on an earlier heuristic approach by Kannan and Wright (1991). Their model assumes a choice process in which choice alternatives are hierarchically organized and, most importantly, allows the segments to differ in the nature of their hierarchical choice processes. The model identifies homogeneous segments in terms of their brand preferences and response to causal factors (e.g. price, and sales promotions), as well as in terms of the type of choice process used by each segment. The authors apply their model to choice data from a household scanner panel and demonstrate that it produces more flexible and realistic patterns of brand competition within each segment. They also demonstrate that their model subsumes the model by Kamakura and Russell (1989) as a special case. A more detailed discussion of this model is given in chapter 7.

A different approach for segmentation based on a nested-logit model was proposed by Bucklin and Gupta (1992) to identify segments on the basis of their brand choices and purchase incidence. Their integrated model provides valuable insights on how consumers respond to price and sales promotions, not only in brand choice but also in category purchase. The authors start with a nested-logit formulation in which brand choices are nested under the decision to make a purchase in the product category. Brand choices are modeled by a multinomial logit model, and the purchase incidence is modeled by a binary logit model in which the probability of purchase depends on the expected value of the maximum utility derived from the choice decision (also called the *inclusive value*). The authors use the finite-mixture of multinomial logits (Kamakura and Russell 1989) in two stages. First, the mixture of multinomial logits is applied to the choice data, yielding segments homogeneous in choice behavior. Once the choice segments are identified and consumers are assigned to a particular segment, a mixture of binary logits is applied to the purchase incidence data of their members, to identify and estimate the segment-level purchase incidence models. Although the model might seem conceptually similar to the finite mixture of nested logits (Kamakura, Kim and Lee 1996), its estimation is more akin to that of the mixture of multinomial logits, applied in two sequential steps in which the segmentation based on purchase incidence is conditional on a prior classification based on the results from the choice-based segmentation. The model formulated by Bucklin and Gupta (1992) combines Guadagni and Little's (1983) loyalty variables with Kamakura and Russell's nonparametric random-effects treatments for consumer heterogeneity. Bucklin, Gupta and Han (1995) also combine the mixture of multinomial logits with brand "loyalty."

Because of time-related factors such as inventory depletion, it is reasonable to expect that the propensity to re-purchase in a product category will change as time since the last purchase increases (Gupta 1991). However, the changes in purchase probability, as well as the impact of marketing variables, are likely to differ across brands and consumers. Wedel et al. (1995) propose a finite-mixture of piecewise exponential hazards that identifies consumer segments that are homogeneous in how they make decisions about when to make a purchase, what brand to choose and how many units to buy in response to weekly changes in marketing variables. To apply their model, they assume that (1) the timing of a shopping trip within a given week does not depend only on the particular product category under study and that (2) purchase quantities can be discretized into standardized *equivalent units* (a common practice in the marketing research industry, where volume is defined in standard units, rather than by each package size). The first assumption allows use of a discrete piecewise hazard function formulation with an exponential hazard within week and varying hazard rates between weeks; the second justifies a Poisson process for purchase volumes.

Chapter 17 Product-specific observable bases: response based segmentation

Wedel et al. (1995) specify nonproportional piecewise exponential hazards for each brand-switching pair, with a hazard rate that includes (1) a Box-Cox baseline, (2) cross-effects of marketing variables for the switching pair, and (3) interactions between the marketing variables and the duration variables to account for non-proportionality in the hazard rate. Even though a simple "memory less" exponential hazard is assumed within a given week, the Box-Cox formulation of the baseline rate across weeks leads to a flexible hazard rate that allows for inventory effects and for decreasing purchase propensities, if needed. The nonproportional component of the hazard rate accounts for the possibility that the impact of marketing variables might change as the time since last purchase increases. Because of the assumed piecewise exponential hazard, estimation of the model is relatively simple as a finite mixture of Poisson regressions with time-varying purchase rates. However, because a first-order Markov process is assumed for brand switches, the model requires a large number of parameters to be estimated for each segment. A more detailed description of this model and empirical results are given in chapter 10.

The same problem of segmenting a market on the basis of brand choice and purchase quantity was tackled by Dillon and Gupta (1996), who used a finite mixture of Poisson regressions for the category volume and multinomial logits for the brand purchases given the category volume. Those authors start with the assumption that the category volume (sum of all brands) purchased by a household follows a Poisson process. Then they use a multinomial model for brand purchases conditional on the number of purchases in the product category. Individual differences in the Poisson purchase rate for the category and choice probabilities given the category volume are considered through a finite mixture, where the conditional (i.e., segment-level) likelihood is built as the joint likelihood of the Poisson and multinomial processes. The model also includes a "stayer/mover" class (Morrison 1966; Colombo and Morrison 1989), an extra latent class of consumers who are not expected ever to switch brands. Except for this additional mover-strayer class, the Dillon and Gupta model is similar to the model proposed by Böckenholt (1993b), described in chapter 7. By modeling the category volume as a Poisson process independent from the multinomial brand choices, the model assumes that the decision on how many units to purchase does not depend on brand choices; factors affecting brand choices can have an impact on category volume only if they are also included in the Poisson model for category volume. In reality, if a consumer's favorite brand is offering an unexpected discount in a given week, he or she is likely to make an unplanned purchase or to buy more than usual, thus affecting category volume. That effect of brand choice on category volume can be represented easily by a system of independent Poisson processes for each brand, which will result into a Poisson process for category volume and in a multinomial process for brand choices conditional on category volume. Such a model can be viewed as a special case of the model proposed by Wedel et al. (1995), with zero-order brand switching and "memory less" proportional hazards.

Despite the richness of information obtained from household scanner panels, more accurate information about sales volume can be obtained from store-tracking panels because of their broader market coverage and higher level of aggregation over consumers. Even though an un-weighted household panel sample might not produce results that match results at the macro level, the richness of information available at the household level provides a better understanding of consumer behavior and characteristics. The possibility that the household sample might not project perfectly into market volumes and response to the marketing mix is acknowledged by Russell and Kamakura (1994), who start with a model of individual purchase behavior to derive a store-level market-share model that considers consumer heterogeneity in brand preferences obtained from a household panel. Their model combines a mixture model of share-of-requirements from a sample of households with a constrained Dirichlet regression of brand shares at the store level. The mixture model applied to the household panel identifies segments of consumers on the basis of their brand preferences. Those segments and their intrinsic preferences for the available brands are then used to constrain the estimation of the Dirichlet regression model at the store level, thus avoiding the unrealistic patterns of brand competition produced by IIA-constrained logit models, and the counter intuitive cross-elasticities often obtained with unconstrained market-share models. A drawback of the procedure is that its validity depends upon the specification of the model in which parameters cannot be different across households and upon the way the marketing mix variables are assumed to evolve over time according to a stationary time series. The advantage is that the procedure is relatively simple and can be used with panel data, which does not contain marketing mix variables.

Conclusion

Because household panels are likely to over/under-represent certain consumer groups, and because tracking panels do not provide the same richness of information on how consumers differ in preferences and response to marketing factors, it is unreasonable to expect aggregate models estimated at the household level to produce results that provide the same managerial insights as those obtained from tracking panels. To draw substantive conclusions and managerial insights from the models discussed here, researchers must concentrate their efforts in two directions: (1) develop models that use information on the characteristics of the sampled households and the population to produce weighted estimates of market response and (2) develop models that integrate the rich information on consumer characteristics, brand preferences and responsiveness to marketing variables from household panels with the broader coverage and more projectable data collected at the store level.

The response-segmentation models discussed here focus on consumers' purchase behavior in one product category in isolation. However, the actual timing of a purchase in most packaged goods categories is determined by the timing of the shopping trip, which depends on many other product categories. By the same token, brand choices are constrained by the choice of store; that is particularly true for private labels. The choice of a store depends on this pricing and promotion strategy over a broad range of brands in many product categories. Given the interdependence among product categories, and between brand choice and store choice, efforts should be made to develop segmentation models based on multiple-category shopping behavior. Such models would be useful to manufacturers, for example in developing and evaluating cross-promotions and in evaluating brand extensions. They should also be of great value to retailers for category management and building store loyalty.

Finally, on the technical side, developments should address the integration of random effect models with mixture models. Note that the mixture model framework advocated by Böckenholt (1993a,b) is a step in that direction, because within-segment heterogeneity is accounted for n these models through the use of compounding distributions that are natural conjugates (i.e. the Poisson and gamma, the binomial and beta, and the multinomial and the Dirichlet). Böckenholt shows that in general fewer segments are needed to represent the data accurately when one accounts for within segment heterogeneity than when it is is neglected. By further integrating mixture and random coefficient models, the limitations of within-segment homogeneity can be further alleviated, thus affording a more appropriate description of consumers' response behavior and improving the predictive power of mixture models. Here, hierarchical Bayes approaches appear to be a fruitful avenue for research (cf. Allenby and Lenk 1994).

18
PRODUCT-SPECIFIC UNOBSERVABLE BASES: CONJOINT ANALYSIS

We start by providing a brief introduction to conjoint analysis. Conjoint segmentation is the most common approach in the class of unobservable product-specific bases. We describe its use for market segmentation, and review the procedures that have been proposed for metric conjoint segmentation. Then we report the results of a recently published Monté Carlo study that demonstrates the relative strengths of a variety of proposed segmentation methods for metric conjoint analysis. We conclude the chapter by discussing segmentation for conjoint choice and rank-order data.

Conjoint Analysis in Marketing

Marketing researchers' interest in conjoint analysis was spawned by the work of Luce and Tukey (1964) on conjoint measurement. In marketing research, the applicability of the approach to the measurement and analysis of multiattribute preferences is now widely acknowledged and such measurement is now commonly known as conjoint analysis. After its introduction, conjoint analysis received great attention in academic research (Green and Srinivasan 1990). A definitive documentation of industry usage of conjoint analysis is contained in the surveys by Wittink and co-authors (Cattin and Wittink 1982; Wittink and Cattin 1989; and Wittink, Vriens and Burhenne 1994). Their surveys indicate that as many as 300 conjoint studies may be completed each year in the United States and Europe. The main applications are in new product evaluation, competitive analysis and market segmentation. We discuss in detail its use for market segmentation. Before discussing conjoint segmentation procedures, we review the basic principles of conjoint analysis, drawing from the recent review papers by Green and Krieger (1991) and Green and Srinivasan (1990).

Conjoint analysis is a decompositional approach for the measurement of preferences. Whereas in compositional preference measurement parameters are elicited directly from the decision maker, in decompositional methods those parameters are derived from the decision makers' holistic evaluative responses (in terms of liking, preference, likelihood of purchasing, stated choices, etc.) to profile descriptions designed by the researcher. Statistical methods are then applied to estimate the contribution of the attributes and levels used in the construction of the profiles. The principal steps in conjoint analysis (Green and Srinivasan 1990), which we discuss in detail below, involve the selection of:

(1) the attributes and levels,
(2) the stimulus set,
(3) the stimulus presentation method,
(4) the data collection method, and the measurement scale,
(5) the preference model and the estimation method, and
(6) a choice simulator.

Choice of the Attributes and Levels

The choice of the attributes in conjoint studies evolves around the following questions: (1) Which attributes should be included? (2) How should the attributes be defined? (3) How many attributes should be included? (4) How many levels of each attribute should be included? and (5) Which levels should be included?

The key determinants of which attributes are to be included in a conjoint study are management uncertainty about the profitability of current product features, and the potential profitability of new product features, but also the need for information on the value of product attributes to consumers and the effects of product attributes of competing brands. The implication is that attributes not of direct importance to managers for product modification or new product development might be included in the conjoint design.

Types of Attributes

Attributes included in the conjoint design are often functional or physical attributes. Examples are the presence of power steering in cars, prices of hotel rooms, and the amount of fluoride in toothpaste. Such a design has obvious advantages for R&D departments, because it directly links the physical characteristics of products to consumer preference or choice. On the other hand, physical characteristics may sometimes be difficult for consumers to use or evaluate. Therefore, the attributes may have to be expressed as benefits, or symbolic or hedonic attributes may need to be included. Examples are the use of attributes such as tooth decay prevention in toothpaste and service standards in hotels. However, such abstract attributes may pose operationalization problems to R&D for product design and modification.

A solution to that problem is provided by Oppewal, Louviére and Timmerman's (1993) hierarchical information integration (HII) approach. The HII approach categorizes physical attributes into several non-overlapping sets on the basis of theory, empirical evidence or demands of the application, such that the sets represent higher level constructs such as "quality," "status," or "value for money." HII designs and administers separate sub-experiments to define the composition of each construct in terms of its attributes. The procedure also develops an overall bridging design, which concatenates the results from the separate sub-designs and overall design into one utility model. Oppewal, Louviére and Timmermans (1994) extend the approach by creating a series of sub-experiments in which the physical attributes defining a specific higher level construct are included together with overall indicators of the other higher level constructs (thus eliminating the need for the bridging designs in the conventional

HII approach). The approach is attractive in the sense that it enables one to cope with large numbers of physical attributes, and includes both physical and abstract attributes in the design of conjoint studies.

Two attributes that warrant special attention in conjoint analysis are brand name and price. Brand name is an important attribute, likely to show interactive effects with other attributes. The inclusion of the brand name enables researchers to assess brand equity from the conjoint results (e.g., Park and Srinivasan 1994). Price poses special problems because it is environmentally correlated with many other attributes: most of the attributes included in the conjoint design come at a specific price. Ignoring that fact may result in unrealistic profiles. Some researchers therefore prefer to include the price of each attribute in the descriptions of the attributes themselves (i.e., to allow the price to depend completely on the descriptions of the other attributes). In that case, all non-price part-worths reflect the presence of price, which cannot be estimated separately.

Number of Attributes

Conjoint analysis in its most common form is restricted to a relatively small number of attributes. That number is limited by the total number of profiles that can be evaluated by respondents without undue fatigue and boredom. Typically, the number is about 30 at maximum, depending on the data collection method. The number of attributes is also limited by the number of attribute descriptions that need to be evaluated in each profile. Green and Srinivasan (1978) recommend use of no more than six attributes in a single profile. The use of fractional factorial designs to construct the profiles from the attributes increases the number of attributes that can be included, but the number of attributes needed in industry studies is often very large, e.g., 40 to 50. That problem has yet not been entirely resolved.

The following options have been considered. First, once the fractional factorial designs have been used to reduce the number of profiles constructed from the attributes, blocking designs may be used to group the alternatives into subsets. The subsets are then presented to different (random) sub-samples. A disadvantage of this approach arises when the data are to be analyzed at the disaggregate level, because insufficient information may be present on the importance of certain attributes. A second solution is provided by bridging procedures (Albaum 1989), which use subsets of attributes in separate experiments. The separate experiments are "bridged" by at least one common attribute. The method appears to be somewhat ad hoc, and its statistical properties are not well understood. A third solution is the use of hybrid conjoint analysis, which uses both self-explicated and full profile data (Green 1984), or adaptive conjoint analysis (ACA, Johnson 1987), which uses self-explicated and paired comparisons data. The self-explicated part of the methods involves eliciting importances on the attributes directly from respondents (a compositional approach). For example, respondents would indicate the importance of price, power steering, engine capacity, etc. on five-point scales. In the hybrid method the utility function is a composite of the data obtained from the self-explicated and the full-profile evaluation tasks. In ACA, a pair of profiles that differ on a limited number of attributes (two to five) is selected for each subject on the basis of the self-stated part-worths. After the respondent provides a preference intensity judgment of the pair, his/her part-worth estimates are updated, and the next pair is selected on the basis of the updated part-worths. ACA can accommodate larger numbers of attributes, but the profile design is no longer orthogonal and the method is partially compositional.

Attribute Levels

Once the attributes are chosen, the researcher needs to determine the levels to be included in the design. Again, there is a tradeoff between the amount of information retrieved from the respondents (which increases with the number of levels of the attributes) and the quality of the responses obtained (which decreases with the number of levels, because of respondent fatigue and boredom). Two-level attributes, for example, reduce the respondents' burden because the number of profiles to be evaluated is smaller than it is when three- or four level attributes are used. However, the use of two levels for an attribute allows the estimation of only linear effects in the attribute, whereas the use of three levels allows the estimation of quadratic effects, and so on. A potential problem in conjoint studies is the "number of levels effect." The results of conjoint studies are often summarized on the basis of attribute importances, defined as the differences between the highest and lowest part-worths. The magnitude of attribute importances, however, is influenced by the number of levels: the addition of intermediate levels (holding the extreme levels constant) tends to increase the importance of an attribute in relation to other attributes (Steenkamp and Wittink 1994). For example, if price of a hotel room in a conjoint design is specified at two levels, $50 and $150, the importance of price relative to the other attributes is less than when price is specified at three levels: $50, $100 and $150. It is therefore advisable to use approximately equal numbers of levels for each of the attributes in conjoint studies. Another issue in choosing the levels in a conjoint study is the operationalization of the levels. Physical levels can be used, as in the assessment of hotel service quality for the attributes check-in time (5 minutes or less/5 to 10 minutes/more than 10 minutes), message service (yes/no), laundry pickup (twice a day/once a day/not at all), and so on. Alternatively, perceived levels can be used. In the example of hotel service quality, the levels of the above three attributes can be operationalized as performing "as expected," "better than expected" or "worse than expected." The HII approach described above uses perceived levels of the attributes (four-point scales) in the operationalization of higher level constructs in conjoint analysis.

Stimulus Set Construction

Orthogonal factorial designs are commonly used to generate all possible combinations of the attributes and their levels. If four two-level attributes are used, for example, a full factorial design will lead to $2^n = 16$ profiles. The resulting profiles are to be used in the evaluation task. An advantage of using full factorial designs is that interactive effects between the attributes can be estimated. However, the number of profiles is often too large for reliable evaluation by respondents. In such situations, orthogonal fractional factorial designs are commonly used (cf. Green and Srinivasan 1990). They allow estimation of the main effects of the attributes, and some of the designs (compromise designs) allow estimation of a limited set of (e.g., first-order) interactive effects of the attributes. If, for example, there are five two-level attributes, a half-replicate of a full factorial design (with 32 profiles) leads to 16 profiles. In such a design, some main and interaction effects are typically confounded. In some applications where the number of profiles resulting from these techniques is still too large, the profiles are grouped into blocks, to create a partially balanced block design. Different groups of respondents evaluate the different blocks of profiles.

Conjoint choice experiments also make extensive use of blocking designs (Louviére 1988). In those experiments consumers make choices among sets of profiles rather than rating or evaluating each in isolation. The sets of profiles are constructed using blocking designs. One can preserve the orthogonality of the design and fix the origin of the utility scale across the different choice sets by using a common or "base alternative," as suggested by Louviére (1988). For example, the 16 profiles from the half-replicate of the 2^5 design can be blocked in eight sets of two profiles (1 and 2, 3 and 4, 5 and 6, etc.) with the addition of a base alternative (such as current brand, or "none"), leading to eight choice sets with three alternatives each.

Problems posed by environmentally correlated attributes can be resolved by creating "super-attributes." Environmentally correlated attributes pose problems because of the occurrence of unrealistic profiles, as explained before for price. The feasible levels of two (or more) attributes can be combined into the levels of a single super-attribute (Green and Srinivasan 1990). For example, two levels of engine capacity (1.5 liters and 2.0 liters) and two levels of fuel consumptions (25 mpg and 30 mpg) can be combined into one two-level attribute (1.5 liters with 30mpg and 2.0 liters with 25mpg).

Stimulus Presentation

Once the profiles have been designed, one must consider how the profiles are to be presented to the respondent. The following presentation methods can be considered.

(1) Paragraph descriptions (texts describing the profile in narrative form),

(2) verbal descriptions (profile cards with terse attribute-level descriptions),

(3) pictorial descriptions (drawings or photographs),

(4) computer aided design (provides pictorial representations of non-existing products), and

(5) physical products.

The use of profile cards with verbal descriptions appears to be by far the most popular. Pictorial materials, make the task more interesting to the respondent, however, and provide an easier and less ambiguous way of conveying information. Such material is virtually indispensable in conjoint studies on appearance and ethical attributes, such as package design and styling. Computer aided design techniques have the same advantages, but may provide even more realistic descriptions of profiles that are costly to produce, as in the case of new products. Whenever feasible, for example in the case of foods, beverages and fragrances, presenting the physical products provides the greatest realism in stimulus presentation.

Data Collection and Measurement Scales

Because of the complexity of conjoint studies, personal interviews are very well suited to obtaining high quality conjoint data. Mail and telephone interviews are less attractive, but peeling-stickers have been used effectively to elicit consumer rank-order statements in mail interviews. Computer-interactive interviewing is attractive, especially for ACA, which was designed for that purpose. Alternative methods for conjoint data collection are:

(1) *The full-profile approach* (respondents evaluate all profiles, based on combinations of all attributes, simultaneously).

(2) *The tradeoff approach* (respondents evaluate combinations of the levels of two attributes).

(3) *The paired comparison approach* (respondents evaluate pairs of profiles consisting of combinations of all attributes).

(4) *The experimental choice approach* (respondents choose one profile from each of a number of choice sets, where each profile is defined on all attributes).

Current data collection in conjoint analysis tends to emphasize the full-profile approach over the tradeoff matrix. The reason appears to be the growing interest in behaviorally oriented constructs such as intentions to buy, likelihood of trying and so on in stead of traditional liking or preference judgments. Full-profile descriptions are therefore more relevant because they provide a complete representation of the product or service. (For similar reasons, the conjoint choice approach is recently gaining popularity). In both the full-profile and the tradeoff approaches, rating or ranking data can be collected. Ratings are commonly assessed on seven- or nine- point scales that have interval-scale properties. However, because of their monadic nature, ratings may be susceptible to response-style biases. Steenkamp and Wittink (1994) used magnitude estimation, which results in scales with ratio-scale properties. The paired comparison approach is used particularly in ACA. In paired comparisons tasks, ratings (graded paired comparisons), rankings or choice measures can be collected.. In conjoint choice tasks, multinomial choice data are collected for each choice set offered to the respondents. Alternatively, respondents may be asked to rank the alternatives in the choice sets. Computer assisted data collection for conjoint choice experiments enables one to collect not only choice data, but also response-time data, which provide additional information on choice-set complexity and Pareto optimality.

Preference Models and Estimation Methods

Four basic models can be used to describe consumers' multiattribute preference functions through conjoint analysis: the vector model, the ideal-point model, the part-worth model and the mixed model. The vector model states that preferences increase with increasing values of the attribute, the ideal-point model assumes that preferences are maximal at the ideal and decrease with movement away from it, and the part-worth model assumes that each level of the attribute has a unique part-utility associated with it. The mixed model combines the other three models. The part-worth model is the most general, but also entails more parameters to estimate. It has nevertheless emerged as the most popular conjoint model. Part-utilities are estimated for each level of each attribute by specifying a dummy variable for each attribute level in the model. The four models are specified in the box below.

A variety of techniques can and have been used for the analysis of conjoint data. The adequacy of the techniques depends on the measurement scale used by consumers in evaluation of the profiles. The three major types of scales in conjoint analysis are: metric, rank-order, and nominal (choice). Metric data are commonly analyzed by ordinary least squares regression. The importances of the attribute levels are estimated by ordinary least squares.

For rank-order preference data, the dominant techniques have been MONANOVA (Kruskal 1965) and LINMAP (Shocker and Srinivasan 1979). The MONANOVA algorithm involves an iteration between two steps. In step t, first the importances of the attribute levels are estimated by ordinary least squares: $\mathbf{B}^t = (\mathbf{X}'\mathbf{X})^{-1}\mathbf{X}'\mathbf{Y}^{t-1}$. In the second step, a monotone transformation is applied to the predicted preferences, such that the order of these predicted preferences conforms to the order of the observed preferences: $\mathbf{Y}^t \sim M(\hat{\mathbf{Y}}^{t-1})$. The monotone transformation M(A) amounts to averaging those pairs of predicted values for which the order is not the same as the order of the corresponding pair of data values. The two steps are alternated until convergence (i.e., until the badness of fit measure *stress* is less than some prespecified value). The MONANOVA algorithm can be applied to both full-profile and tradeoff matrix data. The LINMAP linear programming method also handles those two types of data. As shown by Jain et al. (1979), among others, OLS does a very good job in estimating the part worths, in comparison with the former two non-metric procedures. An alternative way of analyzing rank- order conjoint data is by using the rank explosion rule, which expands the rank orders as a series of successive choices. A multinomial logit model is estimated to the exploded data (Chapman and Staelin 1982; Kamakura, Wedel and Agrawal 1994; see the example provided at the end of this chapter). However, rank-

Part 4 Applied market segmentation

order (and pairwise choice) data have been analyzed only through choice models based on the independence of irrelevant alternatives assumption, that is, multinomial logit models.

Conjoint Preference Models

Notation
$l = 1,...,L_p$ denote the levels of attribute p,
x_{kp} is the level of attribute p for profile k,
y_{nk} is the preference judgment of subject n for profile k,
a_{np} is the ideal of subject n for attribute p,
$f_l(\cdot)$ is an indicator function indicating level l of attribute p.

1. The vector model: $y_{nk} = c_n + \sum_{p=1}^{P} \beta_{np} x_{kp}$

2. The ideal-point model: $y_{nk} = c_n + \sum_{p=1}^{P} \beta_{np} (x_{kp} - a_{np})^2$

3. The part-worth model: $y_{nk} = \sum_{p=1}^{P} \sum_{l=1}^{L_p} \beta_{lp} f_l(x_{kp})$

4. The mixed model, which involves combinations of models 1 through 3.

An alternative is provided by the multinomial probit model, which allows for correlated errors and is therefore not subject to the independence of irrelevant alternatives assumption (e.g., Kamakura and Srivastava 1984). The problems of estimating probit models when choice sets contain more than three alternatives, arising because four and higher dimensional integrals cannot be solved numerically, have been resolved by the use of simulation methods to compute the integrals (McFadden 1989).

Choice Simulations

Once conjoint results (i.e., part-worths) have been obtained, choice simulators can be used to simulate the effects of new product introductions, product modifications, competitive entries and product-line extensions. The output of the simulations typically consists of the proportion of choices received by each product (i.e., their market shares). The simulations may involve single products, competitive products or product-bundles. The choice rules that can be used include (1) maximum utility, (2) Bradley-Terry-Luce, (3) logit and (4) alpha. The alpha rule (Green and Krieger 1993) specifies the choice probability of consumer n for profile k as:

$$P_{kn} = \hat{y}_{kn}^{\alpha} / \sum_{k=1}^{K} \hat{y}_{kn}^{\alpha}.$$

If α = 1 the BTL rule is obtained; if α→∞, the maximum utility rule is obtained (the optimal value of α can be found numerically by maximizing the correspondence of external market shares for a series of current products and those shares as predicted by the model). One must be aware, however, of the heuristic nature of those rules; the part worths do not necessarily translate directly into choice probabilities unless they have been estimated with a probabilistic choice model.

Several procedures are available for post-hoc analyses based on the results of conjoint analysis to find the optimal products that are on the Pareto frontier, that is, are not dominated by other products with respect to return and share (Green and Krieger 1993) or to find the optimal products vis à vis competitors by using game-theoretic procedures (DeSarbo and Choi 1993).

Market Segmentation with Conjoint Analysis

In chapter 2 we saw that the benefits derived from the attributes of a product or service have proved to be one of the most powerful segmentation bases. Therefore, it is not surprising that market segmentation is consistently found to be one of the primary purposes of conjoint analysis in commercial applications in the United States as well as in Europe (Wittink and Cattin 1989; Wittink, Vriens and Burhenne 1994). To describe conjoint segmentation procedures, we distinguish between metric and non-metric (rank-order or choice data) conjoint segmentation. We discuss approaches for the two types of data below. For metric conjoint segmentation we provide the results of a recent extensive Monté Carlo study investigating the relative performance of a large variety of models. First, however, we provide an application of metric conjoint segmentation to a particular situation in which multiple criteria are assessed for each of the profiles.

Application of Conjoint Segmentation with Constant Sum Response Data

We describe the results of a business-to-business segmentation study performed by DeSarbo, Ramaswamy and Chatterjee (1995). They analyzed constant sum data collected for multiple criteria. Constant sum multiple-criteria data may be used in a variety of situations in conjoint analysis, such as for resource allocation decisions, to investigate the importance of evaluation criteria or to examine the relative likelihood of different modes of purchase. The authors applied their approach to investigate industrial buyers' tradeoffs between multiple decision criteria and how they are influenced by the type of product under consideration. We summarize their findings. The study investigated the criteria used by purchasing managers to select suppliers for different types of products. Eight product types were generated in a fractional factorial design, from four two-level attributes:

Part 4 Applied market segmentation

(1) Customization of the product (standard/customized),
(2) complexity of the product (simple/complex),
(3) novelty of the application (standard/novel),
(4) dollar commitment (low/high).

A customized (as opposed to a standard) product is designed to fulfill the function envisaged for it. A complex product is more difficult to evaluate than a simple one. A novel (as opposed to a standard) application is an unfamiliar application of the product. A high (as opposed to a low) dollar commitment is an above average dollar outlay attributable to volume of purchases per annum or high unit cost.

The eight profiles were offered to a sample of 211 purchasing managers, with examples of products implied by the profiles. For each of the eight profiles, the managers allocated 10 points among four supplier-selection criteria according to relative perceived importance: economic costs, functional performance, vendor cooperation and vendor capability. Economic costs included the price paid to the supplier and unanticipated costs. Functional performance consisted of performance characteristics, operating features and product quality, that affected the performance of the functions for which the product was purchased. Vendor cooperation related to the perceived willingness of the supplier to offer services. Vendor capability pertained to the buyer's assessment of the supplier's delivery and production capability. Those constant sum allocations constituted the multiple criterion variables of interest.

The model developed for those data was a mixture Dirichlet regression model. It was a special case of the general model described in chapter 7, with the Dirichlet distribution (Table 6.2) as the distribution of the dependent variable in equation 7.3. On the basis of the CAIC statistic, the authors identified a two-segment solution.

To highlight the differences between the structures of the part-worths for the two segments, the predicted allocations of points for three specific profiles, A, B and C, were computed. Profile A was a base profile, comprising a standard product with simple makeup, standard application and low dollar commitment (scratch pads were used as an example). Profile B represented a customized, complex, novel product with low dollar commitment (e.g., special video monitors). Profile C represented a customized, complex, novel, product with high dollar commitment. The estimates are reported in the Table 18.1.

From Table 18.1 we can see that the expected relative importance of the supplier criteria for profile A does not differ much across the two segments. Both segments stress economic criteria more than non-economic criteria. However, for profile B, in segment 1 the non-economic criteria have higher salience. For segment 2, on the contrary, although the relative salience of functional performance is high, both vendor cooperation and capability have lower salience. Hence, segment 1 is more concerned than segment 2 with reducing risk by trading off economic criteria for non-economic criteria. For profile C, segment 2 attaches greater importance to vendor capability, at the expense of functional performance. Those respondents manage financial risk by attaching greater importance to vendor capability. Hence, even if the procurement involves a customized product with a complex makeup and novel application, if the dollar commitment is high the perceived ability of the supplier to produce and deliver the buyers' requirement is a critical factor in vendor selection.

The segments were profiled with consumer descriptor variables. The respondents in segment 1 were found to be more risk-averse and more loyal to current suppliers. Their greater risk aversion is consistent with their higher emphasis on non-economic criteria and their loyalty is linked to a risk-handling strategy.

Table 18.1: Estimated Allocations for Three Profiles in Two Segments (adapted from DeSarbo et al. 1995)

	Costs	Performance	Cooperation	Capability
Segment 1				
Profile A	4.05	2.13	1.78	2.05
Profile B	1.63	4.16	2.00	2.20
Profile C	1.75	4.16	1.80	2.28
Segment 2				
Profile A	4.02	2.11	1.83	2.05
Profile B	1.91	4.92	1.42	1.76
Profile C	1.95	4.94	1.35	2.21

Market Segmentation with Metric Conjoint Analysis

Both in commercial (e.g., Wittink, Vriens and Burhenne 1994) and academic applications (e.g., Green and Krieger 1991), the segmentation of markets with conjoint analysis has traditionally involved a two-stage approach in which the identification of segments and the estimation of conjoint models are performed separately. However, several alternative techniques have been proposed (i.e., Hagerty 1985; Ogawa 1987; Kamakura 1988; Wedel and Kistemaker 1989; Wedel and Steenkamp 1989; DeSarbo, Oliver and Rangaswamy 1989; De Soete and DeSarbo 1991; DeSarbo et al. 1992). Because the estimation and segmentation stages are integrated, we refer to those approaches as integrated segmentation methods. Note that many of them have already been discussed in chapter 5 (see also Carroll and Green 1995).

Metric conjoint segmentation methods differ in (1) the type partitioning assumed, (2) the algorithms and estimation procedures used, and (3) the criterion being optimized. We discuss those three aspects for each of the proposed conjoint segmentation methods. A summary is provided in Table 18.2 (based on Vriens, Wedel and Wilms 1996).

A-Priori and Post-Hoc Methods Based on Demographics

In a-priori conjoint segmentation (Green and Krieger 1991; Wind 1978) consumers are assigned to segments designated a-priori on the basis of demographic and socioeconomic variables. The conjoint model is estimated within each segment. Alternatively, one can cluster consumers on the basis of demographic and socioeconomic variables to produce segments post-hoc. Subsequently, a conjoint model is estimated within each segment. Both a-priori and post-hoc approaches assume that the bases used in the segmentation stage account for heterogeneity in preferences. For a discussion of the effectiveness of those procedures we refer to chapter 3.

Componential Segmentation

In componential segmentation (Green 1977; Green and DeSarbo 1979), consumer profiles are generated on the basis of sociodemographic characteristics. Respondents matching the profiles are sought from an available sampling frame and each of the selected respondents completes the conjoint task. The componential segmentation model estimates both main effects of the design variables and interactions between the design variables and subject profiles. The success of the componential segmentation method depends on the strength of the associations between consumer background variables and preferences (chapter 2).

Two-Stage Procedures

In the traditional two-stage conjoint segmentation procedure, estimates of the individual-level part-worths β_n are obtained for each subject by using least squares. At the second stage, consumers are clustered into segments on the basis of the similarity of those estimated coefficients (see Green and Krieger 1991). The traditional two-stage approach has the limitation of depending on possibly unreliable individual-level estimates. The fractional factorial designs used in conjoint applications generally leave very few degrees of freedom for estimation at the individual level. The result may be unreliable part-worth estimates, which may in turn cause misclassification. Furthermore, it is not possible to estimate models that are over parameterized at the individual level. The implication is that conjoint experiments in which blocking designs are employed cannot be accommodated. A third limitation results from the fact that the traditional two-stage procedure entails two unrelated steps, in which the procedures optimize different criteria. In the first step the sum of squared errors of the model is minimized whereas in the second step the between-to-within-segment ratio of the variances of the individual part-worths is maximized.

An alternative two-stage procedure has been proposed (Green, Carmone and Wind 1972; Green and Srinivasan 1978). In the first step, individuals are clustered on the basis of the observed preference ratings of the profiles. In the second step, separate conjoint models are estimated across consumers and profiles in each segment. The procedure does not result in possibly unreliable parameter estimates, but models that are over-parameterized at the individual level cannot be accommodated. The procedure also optimizes two distinct criteria in each of the two steps.

Hagerty's Method

Hagerty (1985) has provided a general formulation of the segmentation problem in conjoint analysis. We let P denote the (N × S) matrix representing a general partitioning scheme for assigning consumers to segments, B denote a (K × $\sum L_p$) matrix of regression coefficients and E denote the (K × N) matrix of random errors. The general conjoint segmentation model is formulated as:

$$YP(P'P)^{-1}P' = XB + E. \tag{18.1}$$

Equation 18.1 accommodates individual-level analyses (S = N), segment-level analyses (1 < S < N) and aggregate-level analyses (S = 1). Furthermore, several types of sample partitioning are handled. The common concept behind conjoint segmentation procedures is that by weighting similar subjects together (through P), the variance of the estimates is reduced with respect to individual-level analyses because of the increase in the number of observations available. However, estimating a conjoint model across subjects that are not exactly identical in their true parameters causes a bias in the estimates. Clearly, conjoint segmentation involves a tradeoff between bias and variance of the parameter estimates. Four major forms of the general partitioning matrix P can be distinguished: (1) a nonoverlapping partitioning, (2) an overlapping partitioning, (3) a fuzzy partitioning and (4) a factor analytic partitioning (see chapter 5 for the structure of the matrix P in each of these cases).

The particular method proposed by Hagerty (1985) within his general framework is based on a weighting scheme that optimizes, under a given set of assumptions, the expected mean squared error of prediction in validation samples (EMSEP). That weighting scheme represents an optimal overlapping partitioning obtained by a Q-factor analysis of the between-subjects correlation matrix of preferences and therefore presents a factor-type partitioning of the sample. A possible problem with the method is the interpretation of the factor solution in terms of segments (Stewart 1981). The number of extracted factors need not be an adequate indicator of the number of segments and is limited by the number of independent observations per subject. The procedures proposed to identify segments on the basis of the factor solution result in a loss in predictive accuracy.

Hierarchical and Non-hierarchical Clusterwise Regression

Kamakura (1988) proposes a hierarchical clustering method for simultaneous segmentation and estimation of conjoint models (see chapter 5). The method is based on least-squares estimation, and starts from single-subject clusters that are combined hierarchically to maximize the predictive accuracy index, resulting in nonoverlapping clusters. Its two disadvantages are (1) misclassification at earlier stages in the hierarchical clustering process will carry on to higher levels in the process and (2) models that are over-parameterized at the individual level cannot be estimated.

Wedel and Kistemaker (1989) propose a generalization of clusterwise regression (Späth 1979) for conjoint segmentation. Their method yields nonoverlapping segments obtained by a non-hierarchical clustering procedure. The clusterwise regression methodology proposed by DeSarbo, Oliver and Rangaswamy (1989) is also non-hierarchical and uses simulated annealing. It allows for overlapping clusters and multiple dependent variables. However, the program is limited to small amounts of data. The fuzzy clusterwise regression (FCR) algorithm of Wedel and Steenkamp (1989) permits consumers to be partial members in a number of segments. However, the fuzzy clusterwise regression algorithm depends on the prior specification of a fuzzy weight parameter that affects the degree of overlap among fuzzy clusters. For a more detailed description of those procedures, refer to chapter 5, where they are discussed extensively.

Mixture Regression Approach

DeSarbo, et al. (1992) propose a mixture regression methodology for metric conjoint analysis in which the segments and the conjoint model parameters within those segments are estimated simultaneously, employing mixtures of multivariate normal distributions. In the terminology of chapter 13, the mixture regression model for metric conjoint segmentation is a mixture model that assumes a multivariate normal distribution of the preferences of each subject (Table 6.2). Note that even if no segments are estimated (i.e., $S = 1$), the model is therefore more general than the usual regression model used to estimate the part-worths, because it addresses correlations of the errors on the preferences for different profiles elicited from the same subject. The model assumes the identity link function and is estimated by maximizing the likelihood (equation 7.3), with the EM algorithm described in chapter 7.

Chapter 18 product-specific unobservable bases: conjoint analysis

Table 18.2: Comparison of Conjoint Segmentation Procedures (adapted from Vriens et al. 1996)

Conjoint Segmentation Technique	Possible Types of Partitionings	Possible Partitioning Procedures	Optimization Criterion
A priori using descriptor variables	Nonoverlapping	A priori defined	R^2
Post hoc using descriptor variables	Nonoverlapping	Hierarchical and non-hierarchical	R^2
Componential segmentation	Nonoverlapping	Stepwise regression procedures	R^2
Traditional two-stage conjoint-based segmentation (TTS)	Nonoverlapping	Hierarchical and non-hierarchical	R^2 and W/B
Alternative two-stage conjoint-based segmentation (ATS)	Nonoverlapping	Hierarchical and non-hierarchical	W/B and R^2
Hagerty's (1985) optimal weighting (OW)	Factor solution	Factor analysis	EMSEP
Optimal weighting followed by K-Means (OWKM)	Nonoverlapping	Factor analysis & K-Means	EMSEP, W/B
Kamakura's (1988) method	Nonoverlapping	Hierarchical	EMSEP
DeSarbo et al's (1989) clusterwise regression	Overlapping	Simulated annealing	Likelihood
Wedel & Kistemaker's (1989) clusterwise regression (CR)	Nonoverlapping	Exchange algorithm	Likelihood
Wedel & Steenkamp's (1989, 1991) method (FCR)	Fuzzy	Iteratively weighted east squares	Weighted SSE
DeSarbo, et al. (1992) latent class procedure (LCN)	Probabilistic	EM algorithm	Likelihood

- W/B = the ratio of the within-cluster variance and the between-cluster variance,
- EMSEP = expected mean squared error of prediction.

Part 4 Applied market segmentation

A Monté Carlo Comparison of Metric Conjoint Segmentation Approaches

The large number of conjoint segmentation procedures described here may confuse researchers who seek to identify segments in a market on the basis of a conjoint study. As a result, a procedure is likely to be selected on the basis of availability, coincidental acquaintance or subjective preference rather than on the basis of specific strengths and weaknesses of the methods in relation to a specific application. To alleviate the confusion, Vriens, Wedel and Wilms (1996) performed a detailed comparison of metric conjoint segmentation methods. They performed a Monte Carlo study that assessed the relative performance of the methods empirically on synthetic data. Their results are summarized below.

The Monté Carlo Study

To examine the performance of metric conjoint segmentation methods, Vriens and co-authors designed a Monté Carlo study in which six independent factors (each at two levels) were considered, reflecting sample size, degrees of freedom, type of partitioning and error in the data:

- *Numbers of subjects:* The number of subjects in the sample was varied at two levels: 100 and 200 subjects.

- *Number of profiles:* Simulated preferences were generated from a conjoint design assuming six attributes, each defined on three levels. Given a fixed number of attributes, the number of profiles was 18 or 25.

- *Number of segments:* Two-segment and four-segment conditions were specified. The segments consisted of equal numbers of subjects.

- *Error variance*: The percentages of error variance used were 5% or 35%.

- *Segment homogeneity:* Two conditions were specified, representing "homogeneous" and "diffuse" segments. In the situation of homogeneous segments, there is little aggregation bias if the models recover (0/1) segment memberships exactly.

- *Segment similarity:* To simulate situations of relatively similar versus dissimilar segments, the part-worths in the latter condition were multiplied by two, which results in greater separation of segments.

The study had a full (2^6) factorial design, with 64 datasets of simulated preferences for profiles by a sample of subjects. The segments initially generated were nonoverlapping. Values drawn from a normal distribution with a standard deviation of either .05 (homogeneous) or .10 (diffuse segments) were added to the part-worths to simulate within-segment heterogeneity. The "true" preferences calculated for each subject were based on those coefficients. A random error was added so that the percentages of error variance in "observed" preferences were 5% (low error condition) or 35% (high error condition). Holdout preferences on eight profiles were generated.

The 64 datasets were analyzed with nine methods for metric conjoint segmentation.

(1) TTSWA: the Traditional Two-Stage approach, using Ward's hierarchical clustering procedure
(2) TTSKM: the Traditional Two-Stage approach, using K-Means non-hierarchical clustering
(3) ATSWA: the Alternative Two-Stage approach, using Ward's clustering method
(4) ATSKM: the Alternative Two-Stage approach, using K-Means clustering
(5) OW: Optimal Weighting
(6) OWKM: OW, followed by K-Means clustering on the factor scores
(7) CR: the Clusterwise Regression procedure
(8) FCR: the Fuzzy Clusterwise Regression procedure
(9) LCN: the (Latent Class) Normal mixture regression model

Several of the methods have the problem of convergence to local optima (CR, FCR and LCN). Those methods were applied several times, with different random starts to overcome that problem. For each of the methods, the number of estimated segments (or factors) was set equal to the true number of segments.

The following measures, addressing the most important aspects of the performance of the methods (parameter recovery, goodness of fit, and predictive power) were used to compare their relative performance.

(1) R^2: the percentage of variance accounted for by the conjoint models
(2) RMSE(b): the root mean squared error of the estimated part-worths
(3) RMSE(P): the root mean squared error of the estimated cluster memberships
(4) RMSE(Y): the root mean squared error of the predicted preferences in the out-of-sample

The observations on each of the dependent performance measures were analyzed by analysis of variance. The conjoint segmentation method had an overall significant effect on all performance measures used in the study. The mean performance measures for each of the methods, averaged over the other factors in the study are presented in Table 18.3. In Table 18.4 we report the mean performance measures for each of the levels of the data characteristics. Various interaction effects are statistically significant, the most important being for RMSE(b) and RMSE(P).

Results

With respect to R^2, the two-stage clustering procedures outperform the integrated segmentation procedures. Table 18.4 shows that the fit is lower for a larger number of datapoints, fewer parameters and higher error levels. Optimal weighting and the two-stage clustering procedures provide the best fit among all methods but, given that they do less well with respect to parameter recovery, they may be subject to overfitting.

The mixture regression approach, LCN, is clearly superior in coefficient recovery, as can be seen in Table 18.3. It recovers part-worths almost three times as well as the two-stage clustering procedures. FCR and CR are second best. The optimal weighting procedures outperform the two-stage clustering procedures, but among the latter no significant differences are found. Table 18.3 indicates that for membership recovery (RMSE(P)), the mixture approach, LCN, performs best: it performs twice as well as the two-stage procedures. CR and FCR are second best. The two-stage clustering procedures involving Ward's method (TTSWA, ATSWA) outperform those based on K-means (TTSKM, ATSKM).

Table 18.4 shows that parameter recovery, as measured by RMSE(b), improves for a larger number of subjects and profiles. An increase in the number of segments results in the deterioration of parameter recovery as hypothesized, but it is interesting to note that the effect is much less for the clusterwise and mixture regression methods. Apparently, the performance of those methods is superior, especially for larger numbers of segments, whereas FCR and LCN estimates are four times as good as those of the two-stage clustering procedures. An increase in the error level results in a 50% decrease in the recovery of the parameters, with some minor differences across methods. The recovery of the parameters is somewhat better for homogeneous segments than for diffuse segments.

Parameter recovery improves for all methods with an increasing number of observations (subjects and profiles), and decreases with an increasing number of parameters estimated and higher error levels. The numbers of subjects and profiles do not have a significant effect on segment-membership recovery. Membership recovery deteriorates when the number of segments increases for the two-stage clustering procedures. With increasing error variance, RMSE(P) deteriorates by about a factor five on average. Again, the effect is much larger for the two-stage procedures, and it is smallest for the mixture approach. When segments are homogeneous, dissimilar segment memberships are recovered better. The relative sensitivity of CR, FCR and LCN for diffuse and/or similar segments is less than that of the other methods, however.

From their results, Vriens et al. concluded that LCN, FCR and CR generally outperform the other procedures in coefficient and segment-membership recovery. LCN performs best. The performance of the optimal weighting procedures was not in accordance with expectations. In an additional smaller Monté Carlo study, the authors found that the performance of Kamakura's hierarchical clusterwise regression procedure was superior to the two-stage clustering procedures but inferior to the other integrated procedures, LCN, FCR and RMS.

The performance of the two-stage procedures was found to deteriorate faster than that of the other procedures with increasing numbers of parameters to be estimated, for higher levels of error variance and for less well separated segments. Among the two-stage procedures, the application of Ward's clustering in the second stage tended to perform best.

Table 18.3: Mean Performance Measures of the Nine Conjoint Segmentation Methods[a] (adapted from Vriens et al. 1996)

Method	R^2	Performance Measures		
		RMSE(b)	RMSE(P)	*RMSE(y)*
1. TTSWA	$0.712^{6,8,9}$	0.348	$0.212^{2,3,4}$	1.468^5
2. TTSKM	$0.709^{6,8,9}$	0.355	0.251^4	1.473^5
3. ATSWA	$0.713^{6,8,9}$	0.352	0.243^4	1.466^5
4. ATSKM	$0.7093^{6,8,9}$	0.364	0.284	1.472^5
5. OW	$0.755^{1,2,3,4,5,6,7,8,9}$	$0.335^{2,4}$	*	1.571
6. OWKM	0.702^8	$0.253^{1,2,3,4,5}$	$0.208^{2,3,4}$	1.487^5
7. CR	$0.709^{8,9}$	$0.163^{1,2,3,4,5,6}$	$0.123^{1,2,3,4,6}$	1.470^5
8. FCR	0.691	$0.150^{1,2,3,4,5,6}$	$0.152^{1,2,3,4,6}$	1.471^5
9. LCN	0.695	$0.117^{1,2,3,4,5,6,7,8}$	$0.101^{1,2,3,4,6,8}$	$1.456^{5,6}$

* Not computed for the OW procedure because the factor scores cannot be interpreted as memberships.
[a] Statistically significant differences between means are denoted by superscripts, computed by the least significant difference (LSD) (p < 0.01).

Predictive Accuracy

Table 18.3 shows that the mixture regression procedure LCN results in the lowest mean value of RMSE(y). The ATSWA, CR and FCR procedures perform second best in this respect, although many of the differences are neither significant nor substantial. A surprising result is that OW performs significantly worse than the other procedures. Although the effects for predictive accuracy by and large parallel those for coefficient and segment-membership recovery, differences in predictive accuracy are small and mostly nonsignificant. The predictive accuracy deteriorates for a larger number of segments, for a higher error level, for diffuse and more dissimilar segments (Table 18.4).

An explanation for the absence of differences in predictive accuracy between the methods may be related to the number of segments (and factors) extracted. In the study the true number of segments was retained in all analyses. A larger number of segments may have resulted in a better representation of simulated within-segment heterogeneity and a corresponding improvement in the predictive accuracy of the models. That possibility is confirmed by the predictive accuracy based on individual-level estimates. Individual-level predictions resulted in the lowest mean RMSE(y) among all methods.

The explanation is that within-segment heterogeneity was present in all of the datasets. The individual differences within segments apparently affected the performance of all evaluated conjoint segmentation methods negatively, while the fact that there were at least five degrees of freedom in all synthetic datasets may have favored individual-level predictions. The results in this respect are consistent with those of previous studies (Green and Helsen 1989; Green, Krieger and Schaffer 1993; Montgomery and Wittink 1980). Clearly, the topic of the selection of the number segments in relation to predictive accuracy warrants further study.

The Monté Carlo study revealed that the mixture regression approach performs best. However, the differences with FCR are small and the predictive performance of the methods does not match that of individual level-models. Further, LCN offers the advantage of providing standard errors of the estimated part-worths

Table 18.4: *Mean Performance Measures for Each of the Factors (adapted from Vriens et al. 1996)*

Factor	Performance Measures			
	R^2	RMSE(b)	RMSE(P)	RMSE(y)
Subjects				
100	0.713^1	0.275	0.196	1.484
200	0.708	0.267	0.197	1.479
Profiles				
18	0.711^1	0.274	0.200	1.486
25	0.710	0.268	0.194	1.477
Segments				
2	0.694^1	0.124^1	0.182^1	1.297^1
4	0.727	0.418	0.211	1.666
Error variance				
5%	0.839^1	0.220^1	0.065^1	0.967^1
35%	0.582	0.322	0.329	1.996
Segment homogeneity				
0.05	0.739^1	0.259^1	0.167^1	1.378^1
0.10	0.682	0.283	0.227	1.585
Segment separation				
Low	0.610^1	0.205^1	0.254^1	1.195^1
High	0.761	0.336	0.140	1.769

[1] The difference between the two means is significant (p < 0.01) as indicated by the LSD rule.
[2] The SEDs are 0.0042 (R^2), 0.0110 (RMSE(b)), 0.0168 (RMSE(P)), 0.0090 (%CORCLS), 0.0169 (RMSE(y), and 0.0058 (%1stCH).

Segmentation for Rank-Order and Choice Data

Rank-order preferences are the second most frequently collected data in commercial conjoint studies (Wittink and Cattin, 1989). As mentioned previously, a major advantage of ranks is that they are comparative and therefore less affected by the response-style biases typically found in monadic ratings. Currently, there is much scientific interest in conjoint choice experiments. Choice data may be preferable for predicting actual choice behavior and choice shares (cf, Elrod, Louviere and Davey 1992). In conjoint choice experiments, the direct link between the model estimation and choice simulation alleviates the need to make the arbitrary assumptions of choice simulators. On the other hand, choice models require substantial amounts of data for estimation, and are thus inadequate for individual-level conjoint analysis. Commonly, choice-based conjoint analysis is done at either the aggregate level (i.e., assuming homogeneity of preferences across consumers) or at the segment level. The procedures that have been developed to identify segments from a sample on the basis of rank-order or choice data are conceptually (but not methodologically) similar to those for metric conjoint segmentation. We briefly discuss the procedures below. We describe the mixture regression approach in more detail.

A-Priori and Post-Hoc Approaches to Segmentation

The segmentation procedures described previously for metric conjoint data may, in principle, be applied to rank-order data because OLS regression, on which most of the two-stage and integrated procedures are based, provides good results for rank-order data as well (cf. Jain et al. 1979). However, OLS estimation is not efficient in the statistical sense, as it does not provide the minimum variance of the estimates. Moreover, the OLS estimates are not logically consistent in the sense that predicted preferences do not have an ordinal scale. Segments can be identified for both conjoint rank-order preferences and conjoint choices by using the a-priori or post-hoc approaches to derive segments from demographic, socio-economic data, for example. The part-worths can then be estimated at the segment level by using non-metric procedures such as MONANOVA or LINMAP for rank-order data, or the multinomial logit model for choice data (or rank-order data after rank explosion).

Two-Stage Procedures

Non-metric conjoint data can be grouped into segments by using the traditional and alternative two-stage procedures described above. For the traditional two-stage procedure, non-metric methods, such as MONANOVA and LINMAP can be applied to the data of individual consumers, and the individual-level estimates obtained can subsequently be grouped by using hierarchical or non-hierarchical clustering. Similarly, the alternative two-stage procedure would amount to grouping subjects on the basis of their rank-ordered preferences. Preferably, a hierarchical method employing a similarity measure tailored to rank-orders should be applied (see chapter 5). For each of the segments identified, a non-metric procedure can be applied to estimate the segment-level part-worths. For choice-based conjoint experiments, the traditional two-stage

Part 4 Applied market segmentation

procedures cannot be applied because individual-level estimates obtained with multinomial logistic regression are severely biased and unreliable. The alternative two-stage procedures are applicable. A hierarchical clustering procedure, using a similarity measure tailored to binary data (e.g., Jaccard's coefficient, S2 in Table 5.3, where each choice set represents a variable), can be applied to the individual-level choice data. For each of the segments, a multinomial logit model can be estimated to obtain the part-worths, as long as the number of observations is substantial.

Hierarchical and Non-hierarchical Clusterwise Regression

An integrated conjoint segmentation method that is tailored to the analysis of rank-ordered preferences was proposed by Ogawa (1987). A procedure similar to ridge regression is proposed for estimating individual-level part-worths by using multinomial logit models, and an information theoretic criterion is used to aggregate consumers. The procedure starts from single-subject clusters, which are combined hierarchically to maximize the log-likelihood. The procedure yields non-overlapping clusters. The method is analogous to Kamakura's method for metric conjoint. Because the method employs individual-level estimates, the limitations documented by (e.g.) Elrod, Louviere and Davey (1992) apply, which pertain to instability and bias in the individual-level part-worth estimates. The hierarchical clustering process implies that the method depends in the initial stages on part-worths estimated at the individual level, thereby creating potential misclassification due to unreliable estimates, which may carry on to higher levels in the hierarchical clustering process. To circumvent the computational burden involved in estimating a logit model at each stage of the clustering algorithm, Ogawa uses OLS to estimate part-worths, which results in inefficient estimates.

A non-hierarchical clusterwise regression procedure for multinomial choice data was proposed by Katahira (1987). That procedure (it is discussed in chapter 5) can be applied both to conjoint choice data and conjoint rank-order data (after applying the rank-explosion rule to the data). The procedure estimates a logistic regression model for each segment, and uses an exchange algorithm to allocate consumers into segments in such a way that the likelihood is maximized.

The Mixture Regression Approach for Rank-Order and Choice Data

Kamakura, Wedel and Agrawal (1994) developed a unifying mixture regression model (chapter 7) for segmentation of choice and rank-order data. Their approach assumes that rank orders or choices are based on random utility maximization. We describe their procedure in more detail. More recently, DeSarbo, Ramaswamy and Cohen (1995) proposed a mixture regression model for conjoint choice data. An extension of the mixture regression model, integrating conjoint analysis and multidimensional scaling, was provided by Wedel et al. (1998). Some additional notation is contained in the box below.

Chapter 18 product-specific unobservable bases: conjoint analysis

Notation

Choice data:

t = 1,...,T choice sets,
y_{nkt} = 1 if consumer n chooses profile k at choice occasion t,
0 otherwise.

Rank-order preference data:

t = 1,...,T (T = J),
y_{nkt} = 1 if consumer n's rank order of profile k is greater or equal to t,
0 otherwise.

Along the lines of the GLIMMIX family described in chapter 7, the observed choice variables y_{nkt}, and the rank-ordered preferences are assumed to be independent multinomial (cf, Luce and Suppes 1965), and to arise from a population that is a mixture of S unobserved segments. Given segment s, the choice probability for profile k at (choice replication or rank-order position) t is: $P_{kt|s} = P(U_{kt|s} \geq U_{qt|s})$, where $U_{kt|s}$ is the random utility derived from alternative k at t in segment s. For the choice model consumers are assumed to maximize their utility over the entire choice set, $U_{qt|s} = \max(U_{rt|s}, r \neq k)$, whereas for the preference rankings it holds: $U_{qt|s} = \max(U_{jt|s}, U_{qt-1|s})$; that is, at each rank consumers choose the profile that has maximum utility over the remaining set. The latter is equivalent to the rank-explosion rule derived by Chapman and Staelin (1982), so that the rank orders are modeled as a set of successive first choices. The random utility is assumed to be a function of the attributes:

$$U_{kt|s} = \sum_{p=1}^{P} \sum_{l=1}^{L_p} \beta_{pls} x_{kpl} + \varepsilon_{kts}. \tag{18.2}$$

If the random components, ε_{kts}, are assumed to be independent and identically Weibull distributed, the choice and rank probabilities are respectively

$$P_{kt|s} = \frac{\exp(U_{kt|s})}{\sum_{k=1}^{K} \exp(U_{kt|s})}, \text{ and } P_{kt|s} = \frac{\exp(U_{kt|s})}{\sum_{q \in Q} \exp(U_{qt|s})}, \tag{18.3}$$

where Q is the set of alternatives ranked lower than or equal to t. As an additional feature of the model, the prior probabilities of segment membership are re-

Part 4 Applied market segmentation

parameterized according to a concomitant variables model, as in equation 9.2. The models are estimated by maximizing the likelihood, as explained in chapter 9. Below, we reiterate the application to segmenting consumers on the basis of their rank-order preferences for banking services.

An extension of the procedure was provided by Wedel et al. (1998), who assume that the brand intercepts in a conjoint choice model can be decomposed into latent dimensions and segment-specific ideal points along those dimensions. The utility function further contains a linear combination of the attribute-level dummies, as in equation 11.2 (here part-worths are specified only at the aggregate level). Their model integrates conjoint analysis and multidimensional scaling, which makes it especially suited for product positioning. The model also incorporates the effect of brand familiarity, which moderates the importance of concrete attributes in the conjoint design, and the abstract (latent) dimensions underlying brand equity.

Application of Mixture Logit Regression to Conjoint Segmentation

Kamakura, Wedel and Agrawal (1994) applied their model to a commercial conjoint study on consumers' preferences for bank services. Managers of a large bank wanted information on consumers' concern about four attributes and about segments of consumers with different importances for those attributes. The four attributes (with their levels in parentheses) were:

- MINBAL: minimum balance required to exempt the customer from a monthly service fee ($0, $500, $1000),

- CHECK: amount charged per check issued by the customer (04, 154, 354),

- FEE: monthly service fee charged if the account balance falls below the minimum ($0, $3, $6), and

- ATM: availability and cost of automatic teller machines in a network of supermarkets (not available, free ATM, 75c per transaction).

Two equivalent but distinct sets of nine profiles were created from the attributes and were presented to a random sample of 269 of the banks customers in the form of "peeling stickers" in a mail survey. The customers were instructed to peel off their first choice and stick it to a designated place, then their second choice and so on, until a full ranking was obtained. The effects of the first three attributes were assumed to be linear and modeled with a single dummy variable, whereas the effects of the availability and cost of automatic teller machines were modeled with using two dummy variables. In addition to the conjoint data, the following information for each respondent was obtained from the bank:

- BALANCE: average balance kept in the account during the past six months, earning 5.5% interest,

- NCHECK: number of checks issued per month in the past six months (at no charge), and

- NATM: number of ATM transactions per month (all ATM machines).

Those variables represent the actual past banking behavior of each respondent, and constituted the concomitant variables included in the model to explain segment membership.

Results

The rank orders of the nine stimuli were analyzed with the concomitant variable mixture regression model described before .The model was applied to the data for S = 1 to 5 segments; the CAIC statistic indicated S = 4 segments. Table 18.5 presents the estimated coefficients for the four-segment solution. A likelihood ratio test with a chi-square value of 51.2 with 12 degrees of freedom showed that the contribution of the concomitant variables to fit was statistically significant. (The estimated coefficients $ß_{ks}$ of the model with and without concomitant variables were quite similar.)

Segment 1 comprised 21.1 % of the sample. Customers in this segment were mainly concerned about the amount charged per check (CHECK) and the monthly service fee charged (FEE). Relatively little attention was paid to the minimum balance required (MINBAL) or the availability of automatic teller machines (ATM). The parameters of the sub-model show that customers with high average balance (BALANCE) were more likely to belong to this segment than to others. Further, those customers issued a relatively large number of checks per month (NCHECK), which explains their sensitivity for the amount charged per check.

Consumers in segment 2, 20.9% of the sample, were the most sensitive to the availability of automatic teller machines in supermarkets (ATM, ATM 75¢). Members of this segment assigned a high value to free access to ATMs, and would rather pay 75¢ per ATM transaction than not have access to the machines. That finding is consistent with the effects of the concomitant variables in the sub-model, from which we can observe that subjects in segment 2 were heavy users of teller machines. The remaining attributes, although significant, appear to have less impact than they do in the other segments.

Customers in segment 3, 26.7% of the sample, were somewhat sensitive to minimum balance (MINBAL) and ATM availability. Customers with high average balance (BALANCE) were more likely to belong to this segment than to segments 2 or 4.

Segment 4 consisted of 31.4% of the sample. The coefficient for the minimum balance (MINBAL) required was nearly four times as large as that in segment 3. Although the other attributes also had significant coefficients, consumers in this

Part 4 Applied market segmentation

segment were predominantly concerned about the minimum balance required to prevent a monthly service fee. The coefficients in the sub-model explain this result: consumers with low average balance had the highest probability of belonging to this segment.

Table 18.5: Parameter Estimates of the Rank-Order Conjoint Segmentation Model (adapted from Kamakura et al. 1994)

	Segment 1	Segment 2	Segment 3	Segment 4
Attributes				
MINBAL	-0.320	-1.978*	-2.633*	-10.117*
CHECK	-7.856*	-3.984*	-2.479*	-4.167*
FEE	-0.336*	-0.172*	-0.220*	-0.274*
ATM	0.515*	2.414*	0.818*	0.681*
ATM 75	-0.025	0.736*	-0.707*	-0.357*
Concomitant Variables				
BALANC	0.048*	0.017	0.028[1]	-0.093*
NCHECK	0.023*	-0.030	-0.001	0.008
NATM	-0.049*	0.075*	-0.001	-0.025
Size	0.211	0.209	0.267	0.314

* Statistically significant at $p < 0.05$.

The results of the model have provided information that supports the development of strategies by the bank's managers. The viability of differentiated services targeted at those segments would obviously depend on the marginal revenue generated by each customer and the marginal costs of the service features.

Conclusion

Our review in this chapter shows that a variety of approaches are now available for conjoint analysis and conjoint segmentation. Empirical comparisons and a comprehensive Monté Carlo study have not found substantial differences among the methods in predicting preference ratings and first choices. However, the clusterwise and mixture regression approaches have advantages for market segmentation purposes. First, those approaches produce a better recovery of the true underlying part-worths and a better classification of subjects into segments. Second, those approaches identify segments and estimate their preference functions simultaneously, thus eliminating one step from the traditional two-stage approach. Third, because those approaches do not estimate part-worths at the subject level, they can be applied to conjoint experiments with negative degrees of freedom at the subject level. The mixture regression approach is well suited for choice-based conjoint segmentation, producing predictions of choice shares without the arbitrary choice rules used in traditional conjoint analysis. Finally, with the addition of concomitant variables, the mixture approach results in a hybrid segmentation model that simultaneously identifies benefit segments, estimates the preference function within each segment, and relates segment membership to the background characteristics of each consumer.

PART 5
CONCLUSIONS AND DIRECTIONS FOR FUTURE RESEARCH

In the final part of the book we discuss some of the recent challenges to the mixture model approach forwarded in the marketing literature and formulate directions for future research.

19
CONCLUSIONS: REPRESENTATIONS OF HETEROGENEITY

In this chapter we discuss an issue that has recently received much interest in marketing: is the discrete distribution of heterogeneity, imposed by the assumption of the existence of market segments, adequate?

Continuous Distribution of Heterogeneity versus Market Segments

Most of our discussion in this book is based on the implicit assumption that consumers differ from each other, but only to a certain extent, so that they can be grouped into relatively homogeneous groups or segments. Unobserved heterogeneity has been widely recognized as a critical issue in modeling choice behavior, both from a theoretical and substantive standpoint (DeSarbo et al 1997; Allenby and Rossi 1999). The current state of affairs in both modeling and estimation presents an opportunity to take stock of the basic ideas behind the methods involved, and to identify important debates and issues. In doing so, we will base ourselves on the review in Wedel et al. (1999). Some researchers (Allenby and Lenk 1994; Allenby and Ginter 1995) argue that tastes, brand preferences and response to marketing variables are distributed over the population of consumers according to a continuous (unimodal) distribution rather than assuming a discrete distribution across homogeneous segments. To those authors, market segmentation leads to an artificial partition of the continuous distribution into homogeneous segments that ignores the inherent differences across consumers. Therefore, they advocate the estimation of consumer preferences and response to the marketing mix at the consumer level, applying the recent developments in Bayesian estimation (Allenby and Lenk 1994; Rossi and Allenby 1993).

The most important ways of representing heterogeneity in marketing models currently in use are through either a continuous or a discrete mixture distribution of the parameters. This book has taken the perspective of the discrete mixture distribution, connected to the concept of market segments. For the purpose of exposition, we here adopt a Bayesian formulation. The parameters are not fixed quantities, but random variables. We are interested in the posterior distribution of the individual-level parameters, given the data. Assume that Θ is a set of (hyper) parameters indexing the distribution of individual-level parameters. The posterior distribution of the individual-level parameters can be written as (Allenby and Rossi 1999):

Part 5 Conclusions and directions for future research

$$\pi(\boldsymbol{6},\Theta\mid y)\propto \pi(y\mid\boldsymbol{6})\pi(\boldsymbol{6}\mid\Theta)\pi(\Theta),\tag{19.1}$$

where the three terms after the proportionality sign are the likelihood, the mixing distribution and the prior for Θ, respectively. The mixing distribution is often taken to be the multivariate normal:

$$\pi(\boldsymbol{6}\mid\Theta)=MVN(\mu,\Sigma).\tag{19.2}$$

Frequentist inference, which does not take prior information on the parameters into account (i.e. $\pi(\Theta)$ is omitted), focuses on obtaining point estimates of the (hyper-) parameters given the observed data (the likelihood does not involve a probability measure on the parameters, cf. Lindsey 1996, p.76). Equation (19.1) then involves an integration over the mixing distribution:

$$\pi(\Theta\mid y)=\int \pi(y\mid\theta)\pi(\theta\mid\Theta)\,d\theta.\tag{19.3}$$

The discussion here focuses on the form of the mixing distribution. Two issues are pertinent to the discussion of whether a continuous distribution of consumer parameters or a discrete number of classes is more appropriate to represent heterogeneity for marketing purposes.

Continuous or Discrete

First, the question is whether consumer heterogeneity is better described by a continuous or by a discrete distribution. We are of the opinion that to date, no conclusive evidence has been presented for either of the two arguments. The reason is that the often limited data available from each individual consumer reduces the power of any test comparing the two philosophies. Because consumer-level estimates are often obtained with considerable error, a direct analysis of the estimates does not provide conclusive evidence to support either assumption.

Some have argued that the underlying assumption of a limited number of segments of individuals that are perfectly homogeneous within segments in finite mixture models is overly restrictive (cf. Allenby and Rossi 1999). To those authors, market segmentation in choice modeling leads to an artificial partition of the continuous distribution into homogeneous segments. If the underlying distribution is continuous, then assuming a discrete mixing distribution leads to inconsistent parameter estimates. Assuming a limited number of segments in which all consumers are identical seems too restrictive. Critique levied against the discrete mixture approach to heterogeneity is that its predictive power in hold out samples of alternatives is limited because individual-level estimates are constrained to lie in the convex hull of the segment-level estimates. Research has shown that the assumption of within-segment homogeneity may result in a loss of predictive performance. Because of this, models at the individual-level, or models with continuous heterogeneity distributions have been found to outperform the mixture model approaches (Vriens, Wedel and Wilms 1996, Lenk, DeSarbo, Green and Young 1996). Several approaches have been developed that allow for within-segment heterogeneity by compounding the distribution for the dependent variable, for example a Multinomial distribution for y, with a conjugate heterogeneity distribution, such as the Dirichlet, giving rise to the Dirichlet-Multinomial, that effectively captures over-dispersion of the dependent variable within classes (see for example, Böckenholt 1993).

Assuming a continuous mixing distribution seems to offer several advantages: they seem to characterize the tails of the heterogeneity distribution better and predict individual choice behavior more accurately than finite mixture models (Allenby and Rossi 1999; Allenby, Arora and Ginter 1998). They allow model specification to closely follow an underlying theory of consumer behavior (see for example Allenby, Arora and Ginter 1998). Moreover, individual level estimates of model parameters are easily obtained. But, a continuous distribution of consumer parameters assumes a specific distributional form, such as the unimodal normal distribution. Studies have shown that the estimation of response parameters is sensitive to the specific assumptions made on the form of the heterogeneity distribution, which may result in a loss of predictive accuracy (Heckman and Singer 1984). On the other hand, a continuous heterogeneity distribution can be approximated closely by a discrete one by letting the number of support points of the discrete distribution increase (at the cost of a decrease in the reliability of the parameters).

Both discrete and continuous representations therefore seem to have some disadvantages, and under which conditions one of the two is most appropriate remains an empirical question. As evidence accumulates in the future, empirical generalizations obtained through meta-analysis of published studies may shed light on the issue. More recently, combinations of the discrete and continuous heterogeneity approaches have been developed, that account for both discrete segments and within segment heterogeneity (Allenby, Arora and Ginter 1998; Allenby and Rossi 1998; Lenk and DeSarbo 1999). Table 19.1 summarizes several of the issues discussed above.

ML or MCMC

The two most important ways in which models with heterogeneity have been estimated is through maximizing a likelihood function, and with Bayesian approaches. Both discrete and continuous heterogeneity models can in principle be estimated with either maximum likelihood or Bayesian methods: in fact all models in Table 19.1 can be estimated with both approaches. However, currently many published papers in marketing that utilize a continuous distribution of heterogeneity have relied upon Bayesian methods. The advantage of Bayesian methods lies in obtaining posterior distributions of individual-level parameters, based on the actual distribution of the hyper-parameters is used. ML methods approximate the posterior distribution of hyper-parameters by quadratic approximations to the likelihood

Part 5 Conclusions and directions for future research

around the point estimates, and posterior estimates of individual level parameters can only be obtained by using empirical Bayes estimates, conditioning on the point estimates of the hyper-parameters.

Table 19.1. Comparison of Discrete and Continuous representations of heterogeneity (adapted from Wedel et al. 1999).

Mixing Distribution	Discrete	Continuous	Continuous & Discrete
Example reference	Kamakura & Russell (1989)	Allenby & Lenk (1995)	Allenby, Arora & Ginter (1998)
Typical name	Mixture Model	Random Coefficients Model	Mixture Random Coefficients
Estimation [1]	ML, MCMC	SML, MCMC	MCMC
Individual-level predictions	Constrained in a convex hull	Influenced by aggregate parameters	Influenced by component parameters
Precision of individual – level estimates	Not available	Obtained empirically from iterates of a MCMC	Obtained empirically from iterates of the MCMC
Heterogeneity distributions accommodated [2]	M for θ and compound for y e.g. BB, DM	Arbitrary continuous for θ e.g. N, G, TN	M and arbitrary continuous for θ
Nested models	Aggregate	Aggregate	Aggregate, Discrete and Continuous

[1] ML = Maximum Likelihood, SML = Simulated ML, MCMC = Markov Chain Monte Carlo
[2] BB = Beta-Binomial, G = Gamma, DM = Dirichlet-Multinomial, N = Normal, M = Multinomial TN = Truncated Normal.

Bayesian estimation methods have gained popularity recently because they provide a set of techniques that allow for the development and analysis of complex models. The widely used Monte Carlo Markov Chain (MCMC, e.g. Gelman et al. 1995) methods involve integration over the posterior distribution of the parameters given the data (see equation 19.1) by drawing samples from that distribution. Starting from equation (19.1), this would involve successively drawing samples from the full conditional distributions of the model parameters (Allenby and Lenk 1994). Many applications in marketing involve the Gibbs-sampler as a special case, which can be implemented if expressions for the full conditional distributions of all parameters can be obtained. If that is not the case, powerful alternatives such as the Metropolis-Hastings algorithm are available that, based on a known candidate distribution, involve a rejection-type of sampling method to approximate those posterior distributions. Sample statistics such as the mean, mode and other percentiles, are then computed from the draws to characterize the posterior distribution. Statistical properties of estimates (e.g. precision) and estimates of functions of model parameters are thus easily obtained empirically. As marketing researchers move in the direction of utilizing heterogeneity models to make decisions, this feature of hierarchical Bayes models becomes an important advantage. However, a concern with regard to the use of hierarchical Bayes pertains to the distribution used to characterize heterogeneity, which is determined subjectively by the researcher. A pragmatic fix is a sensitivity analysis with regard to the choice of distribution in order to check model robustness (for model-checking procedures see Allenby and Rossi 1999). From a more dogmatic Bayesian point of view, the subjective choice of the distribution characterizes the analysts' uncertain state of knowledge, which does not need to conform to that of other analysts in this matter.

Under certain conditions (Lindsey 1996, p. 336) the ML and Bayesian approach lead to the same results. For example, for large samples the two approaches converge. In that case the posterior distribution approximates the normal with a covariance matrix which is equal to the inverse of the Hessian evaluated at the maximum likelihood estimates. So, the approaches are equivalent for practical purposes if the database is large, thus providing a pragmatic motivation for continued use of frequentist (ML-based) methods under conditions that occur for many marketing applications. However, in particular for small samples and certain parameterizations, the Bayesian approach, involving MCMC estimation, provides more accurate approximations of the posterior distribution of the parameters.

Managerial relevance

An important issue in the discussion of a continuous versus a discrete distribution of heterogeneity is managerial relevance. In applying models to segmentation, one should recognize that every model is at best a workable approximation of reality. One cannot claim that segments really exist or that the distributional form of unobserved heterogeneity is known. Segmentation is a marketing concept that is used to approximate the condition of market heterogeneity by positing diverse homogeneous groups of customers. It has proven to be a very useful concept to managers, and we conjecture that it will continue to be so for some time. Models that approximate market heterogeneity by a number of unobserved segments have great managerial appeal in many applications. Managers seem comfortable with the idea of market segments, and the models appear to do a good job of identifying useful groups. In other applications, for example in micro-marketing or direct marketing applications, a continuous approximation of customer heterogeneity may be more appropriate, as the purpose is not to find segments but to target individual customers. Hence, the two forms of models can be seen as

Part 5 Conclusions and directions for future research

approaches in which different managerial constraints are implicitly embedded in the statistical procedures.

On a more philosophical level, one might still question whether the assumption of a finite set of homogeneous segments is tenable, or whether the distribution of heterogeneity in populations should be considered continuous and described by some continuous rather than discrete distributions. A promising direction for future research is the development of segmentation models that reach a compromise between the two philosophies, identifying segments but also allowing for a certain degree of heterogeneity within each one of them. That stream of research has already been pursued to a certain extent through the development of finite mixtures of compound distributions, as noted above. Böckenholt (1993a,b) has also shown that accounting for within-segment heterogeneity decreases the effective number of segments needed to describe markets, making such approaches more manageable to marketing managers. A second approach mentioned above is the development of mixture models where there is a continuous distribution of the response coefficients within each segment. These studies tend to reach the same conclusion: if within segment heterogeneity is accounted for, fewer segments are needed to represent the data. From a managerial perspective this is an important finding, since in some segmentation studies the number of segments found is too large to enable development and implementation of efficient and profitable target marketing strategies. It is important to empirically assess the relative contribution of potential sources of heterogeneity, through nested model tests and investigation of predictive validity. Examples are provided by Allenby and Lenk (1994), and Allenby, Arora and Ginter (1998). Such model tests will ultimately allow for empirical generalization of heterogeneity findings, A meta-analysis of published studies could shed light on the question of whether accounting for within-segment heterogeneity through such compound distributions results in a smaller number of segments being identified by e.g. information criteria.

Further effort should also be made to collect evidence for the empirical generalization of the market segmentation concept, so that generalizeable conclusions can be drawn about within-segment heterogeneity and the number of segments, the form of heterogeneity distributions, separation of segments, homogeneity in response parameters, segment stability, and so on. Here the mixture model approach may prove to be a valuable instrument, providing a certain degree of uniformity in the methodology for identifying and testing segments in future research and thus facilitating the execution of meta-analyses on segmentation. The availability of standard user-friendly software is essential to that process. The GLIMMIX and other software tools described in this book provide an important step in making the mixture model methodology accessible. Currently, the GLIMMIX methodology has pervasively made its way into applied marketing research and may show to be marketing science's second great hit, after conjoint analysis.

Chapter 19 Conclusions: representation of heterogeneity

Individual Level versus Segment Level Analysis

The possibility of estimating responses to marketing variables at the consumer level leads to *micro-segmentation* where, taken to the extreme, each consumer may represent a segment. An example is *list segmentation* in direct marketing and database marketing. Once the response to marketing effort is estimated at the consumer level, the manager can target efforts to all consumers for whom the marginal contribution is greater than the marginal cost (Bult and Wansbeek 1995). Although these concepts call for methods that allow for individual differences to a much larger extent than mixture models currently do, Poisson regression has been successfully applied to such problems, not only at the aggregate level (Basu, Basu and Batra 1995) but also at the segment level by using a mixture model approach (Wedel et al. 1993). The application of hierarchical Bayes techniques to a customer database containing each customer's catalogue-ordering history produces individual-level estimates. The customer-level estimates can be used by direct marketers to identify customers with the highest probability of making an order at a given point in time, or to determine the optimal timing for a catalogue drop for each customer.

The rapid growth of new technologies in information (computers, telecommunications), product development (CAD-CAM), production (flexible manufacturing) and distribution (express delivery) is enabling marketers to customize their offerings to very small segments, or even to individual buyers, leading to *mass customization* (Kotler 1989). Such developments as direct marketing, micro-marketing and mass customization seemingly might make the mixture approach to market segmentation obsolete in the near future. However, some authors (Reynolds 1965) contend that much of what appears as individual differences is the result of consumers making random choices across brands or reflects changes in tastes over time. The mixture model approach has already provided some solutions, because customers' response behavior is explicitly considered to be stochastic and, as described in chapter 10, dynamic approaches have been developed that account for changes over time. A major limitation of most of those methods (the latent Markov models described in chapter 10) is that they can be applied only to a limited number of time periods. Extensions of the methods for application to time series beyond two or three points are therefore needed.

More recently, the practice of segmentation and product differentiation has been criticized for leading to runaway brand proliferation. Procter & Gamble, a leading producer of toiletries, has recently come to the conclusion that the world market does not need its 31 varieties of Head & Shoulders shampoo or 52 varieties of its popular Crest toothpaste (*Businessweek*, September 9, 1996). That manufacturer, along with other leading marketers such as Nabisco and Clorox, decided to streamline its product line, scaling back its customization of brands to ever smaller segments. Such events call for more research on the stability of market segments, and on the dynamic nature of market segmentation, leading to brand evolution rather than uncontrolled proliferation. In addition, such events support the use of market research in identifying a limited number of segments that are as homogeneous as possible, for example by the mixture approach.

In Table 19.2 we provide a conceptual comparison of the segment level (mixture model) and individual level (hierarchical Bayes) approaches with respect to technical and managerial issues, summarizing the preceding discussion. It cannot be concluded that one of the methods is to be universally preferred. Both approaches still have some unresolved technical problems, and have substantive areas in marketing where their application best suits the underlying behavioral assumptions and the managerial issues of primary interest.

Table 19.2: Comparison of Segment Level and Individual-Level Approaches

Issues	Segment Level	Individual Level
Methodological		
Heterogeneity assumption	Discrete, all subjects in a segment are identical	Continuous, each subjects has a set of idiosyncratic parameters
Modeling problems	Choosing the number of segments, Model identification, Local optima.	Computational requirements, Lack of criteria for convergence, Lack of software, Choice of prior distributions.
Loss in predictive accuracy	Due to predictions being a convex combination of segment level predictions	Due to parametric assumptions on the form of the heterogeneity distribution
Managerial		
Substantive area of managerial relevance	Market segmentation	Direct- and micro marketing
Supporting developments in marketing practice	Continued importance of market segmentation and scaling back of runaway brand proliferation by marketing management	Trends towards individualization of consumers and mass customization in marketing management

The advantage of segment-level models is that the market segments are often very compelling from a managerial standpoint. Its disadvantage is that they can over-simplify and that of limited predictive validity. Segment-level models are more effective in conveying to managers the overall pattern to heterogeneity, but individual level approaches are more accurate when individual level forecasting is necessary. However, advantages of the individual level approach accrue in particular when one wants to obtain posterior distributions of individual-level parameters. An important consideration is whether one is interested in predicting the future behavior (such as in direct marketing data for financial products) or in obtaining results that are representative for the total market or population, based on a *sample* of individuals (as household-level scanner data on non-durable goods). Models with continuous representation of heterogeneity do better than those with discrete heterogeneity in the former case (especially when estimated with Bayesian methods that allow for individual level estimates to be obtained). However, the issue is unresolved if a sample of individuals is analyzed, since individual level estimates can only be obtained for individuals in the sample and projection to the population is not possible at the individual level. The approaches cannot be seen as roughly equivalent ways of representing consumer segments, but are based on quite different behavioral assumptions, and have different managerial implications. Importance lies in the association of a unique set of managerial implications for each type of heterogeneity assumption. An important issue for future research is to provide theoretical underpinnings of heterogeneity, with the purpose of identifying variables that need to be included in models and to assist researchers in the appropriate model specification. Preferably, models should be based on theory-driven, verifiable assumptions about the underlying process that generates heterogeneity

20
DIRECTIONS FOR FUTURE RESEARCH

In this final chapter we briefly review the past and dwell upon segmentation strategy. We provide an agenda for future research.

The Past

Most of our discussion in this monograph has focused on past developments of bases and methods for market segmentation, and on how they have been applied. We started with a discussion of the main criteria for evaluating market segments, and with a review of the most frequently used bases used to group customers into homogeneous market segments. The review was followed by a detailed discussion of the many methods that can be used to form those homogeneous groups, starting with the traditional clustering techniques, and then moving to the more recent approaches of mixture models, mixture regressions, mixture unfolding and concomitant variable mixture models. Finally, we reviewed some of the application areas that have recently attracted much interest.

The emphasis on criteria and methodology reflects the research efforts in the past, most of which have concentrated on the identification of market segments (Wilkie and Cohen 1977). Early a-priori segmentation approaches started with the researcher's selection of a single partitioning basis (e.g., heavy vs. light users of the product or a demographic characteristic), that would directly define segments. Because it can be very easily applied by managers without the need to collect data, that approach has been -and continues to be- quite popular. More useful managerial insights were obtained by the development of post-hoc segmentation, whereby segments are formed as groups of consumers that are homogeneous along multiple characteristics. In contrast to earlier *descriptive* approaches, which formed segments with homogeneous profiles along observed characteristics, *predictive* segmentation methods enabled marketing researchers to form segments homogeneous on the *relationship* among observed variables and allowed the prediction of some measure of market performance, such as consumer preference or choice. The main concerns in the extensive literature on segmentation research have been related to the particular variables used to define the segments and the statistical approaches used to form those segments. Decisions on the number of segments have been based mostly on statistical fit, and the evaluation of the segments has typically been based on the face validity of the segment profiles and/or on the ability to predict actual or intended behavior.

Segmentation Strategy

The focus on methodology may mislead the reader into believing that segmentation is essentially a marketing research problem. Quite to the contrary, market segmentation *strategy* (Bass, Tigert and Lonsdale 1968) does not entail a mere market condition to be investigated and identified (Dickson and Ginter 1987). Segments are not homogeneous groupings of customers naturally occurring in the marketplace. Market segments are determined by the marketing manager's strategic view of the market. Her/his perspective determines the way homogeneous groups of potential customers are to be identified by marketing research. For different strategic goals, different segments may need to be identified in the same population. Moreover, market segments may result from the activities of the marketing manager, especially in the process of demand function modification (Dickson and Ginter 1987), which is aimed at altering customers' benefit importance or ideals along certain product dimensions. Segments are groups of customers to whom the firm may offer a single marketing mix to maximize its long-term profit goals. The strategic purposes of segmentation determine the bases and methods used in market research; different segments may be identified in the same population of customers in different segmentation studies with different purposes (e.g., new product development, pricing or defining direct marketing targets).

The strategy of market segmentation has its theoretical basis in early microeconomic models of price discrimination that showed how a firm selling a homogeneous good to a heterogeneous market could maximize profits: by selling at differentiated prices, the firm would extract the maximum surplus from the market, charging prices to each group of consumers that optimally match their reservation prices. Thus, the microeconomic approach has taken the *responsiveness* criterion for effective market segmentation as the starting point. That criterion has played an important role, especially in the first steps of the historical development of the market segmentation concept. Nevertheless, even the early days of marketing and economic history, there were other views. Shaw (1912), for example, already emphasized the importance of recognizing economic and social market contours and treating them as separate marketing problems.

However, early economic theory did not account for the firm's limited information and limited ability to target the segments selectively (The *accessibility* criterion, cf. Claycamp and Massy 1968). Nor did it recognize that to be of managerial use segmentation studies should provide information on how segments can be targeted with new products, advertising and the like (the *actionability* criterion), nor that segments may change over time (the *stability* criterion). Market segmentation is one component of a broader strategy of the firm, which Kotler (1989) calls the "four P's of strategic marketing" (not to be confused with the better-known four P's of tactical marketing): *probing* (marketing research), *partitioning* (segmentation), *prioritizing* (selecting the target segments) and *positioning* (pinpointing the competitive options in each target segment).

Most of the research on market segmentation has focused on the probing and partitioning components in isolation, that is, defining segmentation bases and identifying segments without considering managerial constraints, especially in the two other phases of the strategic process. Historically, the question of how to find segments was managers' first most pressing concern. However, now that considerable progress has been made in that area, we should begin to address such issues as how to limit the amount of dissimilarity within a segment only up to the point at which the segments become "unmanageable." Because of discontinuities in the marketing mix due to either environmental constraints (e.g., limited ability to reach different consumers with different media or distribution channels) or institutional constraints (e.g., manufacturing limitations in producing multiple versions of the product), managers' need to consider segment representations that are not necessarily the most efficient in the statistical sense. A statistically "less efficient" segmentation scheme might best fit the environmental and institutional constraints faced by managers (Mahajan and Jain 1978). Segmentation research should be conducted within the context of management of efficient consumer response, whereby the whole chain of production, marketing, distribution logistics and finance is taken into account. Scanner data have been shown to hold promise in that respect. The size and number of segments must be determined on the basis of stability and homogeneity, as well as managerial consideration of the costs of segmentation in relation to efficient consumer response (Wind 1978). Such an approach may provide a potential solution to the classical problem of determining the appropriate number of mixture model segments describing the sample. Information criteria can only be used as heuristics to determine the number of segments, as explained in chapter 6, given that certain conditions for their applicability do not hold in those situations. Promising alternative procedure for determining the number of mixture components was proposed by Böhning, Schlattmann and Lindsay (1992) and applied in a marketing context by, for example, Dillon and Kumar (1994). However, the extension of that procedure to more general situations appears not to be straightforward. In the context of Bayesian estimation promise lies in the application of jump-diffusion methods that consider the number of classes as stochastic unknown parameters to be estimated (Phillips and Smith 1996). The essential characteristic of this approach is that prior probabilities are placed on models with different numbers of segments, and that discrete transitions between those models are allowed in sampling from their posterior distributions. On the other hand, criteria based on manageability, for example based on profit maximization, seem to be quite useful to pursue.

The strategic nature of market segmentation has been considered by several authors in the past (a minority in relation to the extensive research in that area), starting from the seminal work by Smith (1956), who saw market segmentation as "a rational and more precise adjustment of product and market effort to consumer or user requirements." Whereas Smith saw segmentation as "dis-aggregative in its effect," Claycamp and Massy (1968) saw market segmentation as a process of aggregation in which firms build "to a viable segmentation strategy rather than tearing a market apart to find one." Tollefson and Lessig (1978), and many influential scholars after them, also saw market segmentation as an aggregative

Part 5 Conclusions and directions for future research

process. However, the issue of segmentation being aggregative or disaggregative seems to have been cluttered by the techniques that were available to identify segments from marketing data: hierarchical and non-hierarchical clustering. As we argued in chapter 5, although hierarchical clustering may have presented a useful set of heuristics for identifying market segments, they basically derive hierarchical relationships among customers that cannot be motivated from any substantive theory in most markets. Therefore, no theoretical reasons seem to support aggregative or dis-aggregate views of market segmentation.

Claycamp and Massy (1968) show that even in the idealized condition of no scale diseconomies in fitting specialized programs to individual consumers, other constraints will lead firms to market segmentation at a more aggregate level. From analysis of synthetic data, Tollefson and Lessig (1978) concluded that elasticity estimates and other response parameters are ineffective criteria for forming segments. However, in a later study with real data, Elrod and Winer (1982) found exactly the opposite: price elasticities outperformed other segmentation criteria in terms of the profitability of the final segmentation solution. Again, much of the discussion seems to have been influenced by the state of the art of market research at the time. To identify segments on the basis of elasticities, researchers needed first to obtain elasticity estimates at the individual level. Because of data limitations, such elasticity estimates were based on relatively few measurements and therefore were rather unreliable, thus resulting in unstable market segments. Elrod and Winer were the first to obtain individual level estimates with some adequate level of reliability. Later, the advent of scanner-data and mixture models made a great contribution. Elrod and Winer's results provide support for the scanner data based mixture approach Kamakura and Russell (1989), which is essentially a dis-aggregative normative approach in the sense of Smith, as well as for the models derived from it later.

With a few notable exceptions (such as Elrod and Winer 1982), the normative segmentation approaches reviewed above have not been implemented and tested. Moreover, those approaches still focus more on the partitioning aspect of the problem, with an emphasis on the bases and analytical techniques for identifying segments rather than on the manager's problem of designing profitable marketing mixes. An integration of the four P's of strategic marketing is necessary to address the issues that are really relevant to the marketing manager: determining the number of distinctive marketing mixes to be offered and designing each of those marketing mixes in the context of efficient consumer response and competitive environment. Such an integrated approach to market segmentation should go beyond the mere identification of statistically homogeneous groups of consumers. Rather, it should help managers to find the most profitable combination of marketing mixes. The ultimate goal is not to find market segments, but to determine the most effective product line, price, promotion mix, distribution strategy, and so on.

Although a model that integrates all aspects of strategic marketing and produces optimally designed and targeted marketing mixes might be beyond the realm of possibility, the methods presented in this book can be used to move in that direction. Response-based segmentation with the application of GLIMMIX (chapter 7) models can be used to determine the optimal level for the decision variable (e.g., price, sales promotion) for each segment and the number of segments the manager wants to use, assuming a cost structure associated with the decision variable. One of the main obstacles to response-based normative segmentation in the past was the fact that individual demand functions are not directly observable, which precluded the application of aggregative clustering techniques. In GLIMMIX models, segmentation and estimation are done simultaneously, allowing for the estimation of segment-level response functions even when there are not enough degrees of freedom for individual-level estimates. By combining the GLIMMIX model with a concomitant sub-model (see chapter 9) and additional information on the cost associated with reaching consumers defined by the concomitant variables, the researcher will also be able to find optimal targeting segments defined along the concomitant variables. As with the componential segmentation model (Green 1977), emphasis shifts to predicting which person type will respond to what product feature; the question of the effectiveness of the segmentation solution shifts from the methods for identifying segments to the variables that define the person types in the concomitant sub-model.

The integration of segmentation, targeting and positioning is achieved with the application of the STUNMIX model (chapter 8). Suppose a sample of consumers participate in a survey, for example a conjoint choice experiment including brands, and provides information on their life-style and media exposure along with their choices. Application of the STUNMIX model to the choice data would produce a positioning map in which segments of consumers would be placed in the same space as the underlying brand dimensions, so that the implicit worth of each dimension for a given segment would be represented by either projections of the segment onto the attribute vectors (in a vector map) or the distance between the segment and the attribute location (in an ideal-point map) (Note that the work of Russell and Kamakura 1994 and Wedel et al. 1998 already has moved conjoint and choice modeling in that direction). Targeting decisions can be made with the help of a sub-model in which lifestyle and media exposure serve as concomitant variables. The integrated model is similar in purposes to the tandem approach proposed by Green, Krieger and Carroll (1987). Their two-stage approach starts with individual-level conjoint analyses to estimate part-worths and predict shares of choices. Cluster analysis and multidimensional scaling are then used in a second stage to portray conjoint-based results graphically, producing a joint map of attributes, profiles and consumer segments, for positioning decisions (Wind 1978). At that stage, Green and his co-authors needed to translate their conceptual ideas of segmentation strategy into a sequence of steps, which now can be accomplished within a single mixture-model-based framework.

A second model that makes a step in the direction of integrating several aspects of strategic marketing is the international market segmentation model provided by ter Hofstede, Steenkamp and Wedel (1999). The authors propose a methodology to identify cross-national market segments, based on means-end chain (MEC) theory. It is a step towards integrating theory, measurement and analysis in

Part 5 Conclusions and directions for future research

market segmentation. It is based on two pick-any tasks, administered to respondents in several countries, who are asked to identify idiosyncratic relations between product attributes and their benefits, and benefits and values, respectively. The resulting pick-any (0/1) data are analyzed with a binomial mixture model (chapter 6), that describes the links between the MEC-concepts in a probabilistic manner. The model has several distinctive features over the standard binomial mixture model. First, it accounts for the stratified sampling design, often employed in international market segmentation studies where samples are stratified by country, using the procedure described in chapter 6. Second, it deals with differential response tendencies across subjects and countries, by including a response threshold that is assumed to follow a normal distribution. Finally, it uses a concomitant variable specification (chapter 9) to allow segment sizes to vary across countries. The methodology offers the potential for integrating market segmentation, product development and communication strategies by linking product characteristics to consumer benefits and at the segment level. As a segmentation bases, MEC combines the advantage of product-specific and consumer-specific bases. It provides segment-specific estimates of strengths of links that tie the consumer to the product, i.e., that identify cognitive associations between product attributes, benefits of product use, and consumer values at the segment-level. It supports product positioning not only at the level of product attributes, but also includes the benefits derived by segments from those attributes, as well as the values satisfied. This increases the actionability of the identified segments for targeted strategies of international product development and communication. The proposed methodology, integrating international segmentation theory, measurement, and analysis in one framework, provides a step toward a closer fit of marketing research and managerial objectives in the international marketing domain.

The applications suggested above integrate various aspects of the segmentation strategy into a single model, but still separate marketing research from marketing policy. The work presumes an inferential step followed by some form of optimization or decision making. Segments are identified and described before a decision is made, either by the manager or by an optimization algorithm, on which segments should be targeted and how they should be served. An important topic for future research is the integration of marketing research and marketing strategy within a single model that combines inference and profit maximization. With such a model, decisions on the proper number of segments would be based on managerial criteria (expected profits) rather than goodness of fit; inferences would be made directly in terms of the optimal marketing effort rather than mere descriptions of the segments. The mixture models discussed in chapters 6 through 9 are amenable to such extensions. Suppose cost information is readily available and optimal levels of the policy variables can be determined on the basis of response parameters estimated for each segment. Then the optimization algorithm could be implemented in the estimation algorithm to produce direct estimates of the optimal strategy for each segment. Such an extension might also lead to a profit-oriented criterion for the best number of targeted strategies. Admittedly, the details for the integration of inference and policy must be examined more carefully. Nevertheless, such efforts could lead to approaches for normative segmentation that have important practical implications. Moreover, they could lead to more objective and managerially relevant criteria for the optimal number of market segments.

Agenda for Future Research

Market segmentation has been a major research topic in marketing in the 40 years since the concept was introduced by Smith (1956). In a review of the accomplishments in its first 20 years, Wind (1978) concluded that research efforts should focus on:

1. Discovery of new bases for segmentation and assessment of their performance across products, situations and markets.
2. Development of simple and flexible data analysis techniques that combine discrete and continuous variables, and assessment of the conditions under which each technique is most appropriate.
3. Development of new research designs and data collection techniques that place less burden on respondents and better handle incomplete data.
4. Development of normative segmentation approaches that consider the dynamic allocation of marketing effort in the long run.
5. New conceptualizations of the segmentation problem.
6. Integration of marketing segmentation research with the marketing information system and with strategic decision making.
7. External validation studies to determine the effectiveness of the methods and strategies for market segmentation.

Our review of the developments in the 20 years since Wind's (1978) landmark study leads us to conclude that considerable advances have been made on the issues related to the methodology of market segmentation (the first four items listed above).

With respect to segmentation bases (1), we have provided an extensive evaluation in chapter 3 according to criteria for effective segmentation. Product benefits appear to be among the most promising bases. An important conclusion is that several bases need to be used in modern segmentation studies, each according to its own strength. In part 4 we discussed applications of the most promising bases in each of the four main types.

With respect to the segmentation methods (2), the mixture model approach is the major breakthrough. Mixture models as described in part 2 are a flexible tool for market segmentation because they provide a statistical model-building approach to the problem. Their appropriateness can be assessed empirically in each particular application, by using statistical criteria and tests of nested model specifications are available.

Part 5 Conclusions and directions for future research

In the development of alternative data collection methods (3), a major breakthrough is the availability of scanner data (chapters 7 and 17). In addition, conjoint analysis has proven to be a very powerful tool for benefit segmentation (chapter 18), and computer-assisted tailored interviewing methods have substantially reduced respondents' burden. The means-end chain approach described above provides a particular actionable segmentation basis that combines the advantage of product-specific and consumer-specific bases.

The further development of normative segmentation methods (4) remains an important topic for future research. An important first step in that direction has been the development of several mixture regression models for scanner data, in which brand choice is related to marketing instruments at the segment level.

New conceptualizations for market segmentation (5) may be brought about by strategies that focus on individual customers. However, this has heated the debate on the most appropriate forms of representation of consumer heterogeneity as either discrete, continuous or a combination of both.

Much remains to be done in the conceptualization of strategic market segmentation and in the integration of marketing research and strategy. Therefore, we feel that the last three issues raised by Wind are still valid. In the future, research effort should be devoted to those areas, where we think that studies with the following objectives are of particular interest.

1. The further development of the theoretical underpinning of heterogeneity, with the purpose of identifying variables to be included in models and of assisting researchers in appropriate model specification. In particular the development of normative theories of market segmentation that lead to feasible and easily-to-implement solutions for the optimal allocation of marketing effort, require attention. International market segmentation is an important area to be further explored in that respect.

2. Development of models that integrate measurement, segmentation, targeting and positioning and avoid sequential stepwise solutions and decisions. The further integration of measurement, analysis and strategic decision making will lead to decisions based directly on managerial objectives rather than statistical fit. Development of techniques for the efficient estimation of response to the marketing mix at the consumer level in combination with market segmentation, through integration of continuous and discrete distributions of heterogeneity. This will permit empirical validation of the segmentation concept through nested model testing, will improve the implementation of micro-segmentation and enhance predictive performance of models.

3. Empirical testing of the predictive validity segment solutions and the study of the stability of segments over time. Empirical generalizations on segmentation would need to be formulated, and Monte Carlo empirical comparisons done to compare the conditions under which models and

estimation methods provide adequate representations of the complex market conditions facing managers. An understanding of the dynamic nature of preferences and market segments composition is essential for strategies focused on the evolution rather than the proliferation of products and businesses.

References

Aitkin, M., D. Anderson and J. Hinde, (1981), "Statistical Modeling of Data on Teaching Styles" (with discussion), *Journal of the Royal Statistical Society,* A144, 419-461.

Aitkin, M. and D. B. Rubin (1985), "Estimation and Hypothesis Testing in Finite Mixture Ddistributions," *Journal of the Royal Statistical Society,* B47, 67-75.

Akaike, H. (1974), "A New Look at Statistical Model Identification," *IEEE Transactions on Automatic Control,* AC-19, 716-723.

Albaum, G. (1989), "Bridger (Ver. 1.0) and Simgraf (Ver. 1.0)," *Journal of Marketing Research,* 26, 486-488.

Allenby, G. M., N. Arora and J. L. Ginter (1998), "On the Heterogeneity of Demand," *Journal of Marketing Research,* 35, 348-389.

Allenby, G. M. and J. L. Ginter (1995), "Using Extremes to Design Products and Segment Markets," *Journal of Marketing Research,* 32, 392-405.

Allenby, G. M. and P. J. Lenk (1994), "Modeling Household Purchase Behavior with Logistic Normal Regression," *Journal of the American Statistical Association,* 89, 1218-1231.

Allenby, G. M. and P. E. Rossi (1999), "Marketing Models of Heterogeneity," *Journal of Econometrics,* 89, 57-78.

Alpert, L. and R. Gatty (1969), "Product Positioning by Behavioral Life-Style," *Journal of Marketing,* 4, 65-71.

Alwin, D. F. and J. Krosnick (1985), "The Measurement of Values in Surveys: A Comparison of Ratings and Rankings," *Public Opinion Quarterly,* 49, 535-552.

Amemiya, T. (1985), *Advanced Econometrics,* Cambridge, MA: Harvard University Press.

Anderson, E. B. (1980), *Discrete Statistical Models with Social Science Applications,* New York: North Holland.

Anderson, W. T., Jr., E. P. Cox, III and D. G. Fulcher (1976), "Bank Selection Decisions and Market Segmentation," *Journal of Marketing,* 40, 40-45.

Arabie, P. (1977), "Clustering Representations of Group Overlap," *Journal of Mathematical Sociology,*5 113-128.

Arabie, P., J. D. Carroll, W. S. DeSarbo and J. Wind (1981), "Overlapping Clustering: A New Method for Product Positioning," *Journal of Marketing Research,* 18, 310-317.

Arabie, P. and L. Hubert (1994), "Cluster Analysis in Marketing Research," in *Advanced Methods of Marketing Research,* R.P. Bagozzi (ed.), Cambridge: Blackwell, 160-189.

Assael, H. (1970), "Segmenting Markets by Group Purchasing Behavior: An Application of the AID Technique," *Journal of Marketing Research,* 7,

References

153-158.

Assael, H. and A. M. Roscoe, Jr. (1976), "Approaches to Market Segmentation Analysis," *Journal of Marketing*, 40, 67-76.

Baker, M. J. (1988), *Marketing Strategy and Management*, New York: Macmillan Education.

Balakrishnan, P. V., M. C. Cooper, V. S. Jacob and P. A. Lewis (1995), "A Study of the Classification Capabilities of Neural Netwoks Using Unsupervised Learning: A Comparison with K-Means Clustering," *Psychometrika*, 59, 509-524.

Balasubramanian, S. K. and W. A. Kamakura (1989), "Measuring Consumer Attitudes Toward the Marketplace with Tailored Interviews," *Journal of Marketing Research*, 26, 311-326.

Banfield, C. F. and L. C. Bassil (1977), "A Transfer Algorithm for Nonhierarchical Classification," *Applied Statistics*, 26, 206-210.

Basford, K. E. and G. J. McLachlan (1985), "The Mixture Method of Clustering Applied to Three-Way Data," *Journal of Classification*, 2, 109-125.

Bass, F. M., E. A. Pessemier and D. J. Tigert (1969), "A Taxonomy of Magazine Readership Applied to Problems in Marketing Strategy and Media Selection," *Journal of Business*, 42, 337-363.

Bass, F. M., D. J. Tigert and R. T. Lonsdale (1968), "Market Segmentation: Group Versus Individual Behavior," *Journal of Marketing Research*, 5, 264-270.

Basu, A. K., A. Basu and R. Batra (1995), "Modeling the Response Pattern to Direct Marketing Campaigns," *Journal of Marketing Research*, 32, 204-212.

Bearden, W., J. E. Teel, Jr. and R. M. Durand (1978), "Media Usage, Psychographics and Demographic Dimensions of Retail Shoppers," *Journal of Retailing*, 54, 65-74.

Beatty, S. E., L. R. Khale, P. Holmer and S. Misra (1985), "Alternative Measurement Approaches to Consumer Values: The List of Values and the Rokeach Values Survey," *Psychology and Marketing*, 2, 181-200.

Beckenridge, J. N. (1989), "Replicating Cluster Analysis: Method, Consistency and Validity," *Multivariate Behavioral Research*, 24, 147-161.

Becker, B. W. And P. E. Conner (1981), "Personal Values of the Heavy User of Mass Media," *Journal of Advertising Research*, 21, 37-43.

Beckwith, N. E. and M. W. Sasieni (1976), "Criteria for Market Segmentation Studies," *Management Sciences*, 22, 892-908.

Beldo, L. A. (1966), "Market Segmentation and Food Consumption," *Harvard Business Review*, 184-195.

Belk, R. W. (1975), "Situational Variables and Consumer Behavior," *Journal of Consumer Research*, 2, 157-164.

Bezdek, J. C. (1974), "Numerical Taxonomy with Fuzzy Sets," *Journal of Mathematical Biology*, 1, 57-71.

Bezdek, J. C., C. Coray, R. Gunderson and J. Watson (1981a), "Detection and Characterization of Cluster Substructure. I. Linear Structure: Fuzzy c-Lines," *SIAM Journal on Applied Mathematics*, 40, 339-357.

Bezdek, J. C., C. Coray, R. Gunderson and J. Watson (1981b), "Detection and Characterization of Cluster Substructure. II. Fuzzy c-Varieties and Convex Combinations Thereof," *SIAM Journal on Applied Mathematics*, 40, 358-372.

Binder, D. A. (1978), "Bayesian Cluster Analyis," *Biometrica*, 65, 31-38.

Blackwell, R. D. and W. W. Talarzyk (1977), "Lifestyle Retailing: Competition Strategies for the 1980s," *Journal of Retailing*, 59, 7-27.

Blattberg, R. C., T. Buesing, P. Peacock and S. K. Sen (1978), "Identifying the Deal-Prone Segment," *Journal of Marketing Research*, 15, 369-377.

Blattberg, R. C., T. Buesing and S. K. Sen (1980), "Segmentation Strategies for New National Brands," *Journal of Marketing*, 44, 59-67.

Blattberg, R. C. and S. K. Sen (1974), "Market Segmentation Using Models of Multidimensional Purchasing Behavior," *Journal of Marketing*, 38, 17-28.

Blattberg, R. C. and S. K. Sen (1976), "Market Segments and Stochastic Brand Choice Models," *Journal of Marketing Research*, 13, 34-45.

Blattberg, R. C. and K. J. Wisnewski (1989), "Price-Induced Patterns of Competition," *Marketing Science*, 4, 291-309.

Blozan, W. and P. Prabhaker (1984), "Notes on Aggregation Criteria in Market Segmentation," *Journal of Marketing Research*, 21, 332-335.

Böckenholt, U. (1993a), "Estimating Latent Distributions in Recurrent Choice Data," *Psychometrika*, 58, 489-509.

Böckenholt, U. (1993b), "A Latent Class Regression Approach for the Analysis of Recurrent Choice Data," *British Journal of Mathematical and Statistical Psychology*, 46, 95-118.

Böckenholt, U., and I. Böckenholt (1991), "Constrained Latent Class Analysis: Simultaneous Classification and Scaling of Discrete Choice Data," *Psychometrika*, 56, 699-716.

Böckenholt, U., and W. Gaul (1986), "Analysis of Choice Behavior via Probabilistic Ideal Point and Vector Models," *Applied Stochastic Models and Data Analysis*, 2, 202-226.

Böckenholt, U. and R. Langeheine (1996), "Latent Change in Recurrent Choice Data," *Psychometrika* 61, 285-302.

Böhning, D. (1995), "A Review of Reliable Maximum Likelihood Algorithms for Semiparametric Mixture Models," *Journal of Statistical Planning and Inference*, 47, 5-28.

Böhning, D., E. Dietz, R. Schaub, P. Schlatttmann and B.G. Lindsay (1994), " The Distribution of the Likelihood Ratio for Mixtures of Densities from the One-Parameter Exponential Family," *Annals of the Institute of Statistics and Mathematics*, 46, 373-388.

Böhning, D., P. Schlattmann and G. B. Lindsay (1992), " Computer Assisted Analysis of Mixtures C.A.MAN: Statistical Algorithms," *Biometrics*, 48, 283-303.

Bollen, K. A. (1989), *Structural Equations with Latent Variables*, New York: John Wiley & Sons.

Bond, M. H. (1988), "Finding Universal Dimensions of Individual Variation in Multi-cultural Studies of Values: The Rokeach and Chinese Value Surveys," *Journal of Personality and Social Psychology*, 55, 1009-1015.

Bottenberg, R. A. and R. E. Christal (1968), "Grouping Criteria - A Method Which Retains Maximum Predictive Efficiency," *Journal of Experimental Education*, 36, 28-34.

Boyd, H. W. and W. F. Massy (1972), *Marketing Management*, New York: Harcourt Brace Jovanovich.

Boyles, R. A. (1983), "On the Convergence of the EM Algorithm," *Journal of the Royal Statistical Society*, B45, 47-50.

Bozdogan, H. (1994), "Mixture Model Cluster Analysis Using Model Selection Criteria and a New Informational Measure of Complexity," in *Multivariate Statistical Modelling*, 2, H. Bozdogan (ed.), Dordrecht: Kluwer Academic Publishers, 69-113.

Breiman, L., J. H. Friedman, R. A. Olshen and C. J. Stone (1984), *Classification and Regression Trees*, Belmont, CA: Wadsworth International Group.

Brody, R. P. and S. M. Cunningham (1968), "Personality Variables and the Consumer Decision Proccess," *Journal of Marketing Research*, 5, 50-57.

Brookhouse, K. J., R. M. Guion and E. M. Doherty (1986), "Social Desirability Response Bias as as One Source of the Discrepancy Between Subjective Weights and Regression Weights," *Organizational Behavior and Human Decision Processes* 37, 316-328.

Bruno, A. V. and E. Pessemeier (1972), "An Empirical Investigation of the Validity of Selected Attitude and Activity Measures," *Proceedings of the Third Annual Conference of the Association of Consumer Research*, 456-474.

Bucklin, R. and S. Gupta (1992), "Brand Choice, Purchase Incidence and Segmentation: An Integrated Modeling Approach," *Journal of Marketing Research*, 29, 201-215.

Bucklin, R., S. Gupta and S. Han (1995), "A Brand's Eye View of Response Segmentation in Consumer Brand Choice Behavior," *Journal of Marketing Research*, 32, 66-74.

Bucklin, R. E., S. Gupta and S. Siddarth (1991), "Segmenting Purchase Quantity Behavior: A Poisson Regression Mixture Model," working paper, Graduate School of Management, University of California, Los Angeles.

Bult, J. R. and T. Wansbeek (1995), "Optimal Selection for Direct Mail," *Marketing Science*, 14, 378-394.

Bunch, D. S., D. R. Gay and R. E. Welsch (1995), "Algorithm 717, Subroutines for Maximum Likelihood and Quasi-likelihood Estimation of Parameters in Nonlinear Regression Models," *ACM Transactions of Mathematical Software*, 19, 109-130.

Calantone, R. J. and A. G. Sawyer (1978), "The Stability of Benefit Segments," *Journal of Marketing Research*, 15, 395-404.

Calinski, T. and J. Habarasz (1974), "A Dendrite Method for Cluster Analysis," *Communications in Statistics*, 3, 1-17.

Carroll, J. (1976). "Spatial, non-spatial and hybrid models for scaling" *Psychometrika*, 41, 439-463.

References

Carroll, J. D. and A. Chaturvedi (1995), "A General Approach to Clustering and Scaling of Two-way, Three-way or Higer-way Data," *in Geometric Representations of Perceptual Phenomena,* R.D. Luce, M. D'Zmura, D.D. Hoffman, G. Iverson and A.K. Romney (eds). Mahwah, N.J. Erlbaum, 295-318.

Carroll, J. D. and P. Arabie (1983), "INDCLUS: An Individual Differences Generalization of the ADCLUS Model and the MAPCLUS Algorithm," *Psychometrika*, 48, 157-169.

Carroll, J. D. and P. E. Green (1995). "Psychometric Methods in Marketing Research: Part I, Conjoint Analysis," *Journal of Marketing Research*, 32, 385-391.

Carroll, J. D. and P. E. Green, (1997), "Psychometric Methods in Marketing Research: Part II, Multidimensional Scaling," *Journal of Marketing Research*, 34, 193-204.

Carroll, J. D. and S. Pruzansky, (1980), "Discrete and Hybrid Scaling Models," in Similarity and Choice, E.D. Lanterman and H. Feger (eds), Bern: Hans Huber, 108-139.

Casella, G. and E. I. George (1992), "Explaining the Gibbs Sampler," *The American Statistician*, 46, 167-174.

Cattin, P. and D. R. Wittink (1982), "Commercial Use of Conjoint Analysis: A Survey," *Journal of Marketing*, 46, 44-53.

Celeux, G. and G. Soromenho (1996), " An Entropy Based Criterion for Assessing the Number of Clusters in a Mixture Model," *Journal of Classification*, 13, 195-212.

Chamberlain, G. (1980), "Analysis of Covariance with Qualitative Data," *Review of Economic Studies,* 47, 225-238.

Chapman, R. G. and R. Staelin (1982), " Exploiting Rank Ordered Choice Set Data Within the Stochastic Utility Model," *Journal of Marketing Research*, 19, 288-301.

Chaturvedi, A. D., and J. D. Carroll, (1994), "An Alternating Combinatorial Optimization Approach to Fitting the INDCLUS and Generalized INDCLUS Models," *Journal of Classification*, 11, 155-170.

Cheng, R. and G. W. Milligan (1996), "Measuring the Influence of Individual Data Points in a Cluster Analysis," *Joural of Classification*, 13, 315-335.

Chintagunta, P. K. (1993), "Estimating a Multinomial Probit Model of Brand Choice Using the Method of Simulated Moments," *Marketing Science*, 11, 386-407.

Chintagunta, P. K. (1994), "Heterogeneous Logit Implications for Brand Positioning," *Journal of Marketing Research*, 31, 304-311.

Chintagunta, P. K., D. C. Jain and N. J. Vilcassim (1991), "Investigating Heterogeneity in Brand Preferences in Logit Models for Panel Data," *Journal of Marketing Research*, 28, 417-428.

Christal R. E. (1968), "Jan: A Technique for Analyzing Group Judgment," *The Journal of Experimental Education*, 36, 24-27.

Claxton, J. D., J. N. Fry and B. Portis (1974), "A Taxonomy of Pre-purchase Informa-

tion Gathering Patterns," *Jounal of Consumer Research,* 1, 35-42.
Claycamp, H. J. (1965), "Characteristics of Owners of Thrift Deposits in Commercial and Savings and Loan Associations," *Journal of Marketing Research,* 2, 163-170.
Claycamp, H. and W. F. Massy (1968), "A Theory of Market Segmentation," *Journal of Marketing Research,* 5, 388-394.
Clogg, C. (1977), "Unrestricted and Restricted Maximum Likelihood Latent Structure Analysis: A Manual for Users," *Working Paper No. 1977-09,* Population Issues Research Office, Pennsylvania State University.
Cochran W. G. (1957), *Sampling Techniques,* New York: John Wiley & Sons.
Colombo, R. and D. G. Morrison (1989), "A Brand Switching Model with Implications for Marketing Strategies," *Marketing Science,* 8, 89-99.
Cooper, L. and M. Nakanishi (1988), *Market Share Analysis,* Boston: Kluwer Academic Press.
Corter, J. E. (1996), *Tree Models of Similarity,* London: Sage.
Cox, T. F. and M. A.A. Cox (1994), *Multidimensional Scaling,* London: Chapman Hall, London.
Currim, I. S. (1981), "Using Segmentation Approaches for Better Prediction and Understanding from Consumer Mode Choice Models," *Journal of Marketing Research,* 18, 301-309.
Curry, D. J. (1993), *The New Marketing Research Systems,* New York: John Wiley & Sons.
Darden, W. R. and W. D. Perreault (1977), "Classification for Market Segmentation: An Improved Linear Model for Solving Problems of Arbitrary Origin," *Management Science,* 24, 259-271.
Darden, W. and F. Reynolds (1971), "Predicting Opinion Leadership for Men's Apparel Fashion," *Journal of Marketing Research,* 8, 505-508.
Davison, M. L. (1983), *Multidimensional Scaling,* New York: John Wiley & Sons.
Day, N. E. (1969), "Estimating the Components of a Mixture of Two Normal Distributions," *Biometrika,* 56, 463-474.
Dayton, C. M. and G. B. MacReady (1988), "Concomitant Variable Latent Class Models," *Journal of the American Statistical Association,* 83, 173-178.
De Soete, G. (1988), "OWTRE: A Program for Optimal Variable Weighting for Ultrametric and Additive Tree Fitting," *Journal of Classification,* 5, 101-104.
De Soete, G. and W. S. DeSarbo (1991), "A Latent Class Probit Model for Analyzing Pick Any/N Data," *Journal of Classification,* 8, 45-63.
De Soete, G., W. S. DeSarbo and J. D. Carroll (1985), "Optimal Variable Weighting for Hierarchical Clustering: An Alternating Least-Squares Algorithm," *Journal of Classification,* 2, 173-192.
De Soete, G. and W. Heiser (1993), "A Latent Class Unfolding Model for Analyzing Single Stimulus Preference Ratings," *Psychometrika,* 58, 545-566.
De Soete, G. and S. Winsberg (1993), "A Latent Class Vector Model for Preference Ratings," *Journal of Classification,* 10, 195-218.
Deb, P. and P. K. Trivedi (1997), "The Demand for Medical Care by the Elderly: A Finite Mixture Approach," *Journal of Applied Econometrics,* 12, 313-336.

Dempster, A. P., N. M. Laird and Donald B. Rubin (1977), "Maximum Likelihood from Incomplete Data via the EM-Algorithm," Journal *of the Royal Statistical Society*, Series B39, 1-38.

Dennis, J. E. and R. B. Schnabel (1983), *Numerical Methods for Unconstrained Optimization and Nonlinear Equations*, Engelwood Cliffs, NJ: Prentice Hall.

DeSarbo, W. S. (1982), "GENNCLUS: New Models for General Nonhierarchical Clustering Analysis," *Psychometrika*, 47, 449-476.

DeSarbo, W. S., A. Ansari, P. Chintagunta, C. Himmelberg, K. Jedidi, R. Johnson, W.A. Kamakura, P. Lenk, K. Srinivasan and M. Wedel (1997), "Representing Heterogeneity in Consumer Response Models," *Marketing Letters*, 8, 335-348.

DeSarbo, W. S., J. D. Carroll and L. A. Clark (1984), "Synthesized Clustering: A Method for Amalgamating Alternative Clustering Bases with Differential Weighting of Variables," *Psychometrika*, 49, 57-78.

DeSarbo, W. S., and J. Cho (1989), "A Stochastic Multidimensional Scaling Vector Threshold Model for the Spatial Representation of Pick Any/n Data," *Psychometrika*, 56, 105-310.

DeSarbo, W. S. and S. C. Choi (1993), "Game Theoretic Derovations of Competitive Srategies in Conjoint Analysis," *Marketing Letters*, 4, 337-348.

DeSarbo, W. S. and W. L. Cron (1988), "A Maximum Likelihood Methodology for Clusterwise Linear Regression," *Journal of Classification*, 5, 249-282.

DeSarbo, W. S. and D. L. Hoffman (1987), "Constructing MDS Joint Spaces from Binary Choice Data: A Multidimensional Unfolding Threshold Model for Marketing Research," *Journal of Marketing Research*, 24, 40-54.

DeSarbo, W. S., D. J. Howard and K. Jedidi (1990), "MULTICLUS: A New Method for Simultaneously Performing Multidimensional Scaling and Cluster Analysis," *Psychometrika*, 56, 121-136.

DeSarbo, W. S. and K. Jedidi (1995), "The Spatial Representation of Homogeneous Consideration Sets," *Marketing Science*, 14, 326-342.

DeSarbo, W. S., K. J. Jedidi, K. Cool and O. Schendel (1991), "Simultaneous Multidimensional Unfolding and Cluster Analysis: An Investigation of Strategic Groups," *Marketing Letters*, 2, 129-146.

De Sarbo, W. S. and V. Mahajan (1984), "Constrained Classification: The Use of A Priori Information in Cluster Analysis," *Psychometrika*, 49, 57-78.

DeSarbo, W. S., A. K. Manrai and L. A. Manrai (1993), "Non-Spatial Tree Models for the Assessment of Competitive Market Structure: An Integrated Review of the Marketing and Psychometric Literature," in *Handbooks of Operations Research and Management Science: Marketing*, 5, J. Eliashberg and G.L. Lilien (eds.), Amsterdam: North Holland, 193-257.

DeSarbo, W. S., A. K. Manrai and L. A. Manrai (1994), "Latent Class Multidimensional Scaling: A Review of Recent Developments in the Marketing and Psychometric Literature," in *Advanced Methods of Marketing Research*, Richard P. Bagozzi (ed.)., Cambridge: Blacwell, 190-222.

DeSarbo, W.S., R. L. Oliver and A. Rangaswamy (1989), "A Simulated Annealing

References

Methodology for Clusterwise Linear Regression,"*Psychometrika*, 54, 707-736.
DeSarbo, W. S. V. Ramaswamy and R. Chatterjee (1995), "Analyzing Constant-Sum Multiple Criterion Data: A Segment-Level Approach," *Journal of Marketing Research*, 32, 222-232.
DeSarbo, W. S., V. Ramaswamy and S. Cohen (1995), "Market Segmentation with Choice-based Conjoint Analysis," *Marketing Letters*, 6, 137-148.
DeSarbo, W. S., V. Ramaswamy and P. Lenk (1993), "A Latent Class Procedure for the Analysis of Two-Way Compositional Data," *Journal of Classification*, 10, 159-194.
DeSarbo, W. S., V. Ramaswamy , M. Wedel and T. H. A. Bijmolt (1996), "A Spatial Interaction Model for Deriving Joint Space Maps of Bundle Compositions and Market Segments from Pick-Any/J Data: An Application to New Product Options, A *Marketing Letters*,7, 131-145.
DeSarbo, W. S., M. Wedel, M. Vriens and V. Ramaswamy (1992), "Latent Class Metric Conjoint Analysis," *Marketing Letters*, 3, 273-288.
Dhalla, N. K. and W. H. Mahatoo (1976), "Expanding the Scope of Segmentation Research," *Journal of Marketing*, 40, 34-41.
Dichter, E. (1958), "Typology," *Motivational Publications*, 3, 3.
Dickson, P. R. (1982), "Person-Situation: Segmentation's Missing Link," *Journal of Marketing*, 46, 56-64.
Dickson, P. R. and J. L. Ginter (1987), "Market Segmentation, Product Differentiation, and Marketing Strategy," *Journal of Marketing*, 51, 1-11.
Dillon, W. R., and M. Goldstein (1990), *Multivariate Analysis*, New York: John Wiley & Sons.
Dillon, W. R., and S. Gupta (1996), "A Segment-Level Model of Category Volume and Brand Choice," *Marketing Science*, 15, 38-59.
Dillon, W. R., and A. Kumar (1994), "Latent Structure and Other Mixture Models in Marketing: An Integrative Survey and Overview," in *Advanced Methods in Marketing Research*, Richard P. Bagozzi (ed.), Cambridge, MA: Blackwell, 295-351.
Dillon, W. R., A. Kumar and M. Smith de Borrero (1993), "Capturing Individual Differences in Paired Comparisons: An Extended BTL Model Incorporating Descriptor Variables," *Journal of Marketing Research*, 30, 42-51.
Dixon, J. K. (1979), "Pattern Recognition with Missing Data," *IEE Transactions on Systems, Man, and Cybernetics*, SMC9, 617-621.
Dolan, C. V. and H. L. J. van der Maas (1998), "Fitting Multivariate Normal Mixtures Subject to Structural Equation Modeling," *Psychometrika*, 63, 227- 254.
Doyle, P. and I. Fenwick (1975), "The Pitfalls of AID Analysis," *Journal of Marketing Research*, 12, 408-413.
Doyle, P. and P. Hutchinson (1976), "The Identification of Target Markets," *Decision Sciences*, 7, 152-161.
Duffy, D. E. and A. J. Quiroz (1991), "A Permutation Based Algorithm for Block-Clustering," *Journal of Classification*, 8, 65-91.
Dunn, J. C. (1974), "A Fuzzy Relative of the ISODATA Process and Its Use in

Detecting Compact Well-Separated Clusters," *Journal of Cybernetics*, 3, 32-57.
Edwards, A. W. F. and L. L. Cavali Sforza (1965), "A Method for Cluster Analysis, *Biometrics*, 21, 362-375.
Elrod, T. and M. P. Keane (1995), "A Factor-Analytic Probit Model for Representing the Market Structure in Panel Data," *Journal of Marketing Research*, 32, 1-16.
Elrod, T., J. J. Louviere and K. S. Davey (1992), "An Empirical Comparison of Ratings-Based and Choice-Based Models," *Journal of Marketing Research*, 29, 368-377.
Elrod, T. and R. S. Winer (1982), "An Empirical Evaluation of Aggregation Approaches for Developing Market Segments," *Journal of Marketing*, 46, 65-74.
Evans, F. B. (1959), "Psychological and Objective Factors in Prediction of Brand Choice," *Journal of Business*, 17, 340-369.
Everitt, B. S. (1984), "Maximum Likelihood Estimation of the Parameters in a Mixture of Two Univariate Normal Distributions:; A Comparison of Different Algorithms," *The Statistician*, 33, 205-215.
Everitt, B. S. (1992), *Cluster Analysis*, London: Edward Arnold.
Everitt, B. S. and D. J. Hand (1981), *Finite Mixture Distributions*, London: Chapman and Hall.
Fischer, G. W. and D. Nagin (1981), "Random Versus Fixed Coefficient Quantal Choice Models," in *Structural Analysis of Discrete Data with Econometric Application*, Charles Masnki and Daniel McFadden (eds.), Cambridge, MA: The MIT Press.
Fishbein, M. and I. Ajzen (1975), *Belief, Attitude, Intention, and Behavior: An Introduction to Theory and Research*, New York: Addison-Wesley.
Fisher, R. A. (1935), "The Case of Zero Survivors" (Appendix to C.I.Bliss 1935), "*Annals of Applied Biology*, 22, 164-165.
Foekens, E. W. (1995), *Scanner Data Based Marketing Modelling: Empirical Applications*, Capelle a/d IJssel, The Netherlands: Labyrint.
Formann, A. K. (1992), "Linear Logistic Latent Class Analysis for Polytomous Data," *Journal of the American Statistical Association*, 87, 476-486.
Fowlkes, E. B. (1979), "Some Methods for Studying Mixtures of Two Normal (Lognormal) Distributions," *Journal of the American Statistical Association*, 74, 561-575.
Fowlkes, E. B., R. Gnanadesikan, and J. Kettenring (1988), "Variable Selection in Clustering," *Journal of Classification*, 5, 205-228.
Frank, R. E. (1962), "Brand Choice as a Probability Process," *Journal of Business*, 35, 43-56
Frank, R. E. (1967), "Is Brand Loyalty a Useful Basis for Market Segmentation," *Journal of Advertising Research*, 7, 27-33.
Frank, R. E. (1968), "Market Segmentation Research: Findings and Implications," in *The Application of the Sciences to Marketing Management*, F.M. Bass, C.W.

References

King and E.A. Pessemier (eds.), New York: John Wiley & Sons.
Frank R. E. (1972), "Predicting New Product Segments," *Journal of Advertising Research*, 12, 9-13.
Frank, R. E. and P. E. Green (1968), "Numerical Taxonomy in Marketing Analysis: A Review Article," *Journal of Marketing Research*, 5, 83-98.
Frank, R. E., W. F. Massy and Y. Wind (1972), *Market Segmentation*, Englewood Cliffs, NJ: Prentice Hall.
Frank, R. E. and C. E. Strain (1972), "A Segmentation Research Design Using Consumer Panel Data," *Journal of Marketing Research*, 9, 285-390.
GAUSS (1991), *The GAUSS System Version 2.1, User's Manual*, Washington: Aptech Systems Inc.
Gelman, A., J. B. Carlin, H. S. Stern and D. B. Rubin (1995), *Bayesian Data Analysis*, London: Chapman Hall.
Gensch, D. H. (1985), "Empirically Testing a Disaggregate Choice Model for Segments," *Journal of Marketing Research*, 22, 462-467.
Ghose, S. (1998), "Distance representations of Consumer Perceptions: Evaluating Appropriateness by Using Diagnostics," *Journal of Marketing Research*, 35, 137-153.
Ginter, J. L. and E. A. Pessemeier (1978), "Brand Preference Segments," *Journal of Business Research*, 6, 111-131.
Gnanadesikan, R. J. W. Harvey and J. Kettenring (1994), "Mahalanobis Metrics for Cluster Analysis". *Sankhya*.
Gnanadesikan, R., J. R. Kettenring and J. L. Tsao (1995), "Weighting and Selection of Variables for Cluster Analysis," *Journal of Classification*, 12, 113-136.
Gonul, F. and K. Srinivasan (1993), "Modelling Multiple Sources of Heterogeneity in Multinomial Logit Models: Methodological and Managerial Issues," *Marketing Science*, 12, 213-229.
Goodman, L. A. (1974), "Exploratory Latent Structure Analysis Using Both Identifiable and Unidentifiable Models," *Biometrika*, 61, 215-231.
Goodman, L. A. (1974), "The Analysis of Systems of Qualitative Variables When Some of the Variables are Unobservable," *American Journal of Sociology*, 79, 1179-1259.
Gordon, A. D. (1980), *Classification*, London: Chapman and Hall.
Gordon, A. D. and J. T. Henderson (1977), "An Algorithm for Euclidian Sum of Squares Classification," *Biometrics*, 33, 355-362.
Gowda, K. C. and G. Krishna (1978), "Agglomerative Clustering Using the Concept of Mutual Nearest Neighbourhood," *Pattern Recognition*, 10, 105-112.
Gower, J. C. (1971), "A General Coefficient of Similarity and Some of its Properties," *Biometrics*, 27, 857-871.
Gower, J. C. (1974), "Maximal Predictive Classification, *Biometrics*, 30, 643-654.
Green, P. E. (1977), "A New Approach to Market Segmentation," *Business Horizons*, 20, 61-73.
Green, P. J. (1984), "Iteratively Reweighted Least Squares for Maximum Likelihood Estimation, and Some Robust and Resistant Alternatives," *Journal of the*

Royal Statistical Society, B46, 149-192.

Green, P. E. and F. J. Carmone (1977), "Segment Congruence Analysis: A Method for Analyzing Association Among Alternative Bases for Market Segmentation," *Journal of Consumer Research*, 3, 217-222.

Green, P. E., F. J. Carmone and D. P. Wachspress (1976), "Consumer Segmentation via Latent Class Analysis," *Journal of Consumer Research*, 3, 170-174.

Green, P. E., F. J. Carmone and Y. Wind (1972), "Subjective Evaluation Models and Conjoint Measurement," *Behavioral Science*, 17, 288-299.

Green, P. E. and W. S. DeSarbo (1979), "Componential Segmentation in the Analysis of Consumer Trade-Offs," *Journal of Marketing*, 43, 83-91.

Green, P. E. and K. Helsen (1989), "Cross-Validation Assessment of Alternatives to Individual-Level Conjoint Analysis: A Case Study," *Journal of Marketing Research*, 26, 346-350.

Green, P. E. and A. M. Krieger (1991), "Segmenting Markets with Conjoint Analysis," *Journal of Marketing*, 55, 20-31.

Green, P. E. and A. M. Krieger (1993), "Conjoint Analysis with Product Positioning Applications," In: *Handbooks of Operations Research and Management Science, Vol. 5: Marketing*, J. Eliashberg and G.J.Lilien (eds.), North Holland, p.467-516.

Green, P. E., A. M. Krieger and J. D. Carrol (1987), "Conjoint Analysis and Multidimensional Scaling: A Complementary Approach," *Journal of Advertising Research*, 27, 21-27.

Green, P. E., A. Krieger and C. M. Schaffer (1993), "An Empirical Test of Optimal Respondent Weighting in Conjoint Analysis,", *Journal of the Academy of Marketing Science*, 21, 345-351.

Green, P. E. and V. Srinivasan (1978), "Conjoint Analysis in Consumer Research: Issues and Outlook," *Journal of Consumer Research*, 5, 103-123.

Green, P. E. and V. Srinivasan (1990), "Conjoint Analysis in Marketing: New Developments with Implications for Research and Practice," *Journal of Marketing*, 54, 3-19.

Greeno, D. W., N. S. Summers and J. B. Kernan (1973), "Personality and Implicit Behavior Patterns," *Journal of Marketing Research*, 10, 63-69.

Grover, R. and V. Srinivasan (1987), "A Simultaneous Approach to Market Segmentation and Market Structuring," *Journal of Marketing Research*, 24, 139-153.

Grover, R. and V. Srinivasan (1989), "An Approach for Tracking Within-Segment Shifts in Market Shares," *Journal of Marketing Research*, 26, 230-236.

Guadagni, P. and J. Little (1983), "A Logit Model of Brand Choice," *Marketing Science*, 2, 203-238.

Gunderson R. W. (1982), "Choosing the r-Dimension for the FCV Family of Clustering Algorithms," *BIT*, 22, 140-149.

Gunter, B. and A. Furnham (1992), *Consumer Profiles: An Introduction to Psychographics*, London: Routledge.

References

Gupta, S. (1991), "Stochastic Models of Interpurchase Time with Time-Dependent Covariates," *Journal of Marketing Research*, 28, 1-15.

Gupta, S., and P. K. Chintagunta (1994), "On Using Demographic Variables to Determine Segment Membership in Logit Mixture Models," *Journal of Marketing Research*, 31, 128-136.

Gupta, S., P. Chintagunta, A. Kaul and D. R.Wittink (1996), "Do Household Scanner Data Provide Representative Inferences from Brand Choices? A Comparison with Store Data," *Journal of Marketing Research*, 33, 383-398.

Gutman, J. and M. Mills (1982), "Fashion Life-Style, Self-shopping Orientation, and Store Patronage: An Integrative Analysis," *Journal of Retailing*, 58, 48-86.

Hagenaars, J. and R. Luijkx (1990), "LCAG: A Program to Estimate Latent Class Models and Other Loglinear Models with Latent Variables with and without Missing Data," Working Paper #17, Tilburg University, Netherlands.

Hagerty, M. R. (1985), "Improving the Predictive Power of Conjoint Analysis: The Use of Factor Analysis and Cluster Analysis," *Journal of Marketing Research*, 22, 168-184.

Haley, R. I. (1968), "Benefit Segmentation: A Decision-Oriented Research Tool," *Journal of Marketing*, 32, 30-35.

Haley, R. I. (1984), "Benefit Segments: Backwards and Forwards," *Journal of Advertising Research*, 24, 19-25.

Hamilton, J. D. (1991), "A Quasi-Bayesian Approach to Estimating Parameters for Mixtures of Normal Distributions," *Journal of Busioness and Economic Statistics*, 9, 27-39.

Hasselblad, V. (1966), "Estimation of Parameters for a Mixture of Normal Distributions," *Technometrics*, 8, 431-444.

Hasselblad, V. (1969), "Estimation of Finite Mixtures of Distributions from the Exponential Family," *Journal of the American Statistical Association*, 64, 1459-1471.

Hauser, J. R. and G. L. Urban (1977), "A Normative Methodology for Modelling Consumer Response to Innovation," *Operations Research*, 25, 579-617.

Hausman, J. A. and D. A. Wise (1978), "A Conditional Probit Model for Qualitative Choice: Discrete Decisions Recognizing interdependence and Heterogeneous Preferences," *Econometrica*, 46, 403-426.

Heckman, J. J. (1981), "The Incidental Parameters Problem and the Problem of Initial Conditions in Estimating a Discrete Time- Discrete Data Stochastic Process," in: *Structural Analysis of Discrete data with Econometric Applications*, C.F. Manski and D. MacFadden (eds.), Cambridge, MA: MIT-Press, 179-195.

Heckman, J. J. and B. Singer (1984), "A Method for Minimizing the Impact of Distributional Assumptions in Econometric Models for Duration Data," *Econometrica*, 52, 271-320.

Helsen, K., K. Jedidi and W. S. DeSarbo (1993), "A New Approach to Country Segmentation using Multinational Diffusion Patterns," *Journal of Marketing*, 26, 60-71.

Herniter, J.D. and J. MaGee (1961), "Customer Behavior as a Markov Process,"

Operations Research, 9, 105-22.

Hofstede, F. ter, J. B. E. M. Steenkamp, and M. Wedel (1999), "International market segmentation based on consumer-product relations." *Journal of Marketing Research*, 36, 1-17.

Hope, A. C. A. (1968), "A Simplified Monte Carlo Significance Test Procedure," *Journal of the Royal Statistical Society*, B30, 582-558.

Hosmer, D. W. (1974), "Maximum Likelihood Estimates of the Parameters of a Mixture of Two Regression Lines," *Communications in Statistics*, 3, 995-1006.

Howard, J. A. (1985), *Consumer Behavior in Marketing Strategy*, New York: Prentice-Hall International.

Hruschka, H. (1986), "Market Definition and Segmentation Using Fuzzy Clustering Methods," *International Journal of Research in Marketing*, 3, 117-134.

Hubert, L. J. and P. Arabie (1985), "Comparing Partitions," *Journal of Classification*, 2, 193-218.

Hustad, T. P., C. S. Mayer and T. W. Whipple (1975), "Consideration of Context Differences in Product Evaluation and Market Segmentation,'" *Journal of the Academy of Marketing Science*, 3, 34-37.

Jain, A. K., N. K. Malhotra and V. Mahajan (1979), "A Comparison of Internal Validity of Alternative Parameter Estimation Methods in Decompositional Multiattribute Preference Models," *Journal of Marketing Research*, 16, 313-322.

Jain, D. and N. Vilcassim (1991), "Investigating Household Purchase Timing Decisions: A Conditional Hazard Function Approach," *Marketing Science*, 10, 1-23.

Jedidi, K., H. S. Jagpal, W. S. DeSarbo (1997a), "Semi-parametric Structural Equation Models for Response Based Segmentation," *Journal of Classification*, 14, 23-50.

Jedidi, K., H. S. Jagpal, W. S. DeSarbo (1997b), "Finite Mixture Strucutural Equation Models for Response Based Segmentation and Unobserved Heterogeneity," *Marketing Science*, 16, 39-59.

Jedidi, K., V. Ramaswamy, W. S. DeSarbo and M. Wedel (1996), "On Estimating Finite Mixtures of Multivariate Regression and Simultaneous Equation Models," *Journal of Structural Equation Modeling*, 3, 266-289.

Jensen, R. E. (1969), "A Dynamic Programming Algorithm for Cluster Analysis," *Journal of the Operations Research Society of America*, 7, 1034-1056.

Johnson, R. M. (1971), "Market Segmentation: A Strategic Management Tool," *Journal of Marketing Research*, 8, 13-8.

Johnson, R. M. (1987), "Adaptive Conjoint Analysis," in *Proceedings of the Sawtooth Software Conference on Perceptual Mapping, Conjoint Analysis and Computer Interviewing*. Ketchum, ID: Sawtooth Software, Inc., 253-265.

Jones, J. M. and J. T. Landewehr (1988), "Removing Heterogeneity Bias from Logit model Estimation," *Marketing Science*, 7, 41-59.

Jöreskog, K. G. (1971), "Simultaneous factor analysis in several populations,"

Psychometrika, 36, 409-426.
Jöreskog, K. G. (1973), "A General Method for Estimating a Linear Structural Equation System," in: *Structural Equation Models in Mathematical Psychology*, 2, A.S. Goldberger and O.D. Duncan (eds.), New York: Seminar, 85-112.
Jöreskog, K. G. (1978), "Statistical Analysis of Covariance and Correlation Matrices," *Psychometrika*, 43, 443-477.
Kahle, L. R., ed. (1983), *Social Values and Social Change: Adaptation to life in America*, New York: Praeger.
Kahle, L. R., S. E. Beatty and P. Holmer (1986), "Alternative Measurement Approaches to Consumer Values: The List of Values (LOV) and Values and Life Style (VALS)," *Journal of Consumer Research*, 13, 405-409.
Kamakura, W. A. (1988), "A Least Squares Procedure for Benefit Segmentation with Conjoint Experiments," *Journal of Marketing Research*, 25, 157-167.
Kamakura, W. A. (1991), "Estimating Flexible Distributions of Ideal-Points with External Analysis of Preference," *Psychometrika*, 56, 419-448.
Kamakura, W. A., B. Kim and J. Lee (1996), "Modeling Preference and Structural Heterogeneity," *Marketing Science*, 15, 152-172.
Kamakura, W. A. and J. A. Mazzon (1991), "Value Segmentation: A Model for the Measurement of Values and Value Systems," *Journal of Consumer Research*, 18, 208-218.
Kamakura, W. A. and T. P. Novak (1992), "Value-System Segmentation: Exploring the Value of LOV," *Journal of Consumer Research*, 19, 119-132.
Kamakura, W. A. and G. J. Russell (1989), "A Probabilistic Choice Model for Market Segmentation and Elasticity Structure," *Journal of Marketing Research*, 26, 379-390.
Kamakura, W. A. and G. J. Russell (1992), "Measuring Brand Value with Scanner Data," *International Journal of Research in Marketing*, 10, 9-22.
Kamakura, W. A. and R. K. Srivastava (1984), "Predicting Choice Shares Under Conditions of Brand interdependence," *Journal of Marketing Research*, 21, 420-434.
Kamakura, W. A. and R. K. Srivastava (1986), "An Ideal-Point Probabilistic Choice Model for Heterogeneous Preferences," *Marketing Science*, 5, 199-217.
Kamakura W. A. and M. Wedel (1995), "Life Style Segmentation with Tailored Interviewing," *Journal of Marketing Research*, 32, 308-317.
Kamakura W. A. and M. Wedel (1997), "Statistical Datafusion for Cross-Tabulation," *Journal of Marketing Research*, 34, 485-498.
Kamakura, W. A., M. Wedel and J. Agrawal (1994), "Concomitant Variable Latent Class Models for Conjoint Analysis," *International Journal of Research in Marketing*, 11, 451-464.
Kannan, P. K. and G. P. Wright (1991), "Modeling and Testing Structured Markets: A Nested Logit Approach," *Marketing Science*, 10, 58-82.
Kass, G. (1980), "An Exploratory Technique for Investigating Large Quantities of Categorical Data," *Applied Statistics*, 29, 119-127.
Katahira, H. (1987), "A Discrete Choice Model and Heterogeneous Preferences" [in

Japanese], *Journal of Econometrics* (University of Tokyo), 53, 31-45.

Katahira, H. (1990), "Joint Space Market Response Analysis," *Marketing Science*, 36, 13-27.

Kaufman, L. and P. J. Rousseeuw (1990), *Finding Groups in Data*, New York: John Wiley & Sons.

Kernan, J. B. (1968), "Choice Criteria, Decision Behavior, and Personality," *Journal of Marketing Research*, 5, 155-169.

Kiel, G. C. and R. A. Layton (1981), "Dimensions of Consumer Information Seeking Behavior," *Journal of Marketing Research*, 9, 233-239.

Kinnear, T. and J. Taylor (1976), "Psychographics: Some Additional Findings," *Journal of Marketing Research*, 13, 422-425.

Klein, R. W. and R. C. Dubes (1989), "Experiments in Projection and Clustering Using Simulated Annealing," *Pattern Recognition*, 22, 213-220.

Koontz, W. L. G., P. M. Narendra and K. Fukunaga (1975), "A Branch and Bound Clustering Algorithm," *IEE Transactions on Computers*, C-24, 908-915.

Koponen, A. (1960), "Personality Characteristics of Purchasers," *Journal of Advertising Research*, 1, 6-12.

Kotler, P. (1988), *Marketing Management*, Englewood Cliffs, NJ: Prentice-Hall.

Kotler, P. (1989), "From Mass Marketing to Mass Customization," *Planning Review*, (September-October.), 10-47.

Kotrba, R. W. (1966), "The Strategy Selection Chart," *Journal of Marketing*, 30, 89-95.

Krishnamurthi, L.and S. P. Raj (1988), "A Model of Brand Choice and Purchase Quantity Price Sensitivities," *Marketing Science*, 7, 1-20.

Kruskal J. B. (1965). Analysis of Factorial Experiments by Estimating Monotone Transformations of the Data. *Journal of the Royal Statistical Society*, B27, 251-263.

Landon, E. L. (1974), "Self Concept, Ideal Self Concept, and Consumer Purchase Intentions," *Journal of Consumer Research*, 1, 44-51.

Langeheine, R. and J. Rost (1988), *Latent Trait and Latent Class Models*, New York: Plenum Press.

Lastovicka, J. L. (1982), "On the Validation of Lifestyle Traits: A Review and Illustration," *Journal of Marketing Research*, 19, 126-138.

Lastovicka, J. L., J. P. Murray and E. A. Joachimsthaler (1990), "Evaluating the Measurement Validity of Lifestyle Typologies with Qualitative Measures and Multiplicative Factoring," *Journal of Marketing Research*, 27, 11-23.

Lazarfeld, P. F. (1935), "The Art of Asking Why," *National Marketing Review*, 1, 26-38.

Lazer, W. (1963), "Lifestyle Concepts and Marketing," in *Toward Scientific Marketing*, S. Greyser (ed.), Chicago: American Marketing Association, 130.

Leeflang, P. H. S. and A. J. Olivier (1985), "Bias in Consumer Panel and Store Audit Data," *International Journal of Research in Marketing*, 2, 27-41.

Lehman, D. R., W. L. Moore and T. Elrod (1982), "The Development of Distinct

References

Choice Processes Over Time: A Stochastic Modeling Approach," *Journal of Marketing*, 46, 48-59.
Lenk, P. J. and W. S. DeSarbo (1999), "Bayesian Inference for Finite Mixture of Generalized Linear Models With Random Effects," *Psychometrika*, forthcoming.
Lenk, P. J., W. S. DeSarbo, P. E. Green and M. R. Young (1996), "Hierarchical Bayes Conjoint Analysis: Recovery of Partworth Heterogeneity from Reduced Experimental Designs," *Marketing Science*, 15, 152-173.
Lesser, J. A. and M. A. Hughes (1986), "The Generalizability of Psychographic Segmentation Across Geographic Locations," *Journal of Marketing*, 50, 18-27.
Lessig V. P. and J. O. Tollefson (1971), "Market Segmentation Through Numerical Taxonomy," *Journal of Marketing Research*, 8, 480-487.
Lindsay, B. J. and K. Roeder (1992), " Residual Diagnostics in the Mixture Model," *Journal of the American Statistical Association*, 87, 785-795.
Lindsey, J. K. (1996), *Parametric Statistical Inference,* Oxford: Clarendon Press.
Long, S. (1983), *Covariance Structure Models, An Introduction to LISREL*, London: Sage University Press.
Loudon, D. and A. J. Della Bitta (1984), *Consumer Behavior. Concepts and Applications*, London: McGraw-Hill International Editions.
Louis, T. A. (1982), "Finding the Observed Information Matrix When Using the EM Algorithm," *Journal of the Royal Statistical Society*, B44, 226-233.
Louviére, J. J., (1988). Analyzing Decision Making: Metric Conjoint Analysis. Sage University Papers Series on Quantitative Applications in the Social Sciences, No 67. Newbury Park, CA: Sage Publications.
Luce, R. D. and P. Suppes (1965), "Preference, Utility and Subjective Probability," In: *Handbook of Mathematical Psychology* 3, R.D. Luce, R.R. Bush and E. Galanter (eds.), New York: John Wiley & Sons, 249-410.
Luce R. D. and J. W. Tukey (1964), "Simultaneous Conjoint Measurement: A New Type of Fundamental Measurement", *Journal of Mathematical Psychology*, 1, 1-27.
Lutz, J. G. (1977), "The Multivariate Analogue of JAN," *Educational and Psychological Measurement*, 37, 37-45.
MacLachlan, D. L. and J. K. Johansson (1981), "Market Segmentation with Multivariate AID", *Journal of Marketing*, 45, 74-84.
Macnaughton-Smith, P., W. T. Williams, M. B. Dale and L. G. Mockett (1964), "Dissimilarity Analysis: A New Technique of Hierarchical Subdivision," *Nature*, 202, 1034-1035.
MacReady, G. B. and C. M. Dayton (1992), "The Application of Latent Class Models in Adaptive Testing," *Psychometrika*, 57, 71-88.
Madrigal, R. and L. R. Kahle (1994), "Predicting Vacation Activity Preferences on the Basis of Value-System Segmentation," *Journal of Travel Research*, 22-28.
Magdison, J. (1994), "The CAID Approach to Segmentation Modelling: Chi-squares Automatic interaction Detection," *Advanced Methods of Marketing Research*, R.P. Bagozzi (ed.), Cambridge,MA: Blackwell, 118-119.

Mahajan, V. and A. K. Jain (1978), "An Approach to Normative Segmentation," *Journal of Marketing Research*, 15, 338-345.
Maier, J. and J. Saunders (1990), "The Implementation Process of Segmentation in Sales Management," *Journal of Personal Selling & Sales Management*, 10, 39-48.
Manton, K. G., M. A. Woodbury and H. D. Tolley (1994), *Statistical Applications Using Fuzzy Sets*, New York: John Wiley & Sons.
Martin, C. R. and R. L. Wright (1974), Profit-Oriented Data Analysis for Market Segmentation: An Alternative to AID," *Journal of Marketing Research*, 11, 237-242.
Maslow, A. H. (1954), *Motivation and Personality*, New York: Harper.
Massy, W. F. (1966), "Order and Homogeneity of Family Specific Brand-Switching Processes," *Journal of Marketing Research*, 3, 48-53.
Massy, W. F. and R. E. Frank (1965), "Short Term Price and Dealing Effects in Selected Market Segments," *Journal of Marketing Research*, 2, 171-185.
Massy, W. F., R. E. Frank and T. M. Lodahl (1968), *Purchasing Behavior and Personality Attributes*, Philadelphia: University of Pennsylvania Press.
Massy, W. F., D. B. Montgomery and D. Morrison (1970), *Stochastic Models of Buying Behavior*, Cambridge, MA:: MIT Press.
McCann, J. M. (1974), "Market Segment Response to the Marketing Decision Variables," *Journal of Marketing Research*, 11, 399-412.
McConkey, C. W. and W. E. Warren (1987), "Psychographic and Demographic Profiles of State Lottery Ticket Purchasers," *Journal of Consumer Affairs*, 21, 314-327.
McCullagh, P., and J. A. Nelder (1989), *Generalized Linear Models,* London: Chapman Hall.
McDonald, M. and I.Dunbar (1995), *Market Segmentation: A Step-by-Step Aproach to Creating Profitable Market Segments*, Philadelphia: Transatlantic Publications.
McFadden, D. (1989), "A Method of Simulated Moments for the Estimation of Discrete Response Models Without Numerical Integration," *Econometrica*, 57, 995-1026.
McHugh, R. B. (1956), "Efficient Estimation and Local Identification in Latent Class Analysis," *Psychometrika*, 21, 331-347.
McHugh, R. B. (1958), "Note on 'Efficient Estimation and Local Identification in Latent Class Analysis'," *Psychometrika*, 23, 273-274.
McLachlan, G. J. (1982), "The Classification and Mixture Maximum Likelihood Approaches to Cluster Analysis," *In Handbook of Statistics*, 2, P.R. Krishnaiah and L.N. Kanal (eds.), "Amsterdam: North-Holland, 199-208.
McLachlan, G. J. (1987), "On Bootstrapping the Likelihood Ratio Test Statistic for the Number of Components in a Normal Mixture," *The Journal of the Royal Statistical Society*, C36, 318-324.
McLachlan, G. J. (1992), *Discriminant Analysis and Statistical Pattern Recognition*, New York: John Wiley & Sons.

References

McLachlan, G. J. and K. E. Basford (1988), *Mixture Models: Inference and Applications to Clustering*, New York: Marcel Dekker.

McLachlan, G. J. and T. Krishnan (1997) *The EM algorithm and Extensions*. New York: John Wiley & Sons.

Miller, K. E. and J. L. Ginter (1979), "An Investigation of Situational Variation in Brand Choice Behavior and Attitude," *Journal of Marketing Research*, 16, 111-123.

Milligan G. W. (1980), "An Examination of the Effect of Six Types of Error Perturbation on Fifteen Clustering Algorithms," *Psychometrika*, 45, 325-342.

Milligan, G. W. (1989), "A Validation Study of a Variable Weighting Algorithm for Cluster Analysis," *Journal of Classification*, 6, 53-71.

Milligan, G. W. (1994), "Cluster Validation: Results and Implications for Applied Analyses," *Clustering and Classification*, P.Arabie, L. Hubert and G. DeSoete (eds.), River Edge, NJ: World Scientific.

Milligan, G. W. (1995), "Issues in Applied Classification," *CSNA Newsletter*, 36-38.

Milligan, G. W. and M. C. Cooper (1985), "An Examination of Procedures for Determining the Number of Clusters in a Dataset," *Psychometrika*, 50, 159-179.

Milligan, G. W. and M. C. Cooper (1988), "A Study of Variable Standardization," *Journal of Classification*, 5, 181-204.

Mitchell, A. (1983), *The Nine American Life-Styles*, New York: Warner.

Mitchell, S. (1995), "Birds of a Feather Flock Together," *American Demographics*, 40-48.

Monk, D. (1978), "Interviewers' Guide on Social Grading," *Social Grading on National Research Surveys*, JICNARS.

Montgomery, D. B. and A. J. Silk (1972), "Estimating Dynamic Effects of Marketing Communications Expenditures," *Management Science*, 18, 485-501.

Montgomery, D. B. and D. R. Wittink (1980), "The Predictive Validity of Tradeoff Analysis for Alternative Aggregation Schemes,", in *Market Measurement and Analysis*, Proceedings of ORSA/TIMS Special Interest Conference, David B. Montgomery and Dick R. Wittink (eds.), Marketing Science Insitute, 298-308.

Mooijaart, A. and P. G. M. van der Heijden (1992), "The EM Algorithm for Latent Class Analysis with Equality Constraints," *Psychometrika*, 57, 261-270.

Moore, W. L. (1980), "Levels of Aggregation in Conjoint Analysis: An Empirical Comparison," *Journal of Marketing Research*, 17, 516-523.

Morgan, B. T, and A. P. G. Ray (1995), "Non-uniqueness and Inversion in Cluster Analysis," *Applied Statistics*, 44, 117-134.

Moriarty, M. and M. Venkatesan (1978), "Concept Evaluation and Market Segmentation," *Journal of Marketing*, 42, 82-86.

Morrison, D. G. (1966), "Testing Brand Switching Models," *Journal of Marketing Research*, 3, 401-409.

Morrison, D. G. (1973), "Evaluation Market Segmentation Studies: The Properties of R2," *Management Science*, 19, 1213-1221.

Morwitz, V. G. and D. C. Schmittlein (1992), "Using Segmentation to Improve Sales

Forecasts Based on Purchase Intent: Which Intenders Actually Buy?," *Journal of Marketing Research*, 29, 391-405.

Moschis, G. (1976), "Shopping Orientations and Consumer Uses of Information," *Journal of Retailing*, 52, 61-70.

Myers, J. H. (1976), "Benefit Structure Analysis: A New Tool for Product Planning," *Journal of Marketing*, 40, 23-32.

Myers, J. G. and F. M Nicosia (1968), "On the Study of Consumer Typologies," *Journal of Marketing Assosiation*, 107.

Naert, P. and P. Leefang (1978), *Building Implementable Marketing Models*, Leiden: Martinus Nijhoff.

Nakanishi, M. and L. G. Cooper (1982), "Simplified Estimation Procedures for MCI Models," *Marketing Science*, 3, 314-322.

Nelder, J. A. and R. W. M. Wedderburn (1972), "Generalized Linear Models," *Journal of the Royal Statistical Society*, A135, 370-384.

Newcomb, S. (1886), "A Generalized Theory of the Combination of Observations So As To Obtain the Best Result," *American Journal of Mathematics*, 8, 343-366.

Novak, T. P. and B. MacEvoy (1990), "On Comparing Alternative Segmentation Schemes: The List of Values and Values and Life-Styles," *Journal of Consumer Research*, 17, 105-109.

Ogawa, K. (1987), "An Approach to Simultaneous Estimation and Segmentation in Conjoint Analysis," *Marketing Science*, 6 (1), 66-81.

Oppewal, H., J. J. Louviére and H. P. Timmermans (1994), "Modeling Hierarchical Conjoint Processes with Integrated Choice Experiments," *Journal of Marketing Research*, 31, 92-105.

Park, C. S. and V. Srinivasan (1994), "A Survey-Based Method for Measuring and Understanding Brand Equity and Its Extendibility," *Journal of Marketing Research*, 31, 271-288.

Pearson, K. (1894), "Contributions to the Mathematical Theory of Evolution," *Philosophical Transactions*, A185, 71-110.

Penner, L. A. and T. Anh (1977), "A Comparison of American and Vietnamese Value Systems," *Journal of Social Psychology*, 101, 187-204.

Peters, B. C. and H. F. Walker (1978), "An Iterative Procedure for Obtaining Maximum Likelihood Estimates of the Parameters of a Mixture of Normal Distributions," *Journal of Applied Mathematics*, 35, 362-378.

Peterson, R. A. (1972), "Psychographics and Media Exposure," *Journal of Advertising Research*, 12, 17-20.

Phillips, D. B. and A. F. M. Smith (1996), "Bayesian Model Comparison Via Jump Diffusion," in *Markov Chain Monte Carlo in Practice*, W.R. Gilks, S. Richardson and D.J. Spiegelhalter (eds.), London: Chapman & Hall, 215-240.

Piirto, R. (1991), *Beyond Mind Games: The Marketing Power of Psychographics*, Ithaca, NY: American Demographics Books.

Pitts, R. E. and A. G. Woodside (1983), "Personal Value Influences on Consumer Product Class and Brand Preferences," *Journal of Social Psychology*, 10, 233-246.

References

Plat, F. W. and P. S. H. Leeflang (1988), "Decomposing Sales Elasticities on Segmented Markets," *International Journal of Research in Marketing*, 5, 303-315.

Plummer, J. T. (1974), "The Concept of Life Style Segmentation," *Journal of Marketing*, 38, 33-37.

Potter, W. J., E. Forrester, B. Sapolsky and W. Ware (1988), "Segmenting VCR Owners," *Journal of Advertising Research*, 28, 29-37.

Poulsen, C. S. (1990), "Mixed Markov and Latent Markov Modelling Applied to Brand Choice Behavior," *International Journal of Research in Marketing*, 7, 5-19.

Poulsen, C. S. (1993), "Artificial Intelligence Applied to Computer Aided Interviewing," in *Information Based Decision Making in Marketing*, Paris: ESOMAR, 95-108.

Punj, G. and D. W. Stewart (1983), "Cluster Analysis in Marketing Research: Review and Suggestions for Application," *Journal of Marketing Research*, 20, 134-148.

Quandt, R. E. (1972), "A New Approach to Estimating Switching Regressions," *Journal of the American Statistical Association*, 67, 306-310.

Quandt, R. E. and J. B. Ramsey (1972), "A New Approach to Estimating Switching Regressions," *Journal of the Americal Statistical Association*, 73, 730-738.

Quandt, R. E. and J. B. Ramsey (1978), "Estimating Mixtures of Normal Distributions and Switching Regressions," *Journal of the American Statistical Association*, 73, 730-738.

Ramaswamy, V. (1997), "Evolutionary Preference Segmentation with Panel Survey Data: An Application to New Products," *International Journal of Research in Marketing*, 14, 57-80.

Ramaswamy, V., E. W. Anderson and W. S. DeSarbo (1993), "A Disaggregate Negative Binomial Regression Procedure for Count Data Analysis," *Management Science*, 40, 405-417.

Ramaswamy, V. R. Chatterjee and S. H. Cohen (1996), "Joint Segmentation on Distinct Interdependent Bases with Categorical Data," *Journal of Marketing Research*, 33, 337-355.

Ramaswamy, V., R. Chatterjee and S. H. Cohen (1999), "Reply to: A note on Ramaswamy, Chatterjee and Cohen's Latent Joint Segmentation Models," *Journal of Marketing Research*, 36, 115-119.

Ramaswamy, V., W. S. DeSarbo, D. J. Reibstein and W. T. Robinson (1993), "A Latent-Pooling Methodology for Regression Analysis with Limited Time-Series Cross-Sections: A PIMS Data Application," *Marketing Science*, 12, 103-124.

Rao, V. R. and F. W. Winter (1978), "An Application of the Multivariate Probit Model to Market Segmentation and Product Design," *Journal of Marketing Research*, 15, 361-368.

Redner, R. A. and H. F. Walker (1984), "Mixture Densities, Maximum Likelihood and the EM Algorithm," *SIAM Review*, 26, 195-239.

Reynolds, W. H. (1965), "More Sense About Market Segmentation," *Harvard Business Review*, (September-October) 107-114.

Reynolds, F. D. and W. R. Darden (1972), "Intermarket Patronage: A Psychographic Study of Consumer Outshoppers," *Journal of Marketing*, 36, 50-54.

Reynolds T. J. and J. Gutman (1988), "Laddering Theory, Method Analysis and Interpretation," *Journal of Advertising Research*, 11-31.

Robinson, J. (1938), *The Economics of Imperfect Competition*, London: MacMillan.

Rogers, M. (1962), *Diffusion of Innovations*, New York: The Free Press.

Rokeach, M. (1973), *The Nature of Human Values*, New York: The Free Press.

Rokeach, M. and S. J. Ball-Rokeach (1989), "Stability and Change in American Value Priorities 1968-1981," *American Psychologist*, 44, 775-784.

Rosbergen, E. A., F. G. M. Pieters and M. Wedel (1997), "Visual Attention to Advertising: A Segment-Level Analysis," *Journal of Consumer Research*, 24, 305-314.

Rossi, P. and G. M. Allenby (1993), "A Bayesian Approach to Estimating Household Parameters," *Journal of Marketing Research*, 171-182.

Rubin, D. B. (1987), *Multiple Imputations for nonresponse in Surveys*, New York: John Wiley & Sons.

Russell, G. J., R. E. Bucklin and V. Srinivasan (1993), "Identifying Multiple Preference Segments from Own- and Cross-Price Elasticities," *Marketing Letters*, 4, 5-18.

Russell, G. J. and W. A. Kamakura (1994), "Understanding Brand Competition Using Micro and Macro Scanner Data," *Journal of Marketing Research*, 31, 289-303.

Samejima, F. (1969), "Estimation of Latent Ability Using A Response Pattern of Graded Scores," *Psychometrika Monograph Supplement*, 34 (Suppl.17), 1-100.

Sattath, S., and A. Tversky, (1977), "Additive Similarity Trees," *Psychometrika*, 42, 319-345.

Scales, L. E. (1985), *Introduction to Numerical Optimization*, London: Macmillan Publishers.

Schaninger, C. M., V. P. Lessig and D. B. Panton (1980), "The Complementary Use of Multivareate Procedures to Investigate Nonlinear and Interactive Relationships Between Personality and Product Usage," *Journal of Marketing Research*, 17, 119-124.

Schiffman, S. S., M. L. Reynolds, and F. W. Young, (1981). *Introduction to Multidimensional Scaling*. London: Academic Press.

Schwartz, S. H. (1992), "Universals in the Content and Structure of Values: Theoretical Advances and Empirical Tests in 20 Countries," in *Advances in Experimental Social Psychology*, Mark P. Zanna (ed.), New York: Academic Press.

Schwartz, S. and W. Bilsky (1987), "Towards a Universal Psychological Structure of Human Values," *Journal of Personality and Social Psychology*, 53, 550-562.

Schwartz, S. and W. Bilsky (1990), "Towards a Theory of the Universal Content and Structure of Values: Extensions and Cross-cultural Replications," *Journal of Personality and Social Psychology*, 58, 878-891.

References

Scott, A. J. and M. J. Simons (1971), "Clustering method based on likelihood ratio criteria," *Biometrics*, 27, 387-398.

Sethi, S. P. (1971), "Comparative Cluster Analysis for World Markets," *Journal of Marketing Research*, 8, 348-354.

Sewall, M. A. (1978), "Market Segmentation Based on Consumer Ratings of Proposed Product Designs," *Journal of Marketing Research*, 15, 557-564.

Sexton, D. E., Jr. (1974), "A Cluster Analytic Approach to Market Response Function," *Journal of Marketing Research*, 11, 109-114.

Shapiro, B. P and T. V. Bonoma (1984), "How to Segment Industrial Markets," *Harvard Business Review*, 62, 104-110.

Shaw, Arch W. (1912), "Some Problems in Market Distribution," *Quarterly Journal of Economics*, 703-765.

Shepard, R. N. and P. Arabie (1979), "Additive Clustering: Representation of Similarities as Combinations of Discrete Overlapping Properties," *Psychological Review*, 86, 87-123.

Shocker, A. D. and V. Srinivasan (1979), "Multiattribute Approach for Product Concept Evaluation and Generation: A Critical Review," *Journal of Marketing Research*, 16, 159-180.

Singh, J. (1990), "A Typology of Consumer Dissatisfaction Response Styles," *Journal of Retailing*, 66, 57-99.

Singh, J., R.D. Howell and G.K. Rhoads (1990), "Adaptive Designs for Likert-Type Data: An Approach for Implementing Marketing Surveys," *Journal of Marketing Research*, 27, 304-321.

Skinner, C. J. (1989), "Domain Means Regression and Multivariate Analysis," in Analysis of Complex Surveys, C.J. Skinner, D. Holt and T.M.F. Smith (eds), New York: John Wiley & Sons, 59-87.

Skinner, C. J., D. Holt and T. M. F. Smith (1989), *The Analysis of Complex Surveys*, New York: John Wiley & Sons.

Slovic, P. and S. Lichtenstein (1971), "Comparison of Bayesian and Regression Approaches to the Study of Information Processing in Judgment," *Organizational Behavior and Human Performance*, 6, 649-744.

Smith, W. (1956), "Product Differentiation and Market Segmentation as Alternative Marketing Strategies," *Journal of Marketing*, 21, 3-8.

Späth, H. (1979), "Clusterwise Linear Regression," *Computing*, 22, 367-373.

Späth, H. (1981), "Clusterwise Linear Regression," *Computing*, 26, 275.

Späth, H. (1982), "A Fast Algorithm for Clusterwise Linear Regression," *Computing*, 29, 175-181.

Srivastava, R., M. I. Alpert and A. D. Shocker (1984), "A Consumer Oriented Approach for Determining Market Structure," *Journal of Marketing*, 48, 32-45.

Starr, M. K. and J. R. Rubinson (1978), "A Loyalty Group Segmentation Model for Brand Purchasing Simulation," *Journal of Marketing Research*, 15, 378-383.

Steckel, J. H. and W. R. VanHonacker (1988), "A Heterogeneous Conditional Logit Model of Choice," *Journal of Business and Economic Statistics*, 6, 391-398.

Steenkamp, J. E. B. M. and M. Wedel (1991), " Segmenting Retail Markets on Store Image Using a Consumer-Based Methodology," *Journal of Retailing,* 67, 300-320.

Steenkamp, J. E. B. M. and M. Wedel (1993), "Fuzzy Clusterwise Regression in Benefit Segmentation: Application and Investigation into its Validity," *Journal of Business Research,* 26, 237-249.

Steenkamp, J. E. B. M. and D. R. Wittink (1994), "The Metric Quality of Full-Profile Judgements and the Number of Attribute Levels Effect in Conjoint Analysis," *International Journal for Researh in Marketing,* 11, 275-286.

Stewart, D. W. (1981), "The Application and Misapplication of Factor Analysis in Marketing Research," *Journal of Marketing Research,* 18, 51-62.

Stigler, S. M. (1986), *The History of Statistics,* Cambridge, MA: Belknap press.

Stout, R. G. et al. (1977), "Usage Incidence as a Basis for Segmentation," in: *Moving Ahead with Attitude Research,* Wind Y. and Greenberg, M. (eds.), Chicago: American marketing Association, 45-49.

Symons, M. J. (1981), "Clustering Criteria and Multivariate Normal Mixtures," *Biometrics,* 37, 35-43.

Tanner M. A (1993), *"Tools for Statistical Inference,"* New York: Springer-Verlag.

Teel, J. E., W. O. Bearden and R. M. Durand (1979), "Psychographics of radio and television audiences," *Journal of Advertising Research,* 19, 53-56.

Teel, J. E., R. H. Williams and W. O. Bearden (1980), "Correlates of Consumer Susceptibility to Coupons in New Grocery Product Introductions," *Journal of Advertising,* 3, 31-35.

Tigert, D. J. (1969), "A Taxonomy of Magazine Readership Applied to Problems in Marketing Strategy and Media Selection, *Journal of Business,* 42, 357-363.

Titterington, D. M. (1990), "Some Recent Research in the Analysis of Mixture Distributions," *Statistics,* 4, 619-641.

Titterington, D. M., A. F. M. Smith and U. E. Makov (1985), *"Statistical Analysis of Finite Mixture Distributions,"* New York: John Wiley & Sons.

Toler, C. (1975), "The Personal Values of Alcoholics and Addicts," *Journal of Clinical Psychology,* 31, 554-557.

Tollefson, J. O. and P. Lessig (1978), "Aggregation Criteria in Normative Market Segmentation Theory," *Journal of Marketing Research,* 15, 364-355.

Tversky, A. (1977), "Features of Similarity," *Psychological Review,* 84, 327-352.

Twedt, D. W. (1967), "How Does Awareness-Attitude Affect Marketing Strategy?," *Journal of Marketing,* 31, 64-66.

Umesh, U. N. (1987), "Transferability of Preference Models Across Segments and Geographic Areas," *Journal of Marketing,* 51, 59-70.

Urban, G. L. and J. R. Hauser (1980), *Design and Marketing of New Products,* Englewood Cliffs, NJ: Prentice Hall.

Valette-F. (1986), "Les Démarches de Styles de Vie: Concepts, Champs d'Investigations et Problèmes Actuels," *Recherche et Applications en Marketing,* 1, 94-109

van de Pol F. and L. N. de Leeuw (1986), "A Latent-Markov Model to Correct for

References

Measurement Error," *Sociological Methods and Research*, 15, 118-141.
van de Pol, F., R. Langeheine and W. de Jong (1991), *PANMARK User's Manual*, Report Number 8493-89-M1-2, Netherlands Central Bureau of Statistics, Department of Statistical Methods, P.O. Box 959, 2270 AZ Voorburg, Netherlands.
van Duÿn, M. A. J. and U. Böckenholt (1995), "Mixture Models for the Analysis of Repeated Count Data," *Applied Statistics*, 44, 473-485.
van Raaij and T. M. M. Verhallen (1994), "Domain -Specific market segmentation," *European Journal of Marketing*, 28, 49-66.
Vermunt, J. K. (1993), "LEM: Log-Linear and Event History Analysis with Missing Data Using the EM Algorithm," Work Paper 93.09015/7, Methodology Department, Tilburg University, Tilburg, Netherlands.
Vilcassim, N. and D. C. Jain (1991), "Modeling Purchase Timing and Brand Switching Behavior Incorporating Explanatory Variables and Unobserved Heterogeneity," *Journal of Marketing Research*, 28, 29-41.
Vinson, D. E., J. E.Scott and L. M. Lamont (1977), "The Role of Personal Values in Marketing and Consumer Behavior," *Journal of Marketing*, 41, 44-50.
Vriens M, M. Wedel and T. Wilms (1996), "Metric Conjoint Segmentation Methods: A Monte Carlo Comparison," *Journal of Marketing Research*, 32, 73-85.
Walker, S., and P. Damien (1999), "A note on Ramaswamy, Chatterjee and Cohen's Latent Joint Segementation Models," *Journal of Marketing Research*, 36, 113-114.
Wang, P. I. M., Cockburn and M. L. Puterman (1998), "Analysis of Latent Data – A Mixed Poisson Regression Model Approach," *Journal of Business and Economic Statistics*, 16, 27-41.
Ward, J. (1963), "Hierarchical Grouping to Optimize an Objective Function," *Journal of the American Statistical Association*, 58, 236-244.
Wedel M. (1995), "Market Segmentation Research: A Review of Bases and Methods with Special Emphasis on Latent Class Models," in: *Information Based decision Making in Marketing*, Paris: ESOMAR, 191-221.
Wedel, M. (1996), "The identification of Sensory Dimensions of Food Productsfrom Scanner Data using the STUNMIX Methodology," in: *Agricultural Marketing and Consumer Behavior in a Changing World*, B Wierenga, A. Van Tilburg, K. Grunert J.E.B.M. Steenkamp and M Wedel (eds.), Dordrecht, Kluwer.
Wedel, M. (1999) "Computing the Standard Errors of Mixture Model Parameter Estimates with EM," *SOM Working Paper*, Faculty of Economics University of Groningen.
Wedel, M. (1999). Concomitant variables in mixture models. *SOM Working paper*, Faculty of Economics, University of Groningen, Netherlands.
Wedel, M. and H. A. Bijmolt (1998), "Mixed Tree and Spatial Representations of Dissimilarity Judgements," *Center working Paper*, Tilburg University, Netherlands.
Wedel, M. and W. S. DeSarbo (1993), "A Latent Class Binomial Logit Methodology for the Analysis of Paired Comparison Choice Data: An Application Reinvestigating the Determinants of Perceived Risk," *Decision Sciences*, 24,

1157-1170.
Wedel, M. and W. S. DeSarbo (1994), "A Review of Latent Class Regression Models and their Applications," in *Advanced Methods for Marketing Research*, Richard P. Bagozzi (ed.), 353-388.
Wedel, M., and W. S. DeSarbo (1995), "A Mixture Likelihood Approach for Generalized Linear Models," *Journal of Classification*, 12, 1-35.
Wedel, M. And W. S. DeSarbo (1996), "An Exponential-Family Multidimensional Scaling Mixture Methodology," *Journal of Business and Economic Statistics*, 14, 447-459.
Wedel, M. and W. S. DeSarbo (1998), "Mixtures of Constrained Ultrametric Trees," *Psychometrika*, 63, 419-444.
Wedel, M., W. S. DeSarbo, J. R.Bult and V. Ramaswamy (1993), "A Latent Class Poisson Regression Model for Heterogeneous Count Data," *Journal of Applied Econometrics*, 8, 397-411.
Wedel, M., W. A. Kamakura, N. Arora, A. Bemmaor, J. Chiang, T. Elrod, R. Johnson, P. Lenk, S. Neslin, C. Poulsen, (1999), "Discrete and Continuous Representations of Unobserved Heterogeneity in Choice Modeling," *Marketing Letters*, 10, 217-230.
Wedel M., W. A. Kamakura, W.S. DeSarbo and F. ter Hofstede (1995), "Implications for Asymmetry, Nonproportionality and Heterogeneity in Brand Switching from Piece-Wise Exponential Hazard Models," *Journal of Marketing Research*, 32, 457-462.
Wedel, M. and C. Kistemaker (1989), "Consumer Benefit Segmentation using Clusterwise Linear Regression," *International Journal of Research in Marketing*, 6, 45-49.
Wedel, M., and J. B. E. M. Steenkamp (1989), "Fuzzy Clusterwise Regression Approach to Benefit Segmentation," *International Journal of Research in Marketing*, 6, 241-258.
Wedel, M. and J. B. Steenkamp (1991), "A Clusterwise Regression Method for Simultaneous Fuzzy Market Structuring and Benefit Segmentation," *Journal of Marketing Research*, 28, 385-396.
Wedel, M., F. ter Hofstede and J. E. B. M. Steenkamp (1998), "Mixture model analysis of complex samples," *Journal of Classification*, 15, 225-244.
Wedel, M., M. Vriens, T. H. A. Bijmolt, W. Krijnen and P. S. H. Leeflang (1998), "Including Abstract Attributes and Brand Familiarity in Conjoint Choice Experiments," *International Journal of Research in Marketing*, 15, 71-78.
Weiss, M. J. (1988), *Clustering of America*, New York: Tilden Press.
Wells, W. D. (1974), *Life-style and Psychographics*, Chicago: American Marketing Association.
Wells, W. D. (1975), "Psychographics: A Critical Review," *Journal of Marketing Research*, 12, 196-213.
Wells, W. D. and A. D. Beard (1974), "Personality and Consumer Behavior," in *Consumer Behavior: Theoretical Sources*, S. Ward and T. S. Robertson

References

(eds), Englewood Cliffs, NJ: Prentice Hall.

Wells, W. D. and D. Tigert (1971), "Activities, Interests and Opinions," *Journal of Advertising Research*, 11, 27-35.

White, H. (1982), "Maximum Likelihood Estimation of Misspecified Models," *Econometrica*, 50, 25.

Wildt, A. R. (1976), "On Evaluating Market Segmentation Studies and the Properties of R2," *Management Science*, 22, 904-908.

Wildt, A. R. and J. M. McCann (1980), "A Regression Model for Market Segmentation Studies," *Journal of Marketing Research*, 17, 335-340.

Wilkie, W. L. (1970), "An Empirical Analysis of Alternative Basis for Market Segmentation," unpublished doctoral dissertation, Stanford University.

Wilkie, W. L. and J. B. Cohen (1977),"An Overview of Market Segmentation: Behavioral Concepts and Research Approaches," *Marketing Science Institute Working Paper*.

Wind, Y. (1978), "Issues and Advances in Segmentation Research," *Journal of Marketing Research*, 15, 317-337.

Winter, F. W. (1979), "A Cost-Benefit Approach to Market Segmentation," *Journal of Marketing*, 43, 103-111.

Wishart, D. (1987), *Clustan User Manual*, 4th ed., Computing Laboratory, University of St. Andrews.

Wittink, D. R. and P. Cattin (1989), "Commercial Use of Conjoint Analysis: An Update," *Journal of Marketing*, 53, 91-96.

Wittink, D. R., M. Vriens and V. Burhenne (1994), "Commercial Use of Conjoint Analysis in Europe: Results and Critical Reflections," *International Journal of Research in Marketing*, 11, 41-52.

Wolfe, J. H. (1970), "Pattern Clustering by Multivariate Mixture Analysis," *Multivariate Behavioral Research*, 5, 329-350.

Wu, C. F. J. (1983), On the Convergence Properties of the EM Algorithm, *Annals of Statistics*, 11, 95-103.

Yankelovich, D. (1964), "New Criteria for Market Segmentation," *Harvard Business Review*, 42, 83-90.

Zadeh, L. A. (1965), "Fuzzy sets," *Information and Control*, 8, 338-353.

Zenor, M. J. and R. K. Srivastava (1993), "Inferring Market Structure with Aggregate Data: A Latent-Segment Logit Approach," *Journal of Marketing Research*, 30, 369-379.

Ziff, R (1971), "Psychographics for Market Segmentation," *Journal of Advertising Research*, 11, 3-9.

Index

A
ACORN, 247; 248; 250; 252; 256
agglomerative method, 44
Agrawal, J., 26; 102; 118; 119; 146;
 152; 157; 158; 301; 316; 318;
 358
AIC, 92; 93; 102; 134; 136 152; 169;
 173; 181; 186; 287;
AID, 17; 23; 29; 72;
AIO, 15; 260; 261; 268; 272
Aitkin, M., 90; 91; 345
Ajzen, I., 14; 15; 353
Akaike, H., 56; 92; 345
Akaike information criterion, 56; 92;
Albaum, G., 297; 345
Allenby, G., 284; 293; 325; 326; 327;
 328; 329; 330; 345; 365
Alpert, L., 11; 260; 345; 366
Alwin, D.F., 262; 345
Amemiya, T., 283; 345
Anderson, D., 91; 345
Anderson, E.B., 134; 345
Anderson, E.W., 116; 364
Anderson, W., 72; 345
Anh, T., 262; 363
Arabie, P., 19; 20; 58; 60; 65; 255;
 345; 349; 357; 366
Arora, 327; 328; 330; 345; 369
Assael, H., 23; 27; 72; 345; 346
average linkage, 48; 49; 50; 59; 72

B
Baker, M., 4; 7; 346
Balakrishnan, P., 24; 54; 346
Balasubramanian, S., 195; 197; 198;
 346
Ball-Rokeach, S., 262; 264; 270;
 276; 365

Banfield, C., 53; 55; 88; 346
Basford, K., 21; 78; 80; 81; 88; 89;
 90; 346; 362
Bass, F., 22; 43; 72; 268; 336; 346
Bassil, L., 53; 55; 88; 346
Basu, A., 331; 346
Basu, A.K., 331; 346
Batra, R., 331; 346
Bayes' theorem, 88; 238
Bayesian information criterion, 92
Beard, A., 268; 369
Bearden, W., 260; 268; 346; 367
Beatty, S., 12; 13; 265; 266; 270;
 346; 358
Beckenridge, J., 59; 346
Becker, B., 264; 269; 346
Beckwith, N., 22; 346
Beldo, L., 15; 346
Belk, R., 10; 11; 346
benefit segmentation, 15; 20; 24; 31;
 73; 167; 342
beta-binomial, 79
Bezdek, J., 20; 65; 66; 67; 346; 347
BIC, 92; 149; 181; 186; 190; 191;
 192; 226; 227; 228
Bijmolt, T., 238; 352; 368; 369
Bilsky, W., 12; 262; 264; 270; 365
Binder, D., 53; 347
binomial, 77; 79; 85; 91; 93; 98;
 106; 107; 110; 112; 113; 114;
 115; 116; 117; 118; 126; 128;
 129; 131; 132; 136; 137; 138;
 140; 141; 150; 158; 170; 176;
 181; 186; 192; 293; 340
Blackwell, R., 260; 269; 347
Blattberg, R., 27; 282; 347
Blozan, W., 14; 347
Böckenholt, U., XXII; 97; 98; 116;
 117; 118; 126; 133; 140; 141;

Index

159; 169; 175; 176; 291; 293; 327; 330; 347; 368
Böckenholt, I., 133; 140; 141; 347
Böhning, D., 80; 91; 178; 337; 347
Bollen, K., 217; 347
Bond, M., 262; 347
Bonoma, T., 18; 366
Bottenberg, R., 25; 348
Boyd, H., 10; 348
Boyles, R., 81; 348
Bozdogan, H., 92; 348
brand loyalty, 10; 11; 22; 283
brand-loyal, 165; 173
Breiman, L., 24; 348
Brody, R., 12; 348
Brookhouse, K., 25; 348
Bruno, A., 13; 348
BTL, 97; 146; 147; 148
Bucklin, R., 28; 115; 116; 118; 282; 290; 348; 365
Buesing, T., 27; 347
Bult, J., 331; 348; 369
Bunch, D., 185; 348
Burhenne, V., 295, 303, 305; 370

C

CAIC, 92; 93; 117; 127; 134; 136; 152; 164; 169;; 186; 190; 223; 226; 227; 228; 236; 269; 304; 319
Calantone, R., 15; 72; 159; 167; 348
Calinski, T., 54; 55; 61; 348
Carmone, F., XXI,; 18; 21; 76; 96; 97; 306; 355
Carroll, J., 20; 57; 58; 65; 231; 238; 305; 339; 345; 348; 349; 350; 351; 355
CART, 17; 23
Casella, G., 201; 349
Cattin, P., 15; 295; 303; 315; 349; 370
Cavali Sforza, L., 50; 353
Celeux, G., 93; 349
census block groups, 241; 244; 247; 254
census blocks, 254
census tracts, 254
centroid linkage, 48; 49; 59
CHAID, 23; 73
Chamberlain, G., 284; 349
Chapman, R., 301; 317; 349
Chatterjee, R., 145; 189; 191; 194; 303; 352; 364
Chaturvedi, A., 238; 349
Cheng, R., 58; 349
Chintagunta, P., 23; 26; 28; 115; 118; 140; 141; 146; 147; 149; 150; 156; 158; 284; 285; 288; 349; 351; 356
Cho, J., 126; 351
Choi, S., 303; 351
choice-based conjoint, 119; 315; 321;
Christal, R., 25; 348; 349
Clark, L., 20; 58; 351
Claxton, J., 72; 349
Claycamp, H, 12; 14; 26; 27; 28; 336; 338; 350
Clogg, C., 178; 350
clout index, 288
cluster centroid, 43
cluster sample, 62
clusterwise logistic regression, 55
clusterwise logit, 26
clusterwise regression, 25; 29; 32; 39; 40; 51; 55; 56; 65; 66; 67; 68; 101; 107; 307; 308; 309; 311; 312; 316
Cochran, W., 60; 64; 350
Cockburn, 158; 368
Cohen, S., 6; 7; 13; 14; 16; 22; 119; 189; 191; 194; 316; 335; 352; 364; 370
Colombo, R., 291; 350
complete linkage, 19; 48; 49; 57; 59; 72
componential segmentation, 24; 306; 339
concomitant variables, 26; 31; 115; 119; 140; 145; 147; 149; 150;

151; 152; 153; 155; 156; 201; 318; 319; 321
confirmatory factor analysis, 43; 217; 221
conjoint analysis, XIX; XX; 15; 23; 24; 25; 31; 34; 113; 117; 119; 239; 295; 297; 298; 300; 301; 303; 305; 307; 308; 315; 316; 318; 320; 330; 339; 342
Conner, P., 269; 346
consistent Akaike information criterion, 92
constant-sum, 119
convergence method, 53
Cooper, L., 19; 54; 57; 279; 282; 287; 288; 350; 362; 363
Corter, J., 51; 233; 234; 350
Cox, E., 72; 345
Cox, M., 231; 350
Cox, T., 231; 350
Cron, W., 6; 102; 103; 113; 118; 351
cross-tabulation, 17; 18
Cunningham, S., 12; 348
Currim, I., 15; 22; 25; 73; 350
Curry, D., 244; 350

D
Damien, P., 175; 194; 368
Darden, W., 18; 260; 350; 365
datafusion, 256
Davey, K., 315; 316; 353
Davison, M., 125; 350
Day, G., 80; 350
Dayton, C., 115; 146; 151; 198; 288; 350; 360
de Jong, W., 178; 368
de Leeuw, L., 169; 367
De Soete, G., 20; 57; 58; 91; 113; 115; 118; 134; 138; 141; 305; 350
Deb, P., 93; 116; 118; 350
Della Bitta, A., 4; 7; 10; 360
Dempster, A., 80; 81; 120; 351
Dennis, J., 85; 351

DeSarbo, W., 6; 20; 24; 25; 26; 51; 57; 58; 65; 78; 80; 91; 102; 103; 106; 110; 113; 114; 115; 116; 118; 119; 126; 127; 128; 129; 130; 131; 134; 136; 137; 138; 139; 140; 141; 145; 152; 156; 217; 223; 225; 226; 233; 235; 236; 237; 303; 305; 306; 309; 316; 325; 327; 345; 350; 351; 352; 355; 356; 357; 360; 364; 368; 369
Dhalla, N., 14; 27; 352
Dichter, E., 12; 259; 352
Dickson, P., 3; 10; 11; 13; 15; 26; 336; 352
Dillon, W., 42; 75; 76; 78; 88; 90; 91; 96; 97; 115; 118; 146; 147; 148; 151; 152; 158; 291; 337; 352
direct marketing, 4; 6; 34; 116; 258; 329; 331; 332; 333
Dirichlet, 79; 80; 82; 83; 97; 98; 115; 116; 117; 118; 119; 127; 138; 140; 141; 292; 293; 304; 327
Dirichlet multinomial, 79; 329
Dirichlet regression, 115; 292; 304
discriminant analysis, 23; 29; 43; 56; 212
divisive methods, 44; 50
Dixon, J., 59; 352
Doherty, E., 25; 348
Dolan, C, 223; 352
Doyle, P., 24; 352
Dubes, R., 54; 359
Duffy, D., 56; 352
Dunbar, I., 5; 8; 9; 361
Dunn, J., 20; 65; 66; 352
Durand, R., 260; 346; 367
dynamic segmentation, XX; 159; 176

E
Edwards, A., 50; 353
Elrod, T., 14; 21; 23; 27; 28; 96; 97;

315; 316; 338; 353; 359; 369
EM, 75; 80; 81; 84; 85; 86 87; 88;
    90; 94; 108; 109; 110; 111; 113;
    120; 133; 136; 142; 152; 161;
    168; 169; 179; 180; 185; 186;
    190; 226; 229; 236; 258; 287;
    308; 309
endogeneous variable, 218
entropy, 90; 92; 93; 127; 165; 186;
    200; 202; 211; 226; 286
entropy of classification, 200; 202;
    211; 212
entropy statistic, 92; 165; 226
Evans, F., 12; 353
Everitt, B., 39; 42; 44; 49; 54; 56;
    57; 78; 81; 353
exchange algorithm, 53; 55; 56; 316
expectation-maximization, 80
experimental choice approach, 300
exponential family, 79; 80; 82; 85;
    90; 91; 93; 96; 98; 99; 106;
    107; 109; 113; 120; 125; 126;
    127; 131; 132; 142; 150; 178;
    181; 189
exponential gamma, 79
external analyses, 125; 132
E-step, 84; 85; 88; 90; 108; 112; 121;
    133; 142; 144; 169

F
Fenwick, I., 24; 352
finite mixture, XIX; XX; XXI;
    XXII; 35; 37; 75; 78; 81; 102;
    103; 107; 190; 226; 284; 285;
    286; 290; 291; 327; 345
first-order Markov, 96; 161; 175
Fischer, G., 282; 353
Fishbein, M., 14; 15; 353
Fisher, R., 85; 89; 106; 109; 200; 353
Foekens, E., 277; 280; 353
Forgy's method, 53
Formann, A., 146; 353
Fowlkes, E., 57; 58; 80; 353
fractional factorial design, 303

Frank, R., 4; 5; 6; 7; 10; 11;12; 13;
    14; 15; 19; 23; 26; 27; 28; 178;
    278; 353; 354; 361
Fry, J., 72; 349
Fukunaga, K., 54; 359
Fulcher, D., 72; 345
full factorial design, 298
full-profile approach, 300
Furnham, A., 8; 11; 12; 268; 355
fuzzy clustering, 19; 20; 32; 37; 39;
    41; 65; 66
fuzzy c-means, 20; 66
fuzzy c-varieties, 20; 65

G
Gatty, R., 260; 345
Gaul, W., 126; 347
Gay, D., 185; 348
Gelman, A., 328; 354
generalized EM, 258
generalized linear model, 106; 107;
    108; 122; 123
generalized linear regression, 101
Gensch, D., 23; 354
George, E., 201; 349
geo-demographic segmentation,
    XX; 4; 34; 241; 243; 248;
    254; 256
Geo-Marktprofiel, 248
Ghose, S., 231; 238; 354
Ginter, J., 3; 11; 15;26; 325; 327;
    328; 330; 336; 345; 352; 354;
    362
GLIMMIX, XX; 35; 106; 108;
    110; 112; 114; 117; 118; 120;
    125; 126; 131; 132; 145; 150;
    152; 156; 158; 160; 176; 178;
    181; 182; 285; 296; 288; 317;
    330; 338; 339
Gnanadesikan, R., 56; 57; 58; 353;
    354
Goldstein, M., 42; 352
Gonul, F., 285; 354
Goodman, L., 21; 279; 354

Gordon, A., 39; 42; 49; 54; 354
Gowda, K., 50; 354
Gower, J., 46; 47; 53; 354
Green, P.E., XXI; XXII; 15; 17; 18; 19; 21; 24; 51; 57; 76; 77; 78; 96; 97; 231; 295; 299; 302; 303; 305; 306; 314; 327; 339; 349; 354; 355; 360
Green, P.J., 297; 354
Greeno, D., 20; 72; 355
Grover, R., XXI; 21; 96; 97; 172; 179; 279; 355
Guadagni, P., 10; 283; 290; 355
Guion, R., 25; 348
Gunderson, R., 67; 346; 347; 355
Gunter, B., 8; 11; 12; 268; 355
Gupta, S., 26; 28; 115; 116; 118; 146; 147; 149; 150; 156; 158; 280; 288; 290; 291; 348; 352; 356
Gutman, J., 260; 261; 356; 365

H

Habarasz, J., 54; 55; 61; 348
Hagenaars, J., 178; 356
Hagerty, M., 25; 305; 307; 309; 356
Haley, R., 15; 356
Hamilton, J., 88; 356
Han, S., 290; 348
Hand, D., 78; 353
Harvey, J., 57; 354
Hasselblad, V., 80; 81; 356
Hauser, J., 15; 25; 356; 367
Hausman, J., 282; 284; 285; 356
hazard model, 162; 163; 164; 167; 180
Heckman, J., 283; 284; 285; 327; 356
Heiser, W., 138; 141; 350
Helsen, K., 51; 113; 118; 314; 355; 356
Henderson, J., 54; 354
Herniter, J., 278; 282; 356
heterogeneous conditional logit, 285
hierarchical clustering, 20; 27; 43; 48; 50; 52; 54; 59; 232; 307; 308; 311; 315; 316; 338
Hinde, J., 91; 345
Hoffman, D.,126; 351
Hofstede, F. ter, XXII; 94; 95; 96; 261; 339; 357; 369
Holmer, P., 12; 13; 346; 358
Holt, D., 60; 366
Hope, A., 91; 357
Hosmer, D., 102; 357
household panel, 280; 281; 292
Howard, J., 14; 138; 141; 351; 357
Howell, R., 195; 366
Hruschka, H., 19; 20; 65; 66; 73; 357
Hubert, L., 19; 58; 60; 255; 345; 357
Hughes, M., 269; 360
Hustad, T., 11; 357
Hutchinson, P., 24; 352
hybrid segmentation, 18; 62; 321

I

ICOMP, 92; 93; 129; 134; 136; 152; 169
ideal-point model, 125; 129; 132; 136; 138; 301; 302
identification, XXI; 3; 5; 12; 15; 25; 34; 39; 66; 75; 88; 90; 91; 96; 98; 101; 109; 129; 134; 138; 141; 147; 149; 151; 152; 168; 169; 175; 176; 177; 179; 190; 200; 203; 220; 221; 224; 226; 227; 245; 258; 262; 264; 276; 279; 305; 345
IIA, 101; 104; 105; 280; 284; 292
independence from irrelevant alternatives, 104
information matrix, 88; 89; 92; 109; 200
instrumental values, 12; 262; 264; 266
internal analyses, 125; 126
inverse Gaussian, 79; 80; 106; 107; 126; 131; 132
inversion, 59

item response theory, 195

J
Jagpal, H., 217; 223; 225; 226; 357
Jain, A., 28; 301; 315; 337; 357; 361
Jain, D., 163; 167; 284; 285; 349; 357; 368
Jancey's method, 53
Jedidi, K., 113; 118; 138; 141; 217; 218; 219; 223; 224; 225; 226; 227; 228; 229; 351; 356; 357
Jensen, R., 54; 357
Joachimsthaler, E., 13; 14; 359
Johansson, J., 23; 360
Johnson, R., 5, 297; 351; 357; 369
joint segmentation, XX; 187; 189; 190; 191; 192; 194; 256;
Jones, J., 284; 357
Jöreskog, K., 217; 221; 222; 357; 358

K
Kahle, L., 12; 13; 265; 266; 270; 346; 358; 360
Kamakura, W., XXII; 6; 12; 21; 25; 26; 27; 28; 31; 51; 55; 97; 98; 102; 103; 105; 115; 118; 119; 133; 146; 149; 152; 153; 156; 157; 158; 159; 167; 168; 176; 195; 197; 198; 199; 203; 205; 207; 210; 212; 213; 214; 256; 261; 262; 264; 266; 269; 270; 271; 272; 273; 274; 275; 280; 284; 285; 286; 288; 289; 290; 292; 301; 302; 305; 307; 309; 312; 316; 318; 320; 328; 338; 339; 346; 351; 358; 365; 369
Kannan, P., 289; 358
Kass, G., 23; 358
Katahira, H., 55; 56; 316; 358; 359
Kaufman, L., 42; 49; 359
Keane, M., 23; 353
Kernan, J., 20; 72; 355; 359
Kettenring, J., 56; 57; 58; 353; 354

Kiel, G., 72; 359
Kim, B., 28; 102; 103; 118; 159; 168; 289; 290; 358
Kinnear, T., 260; 359
Kistemaker, C., 14; 25; 27; 55; 56; 65; 305; 308; 309; 369
Klein, R., 54; 359
Koontz, W., 54; 359
Koponen, A., 12; 259; 359
Kotler, P., 4; 7; 331; 336; 359
Kotrba, R., 5; 359
Krieger, A., 15; 295; 302; 303; 305; 306; 314; 355
Krijnen, W., XXII; 369
Krishna, P., 50; 354
Krishnan, T., 90; 362
Krishnamurthi, L., 283; 359
Krosnick, J., 262; 345
Kruskal, J., 301; 359
Kumar, V., 75; 76; 78; 88; 90; 91; 96; 97; 115; 118; 146; 147; 148; 152; 158; 337; 352
K-means, 19; 20; 24; 43; 50; 52; 55; 57; 58; 66; 72; 72; 85; 88; 101; 110; 112; 114; 223; 229; 312

L
Laird, N., 80; 81; 351
Lamont, L., 261; 368
Landewehr, J., 284; 357
Landon, E., 72; 359
Langeheine, R., 78; 169; 175; 176; 178; 347; 359; 368
Lastovicka, J., 13; 14; 359
latent change, 160; 167; 175; 176
latent Markov, 168; 169; 170; 171; 179; 189; 190; 194; 331
latent trait theory, 195
Layton, R., 72; 359
Lazarfeld, P., 12; 359
Lazer, W., 13; 203; 359
Lee, J., 28; 102; 103; 118; 159; 168; 289; 290; 358

Leeflang, P., 27; 280; 359; 363; 364; 369
Lehman, D., 21; 96; 97; 359
Lenk, P., 127; 140; 141; 293; 325; 327; 328; 330; 345; 351; 352; 360; 369
Lesser, J., 269; 360
Lessig, V., 14; 27; 28; 72; 337; 338; 360; 365; 367
Lichtenstein, S., 25; 366
life-style segmentation, XX; 34; 195; 203; 214
likelihood ratio, 91; 92; 134; 136; 149; 152; 169; 221; 319
Lilien, G., XXII
Lindsay, B., 93; 178; 337; 347; 360
Lindsey, J., 326; 329; 360
link function, 80; 107; 109; 119; 123; 131; 133; 308
LINMAP, 303; 315
LISREL, XX; 34; 180; 217; 221; 229
list segmentation, 331
Little, J., 10; 14; 15; 283; 290; 355
local optima, 53; 54; 65; 67; 75; 81; 88; 99; 110; 112; 228; 311
Lodahl, T., 12; 361
logit, 17; 23; 26; 27; 31; 56; 104; 106; 107; 115; 117; 131; 145; 151; 167; 176; 180; 186; 264; 266; 279; 280; 282; 283; 285; 286; 287; 289; 290; 292; 301; 302; 315; 316
log-linear models, 17; 180
Long, S., 217; 360
Lonsdale, T., 22; 336; 346
Loudon, D., 4; 7; 10; 360
Louis, T., 88; 89; 360
Louviére, J., 315; 316; 353; 360; 363
LOV, 12; 13; 21; 97; 265; 266; 270; 276
Luce, R., 96; 146; 295; 302; 317; 360
Luijkx, R., 178; 356
Lutz, J., 25; 360

M

MacEvoy, B., 12; 13; 266; 363
MacLachlan, D., 23; 360
MacNaughton-Smith, P., 50; 360
MacReady, G., 115; 146; 151; 198; 350; 360
Madrigal, R., 270; 360
Magdison, J., 24; 73; 360
MaGee, J., 278; 282; 356
Mahajan, V., 20; 28; 65; 337; 351; 357; 361
Mahalanobis distance, 46; 47; 53
Mahatoo, W., 14; 27; 352
Maier, J., 18; 73; 361
Makov, U., 78; 81; 85; 88; 90; 121; 367
manifest change, 159; 160; 176
Manrai, A.K., 20; 26; 51; 126; 138; 233; 235; 351
Manrai, L.A., 20; 26; 51; 126; 138; 233; 235; 351
Manton, K., 66; 68; 69; 361
Martin, C., 23; 361
Maslow, A., 12; 13; 361
mass customization, 4; 331; 332;
Massy, W., 4; 5; 6; 7; 10; 11; 12; 13; 14; 15; 23; 26; 27; 28; 278; 336; 338; 348; 350; 354; 361
maximum likelihood, 24; 68; 75; 80; 85; 88; 89; 94; 126; 179; 180; 195; 196; 221; 226; 232; 256; 284; 287; 327
Mayer, C., 11; 357
Mazzon, J., 12; 21; 97; 98; 118; 262; 264; 269; 270; 358
McCann, J., 10; 11; 22; 27; 361; 370
McConkey, C., 260; 268; 361
McCullagh, P., 80; 120; 122; 123; 124; 143; 361
McDonald, M., 5; 8; 9; 361
McFadden, D., 285; 302; 361
McHugh, R., 80; 361
McLachlan, G., 21; 78; 80; 81; 88; 89; 90; 91; 346; 361; 362
measurement model, 180; 219; 220;

Index

223; 224; 226; 227; 228
median linkage, 48; 49
metric conjoint segmentation, 295; 303; 308; 310; 311; 315
micro-marketing, 329; 332
micro-segmentation, 331; 342
Miller, K., 11; 362
Milligan, G., 19; 49; 53; 54; 56; 57; 58; 59; 349; 362
Mills, M., 260; 356
minimum variance linkage, 19; 49
Minkowske metric, 46
missing data, 84; 120; 142; 179;
Mitchell, A., 13; 362
Mitchell, S., 241; 243; 362
mixture MDS, 26; 29
mixture regression, 17; 18; 26; 28; 29; 32; 33; 34; 86; 93; 96; 99; 102; 103; 106; 108; 109; 110; 112; 113; 115; 116; 117; 119; 132; 145; 146; 150; 151; 156; 167; 168; 169; 181; 185; 308; 311; 312; 313; 314; 315; 316; 319; 320; 335; 342
mixture unfolding, 33; 37; 96; 125; 131; 138; 140; 145; 146; 150; 151; 156; 167; 168; 255; 335
modified AIC, 92; 173
Monk, D., 8; 362
MONANOVA, 301; 315
Montgomery, D., 43; 72; 278; 314; 361; 362
Mooijaart, A., 81; 362
Moore, W., 21; 25; 96; 97; 359; 362
Morgan, B., 59; 362
Moriarty, M., 15; 20; 25; 72; 362
Morrison, D., 278; 291; 350; 361; 362
Morwitz, V., 189; 362
Moschis, G., 260; 363
multinomial, 21; 23; 55; 56; 68; 75; 77; 79; 80; 82; 83; 96; 97; 98; 102; 104; 106; 107; 113; 115; 116; 117; 118; 119; 120; 138; 140; 141; 149; 150; 152; 156; 167; 175; 176; 266; 280; 283; 285; 286; 287; 289; 290; 291; 293; 301; 302; 315; 316; 317; 327
multionmial mixture, 98; 117
Murray, J., 13; 14; 359
Myers, J.G., 72; 363
Myers, J.H., 15; 363
M-step, 84; 85; 86; 89; 108; 109; 113; 121; 133; 142; 152; 169; 180; 185; 226; 236; 287

N

Naert, P., 27; 363
Nagin, D., 282; 353
Nakanishi, M., 279; 282; 287; 288; 350; 363
Narendra, P., 54; 359
nearest neighbor, 49
negative binomial, 79; 97; 106; 126
Nelder, J., 80; 106; 109; 120; 122; 123; 124; 143; 361; 363
nested logit, 102; 103; 118; 176; 289
neural network, 24
Newcomb, S., 75; 81; 363
Newton-Raphson, 80; 85; 124
Nicosia, F., 72; 363
Nielsen, A.C., 147; 163; 281; 243; 258; 281
nonhierarchical clustering, 20; 52; 54; 338
nonoverlapping clustering, 19; 32; 41; 56
non-response, 214
non-uniqueness, 49; 59
normative segmentation, 26; 31; 102; 338; 339; 342
normed entropy criterion, 93
Novak, T., 12; 13; 21; 97; 98; 133; 266; 270; 358; 363

O

Ogawa, K., 25; 305; 316; 363

Oliver, R., 25; 65; 305; 308; 351
Olivier, A., 280; 359
Oppewal, H., 296; 363
overlapping clustering, 19; 20; 37; 64

P

paired comparison, 115; 147; 300
Panton, D., 72; 365
Park, C., 297; 363
Pearson, K., 47; 75; 180; 363
Penner, L., 262; 363
Perreault, W., 18; 350
Pessemeier, E., 13; 15; 43; 346; 348; 354
Peters, B., 88; 363
Peterson, R., 260; 268; 363
Phillips, D., 337; 363
pick-any data, 126; 128; 132; 140
piecewise exponential hazard, 291
Pieters, F., 117; 365
Piirto, R., 267; 268; 276; 363
Pitts, R., 264; 269; 363
Plat, F., 27; 364
Plummer, J., 13; 260; 364
Poisson, 75; 79; 82; 83; 85; 86; 87; 90; 91; 93; 96; 98; 102; 106; 107; 110; 112; 114; 116; 117; 118; 126; 131; 132; 136; 137; 138; 140; 158; 164; 175; 176; 181; 186; 290; 291; 293
Poisson mixture, 96; 116
Poisson regression, 331
Portis, B., 72; 349
post-hoc segmentation, 27; 335
Potter, W., 268; 364
Poulsen, C., 96; 97; 161; 162; 163; 169; 170; 171; 172; 176; 179; 198; 279; 364; 369
Prabhaker, P., 14; 347
principal component analysis, 242; 247
PRIZM, 242; 243; 244; 245; 246; 247; 256
Pruzansky, S., 238; 349

pseudo-maximum likelihood, 94
psychographics, 12; 14; 16; 20; 22; 259; 260; 268
Punj, G., 19; 32; 47; 50; 54; 69; 364
purchase frequency, 11; 97; 98; 116; 117; 141; 164
purchase incidence, 98; 101; 115; 118; 175; 279; 290
Puterman, M., 158; 368

Q

Quandt, R., 102; 364
Quiroz, A., 56; 352
Q-type factor analysis, 18

R

Raj, S., 283; 359
Ramaswamy, V., XXII; 113; 116; 118; 119; 127; 140; 141; 145; 169; 171; 173; 174; 175; 176; 179; 189; 190; 191; 192; 194; 303; 316; 352; 357; 364; 369
Ramsey, J., 102; 364
Rangaswamy, A., 25; 65; 305; 308; 351
Rao, V., 23; 45; 46; 325; 364
Ray, A., 59; 362
recurrent choice, 175
Redner, R., 81; 364
repeat purchase, 164; 165
response based segmentation, XX
response segmentation, 27
Reynolds, F., 260; 350; 365
Reynolds, M., 232; 365
Reynolds, T.J., 261; 365
Reynolds, W.H., 5; 331; 364
Rhoads, G., 195; 366
Robinson, J., 3; 365
Roeder, K., 93; 360
Rogers, M., 10; 46; 365
Rokeach, M., 12; 261; 262; 263; 264; 265; 266; 269; 270; 271; 276; 365

Index

Rosbergen, E., 117; 118; 365
Roscoe, A., 23; 27; 72; 346
Rossi, P., 284; 325; 326; 327; 329; 345; 365
Rost, J., 78; 359
Rousseeuw, P., 42; 49; 359
Rubin, D.B., 58; 81; 90; 345; 351; 354; 365
Rubinson, J., 27; 366
Russell, G., XXII; 6; 26; 27; 28; 31; 45; 102; 103; 115; 118; 149; 167; 280; 282; 285; 286; 288; 289; 290; 292; 328; 338; 339; 358; 365

S

Samejima, F., 198; 365
Sasieni, M., 22; 346
Sattath, S., 231; 365
Saunders, J., 18; 73; 361
Sawyer, A., 15; 72; 159; 167; 348
Scales, L., 85; 365
Schaninger, C., 72; 365
Schaffer, C., 314; 355
Schiffman, S., 232; 233; 365
Schlattmann, P., 337; 347
Schmittlein, D., 189; 362
Schnabel, R., 85; 351
Schwartz, S., 12; 262; 264; 270; 365
Scott, A., 53; 366
Scott, J., 261; 368
segment membership, 21; 22; 24; 26; 55; 65; 85; 90; 115; 125; 127; 133; 137; 145; 159; 160; 170; 173; 199; 201; 202; 225; 227; 228; 286; 288; 317; 319; 321
segment profiling, 145
Sen, S., 27; 347
Sethi, S., 72; 366
Sewall, M., 15; 366
Sexton, D., 11; 14; 27; 366
Shapiro, B., 18; 366
Shaw, A., 336; 366
Shepard, R., 19; 20; 64; 366

Shocker, A., 11; 301; 366
Siddarth, S., 116; 118; 348
Silk, A., 43; 72; 362
Simons, M., 53; 366
simple random sample, 60; 62; 95
simulated annealing, 54; 65; 308
Singer, B., 285; 327; 356
Singh, J., 42; 73; 195; 198; 366
single linkage, 19; 48; 49; 58; 59
single-source database, 280
singular-value decomposition, 98
Skinner, C., 60; 94; 366
Slovic, P., 25; 366
Smith, A., 78; 81; 85; 88; 90; 121; 363; 367
Smith, T., 60; 366
Smith, W., 3; 337; 338; 341; 366
Smith de Borrero, M., 146; 148; 152; 158; 352
Soromenho, G., 93; 349
Späth, H., 25; 55; 308; 366
splinter average, 50
spurious state dependence, 283
Srinivasan, V., XXI; 15; 21; 96; 97; 179; 279; 282; 285; 295; 297; 299; 301; 306; 351; 354; 355; 363; 365; 366
Srivastava, R., 11; 282; 284; 302; 358; 366; 370
Staelin, R., 301; 317; 349
Starr, M., 27; 366
stationary equations, 122; 143
Steckel, J., 285; 366
Steenkamp, J., 25; 65; 66; 68; 73; 94; 95; 96; 261; 298; 300; 305; 308; 309; 339; 357; 367; 369
Stewart, D., 19; 25; 32; 47; 50; 54; 69; 307; 364; 367
Stigler, S., 106; 367
stochastic mixture unfolding model, 131
store patronage, 10
Stout, R., 11; 367
Strain, C., 27; 354
stratified sample, 62

structural equation model, 187; 217; 218; 219; 220; 221; 222; 223; 224; 225; 226; 227; 229
structural model, 180; 220; 224; 226; 228
structural state dependence, 283
STUNMIX, 126; 127; 131; 132; 133; 134; 135; 136; 137; 138; 139; 140; 141; 142; 145; 150; 152; 156; 158; 160; 161; 176; 177; 255; 258; 288; 339
Summers, N., 20; 355
Suppes, P., 317; 360
Symons, M., 80; 367

T

tailored interviewing, XX; 31; 34; 195; 196; 198; 199; 200; 202; 203; 209; 211; 214; 215
Talarzyk, W., 269; 347
tandem approach, 255
Tanner, M., 84; 367
Taylor, J., 260; 359
Teel, J., 260; 268; 346; 367
terminal values, 12; 98; 262; 263; 264; 265; 270; 271
Tigert, D., 13; 22; 43; 72; 260; 269; 336; 346; 367; 370
Timmermans, H., 296; 363
Titterington, D., 78; 81; 85; 88; 90; 121; 367
Toler, C., 264; 367
Tollefson, J., 14; 27; 28; 337; 338; 360; 367
Tolley, H., 66; 68; 69; 361
tracking panel, 279; 281
tradeoff approach, 300
Trivedi, P., 93; 116; 118; 350
Tsao, I., 56; 58; 354
Tukey, J., 295; 360
Tversky, A., 231; 365; 367
Twedt, D., 10; 367

U

Umesh, U., 22; 367
unobserved heterogeneity, 119; 163; 180; 278; 283; 284;
Urban, G., 15; 25; 356; 367
usage situation, 10; 11; 18; 23; 73

V

VALS, 266; 267; 276
Valette-Florence., 261; 367
value systems, 12; 21; 98; 262; 264; 266; 269; 270; 276
van de Pol, F., 169; 178; 367; 368
van der Heijden, P., 81; 362
van der Maas, H., 223; 352
van Duÿn, M., 118; 368
van Raaij, F., 259; 368
VanHonacker, W., 285; 366
vector model, 125; 127; 131; 134; 136; 137; 138; 143; 301; 302
Venkatesan, M., 15; 20; 25; 72; 362
Verhallen, T., 259; 368
Vermunt, J., 21; 178; 179; 368
Vilcassim, N., 163; 167; 284; 285; 349; 357; 368
Vinson, D., 261; 368
Vriens, M., 295; 303; 305; 309; 310; 312; 313; 314; 327; 352; 368; 369; 370
vulnerability index, 288

W

Wachspress, D., XXI; 21; 96; 97; 355
Walker, H., 81; 88; 194; 363
Walker, S. 175; 364; 368
Wang, P., 158; 368
Wansbeek, T., 331; 348
Ward, J., 43; 50; 51; 61; 66; 368
Warren, W., 260; 268; 361
Wedderburn, R., 106; 123; 363
Wedel, M., XXII; 14; 25; 26; 27; 55; 56; 65; 66; 68; 73; 78; 80; 90; 94; 95; 96; 102; 106; 110; 114;

115; 116; 117; 118; 119; 126;
128; 129; 130; 131; 134; 136;
137; 138; 139; 140; 141; 145;
146; 152; 154; 155; 156; 157;
158; 162; 163; 165; 176; 181;
198; 199; 203; 205; 207; 210;
212; 213; 214; 236; 237; 238;
256; 261; 290; 291; 301; 305;
308; 309; 310; 316; 318; 325;
327; 328; 331; 339; 351; 352;
357; 358; 365; 367; 368; 369
Weiss, M., 242; 369
Wells, W., 13; 14; 259; 260; 268; 269; 369; 370
Welsch, R., 185; 348
Whipple, T., 11; 357
White, H., 89; 94; 370
Wildt, A., 22; 27; 370
Wilkie, W., 6; 7; 13; 14; 15; 16; 22; 335; 370
Wilms, T., 305; 310; 327; 368
Wind, Y., 4; 5; 6; 7; 10; 11; 13; 14; 15; 17; 23; 28; 35; 159; 306; 337; 339; 341; 342; 345; 354; 355; 370
Winer, R., 14; 27; 28; 338; 353
Winsberg, S., 134; 138; 141; 350
Winter, F., 23; 28; 364; 370
Wise, D., 282; 284; 285; 356
Wishart, D., 45; 370
Wisnewski, K., 282; 347
within segment heterogeneity, 79; 98; 99; 116; 117; 119; 136; 137; 286; 293; 311; 313; 327; 330
Wittink, D., 15; 295; 298; 300; 303; 305; 314; 315; 349; 356; 362; 367; 370
Wolfe, J., 80; 81; 370
Woodbury, M., 66; 68; 69; 361
Woodside, A., 264; 269; 363
Wright, G., 289; 358
Wright, R., 23; 361
Wu, C., 81; 370

Y

Yankelovich, D., 14; 15; 370
Young, F., 232; 365
Young, M., 327; 360

Z

Zadeh, L., 20; 65; 370
Zenor, M., 282; 370
Ziff, R., 13; 260; 269; 370
ZIP+4, 32; 241; 256; 257